Palgrave Studies in Pragmatics, Language and (

Series Editors: **Noël Burton-Roberts** and **Robyn** (

Series Advisors: **Kent Bach, Anne Bezuidenhc Glucksberg, Francesca Happé, François Recanati, Deirdre Wilson**

Palgrave Studies in Pragmatics, Language and Cognition is a new series of high quality research monographs and edited collections of essays focusing on the human pragmatic capacity and its interaction with natural language semantics and other faculties of mind. A central interest is the interface of pragmatics with the linguistic system(s), with the 'theory of mind' capacity and with other mental reasoning and general problem-solving capacities. Work of a social or cultural anthropological kind will be included if firmly embedded in a cognitive framework. Given the interdisciplinarity of the focal issues, relevant research will come from linguistics, philosophy of language, theoretical and experimental pragmatics, psychology and child development. The series will aim to reflect all kinds of research in the relevant fields – conceptual, analytical and experimental.

Titles include:

Anton Benz, Gerhard Jäger and Robert van Rooij (*editors*)
GAME THEORY AND PRAGMATICS

Reinhard Blutner and Henk Zeevat (*editors*)
OPTIMALITY THEORY AND PRAGMATICS

Corinne Iten
LINGUISTIC MEANING, TRUTH CONDITIONS AND RELEVANCE
The Case of Concessives

Ira Noveck and Dan Sperber (*editors*)
EXPERIMENTAL PRAGMATICS

Forthcoming titles:

María J. Frápolli (*editor*)
SAYING, MEANING AND REFERRING
Essays on François Recanati's Philosophy of Language

Ulrich Sauerland and Penka Stateva
PRESUPPOSITION AND IMPLICATURE IN COMPOSITIONAL SEMANTICS

Hans-Christian Schmitz
OPTIMAL ACCENTUATION AND ACTIVE INTERPRETATION

Christoph Unger
ON THE COGNITIVE ROLE OF GENRE

Palgrave Studies in Pragmatics, Language and Cognition Series
Series Standing Order ISBN 0-333-99010-2 Hardback   0-333-98584-2 Paperback
(*outside North America only*)

You can receive future titles in this series as they are published by placing a standing order. Please contact your bookseller or, in case of difficulty, write to us at the address below with your name and address, the title of the series and the ISBN quoted above.

Customer Services Department, Macmillan Distribution Ltd, Houndmills, Basingstoke, Hampshire RG21 6XS, England

Experimental Pragmatics

Edited by

Ira A. Noveck and Dan Sperber

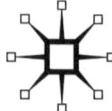

Introduction, selection and editorial matter © Ira A. Noveck and Dan Sperber 2004
All chapters © Palgrave Macmillan Ltd 2004
All rights reserved. No reproduction, copy or transmission of this publication may be made without written permission.

No paragraph of this publication may be reproduced, copied or transmitted save with written permission or in accordance with the provisions of the Copyright, Designs and Patents Act 1988, or under the terms of any licence permitting limited copying issued by the Copyright Licensing Agency, 90 Tottenham Court Road, London W1T 4LP.

Any person who does any unauthorised act in relation to this publication may be liable to criminal prosecution and civil claims for damages.

The authors have asserted their rights to be identified as the authors of this work in accordance with the Copyright, Designs and Patents Act 1988.

First published in hardcover 2004

First published in paperback 2006 by
PALGRAVE MACMILLAN
Houndmills, Basingstoke, Hampshire RG21 6XS and
175 Fifth Avenue, New York, N.Y. 10010
Companies and representatives throughout the world

PALGRAVE MACMILLAN is the global academic imprint of the Palgrave Macmillan division of St. Martin's Press, LLC and of Palgrave Macmillan Ltd. Macmillan® is a registered trademark in the United States, United Kingdom and other countries. Palgrave is a registered trademark in the European Union and other countries.

ISBN-13: 978–1–4039–0350–1 hardback
ISBN-10: 1–4039–0350–6 hardback
ISBN-13: 978–1–4039–0351–8 paperback
ISBN-10: 1–4039–0351–4 paperback

This book is printed on paper suitable for recycling and made from fully managed and sustained forest sources.

A catalogue record for this book is available from the British Library.

Library of Congress Cataloging-in-Publication Data
Experimental pragmatics / edited by Ira A. Noveck and Dan Sperber.
    p. cm. — (Palgrave studies in pragmatics, language and cognition)
    Includes bibliographical references and index.
    ISBN 1–4039–0350–6 (cloth)
    1. Pragmatics. I. Noveck, Ira A., 1962– II. Sperber, Dan. III. Series.

P99.4.P72E93 2004
306.44—dc22
                                                   2004042092

10 9 8 7 6 5 4 3 2 1
15 14 13 12 11 10 09 08 07 06

Printed and bound in Great Britain by
Antony Rowe Ltd, Chippenham and Eastbourne

Contents

Acknowledgements   vii

Contributors   viii

1 Introduction   1
  Dan Sperber and Ira A. Noveck

Part I   Pioneering Approaches

2 Changing Ideas about Reference   25
  Herbert H. Clark and Adrian Bangerter

3 Psycholinguistic Experiments and Linguistic-Pragmatics   50
  Raymond W. Gibbs, Jr.

4 On the Automaticity of Pragmatic Processes: a Modular Proposal   72
  Sam Glucksberg

5 Reasoning, Judgement and Pragmatics   94
  Guy Politzer

6 Exploring Quantifiers: Pragmatics Meets the Psychology
  of Comprehension   116
  A. J. Sanford and Linda M. Moxey

Part II   Current Issues in Experimental Pragmatics

7 Testing the Cognitive and Communicative
  Principles of Relevance   141
  Jean-Baptiste Van der Henst and Dan Sperber

8 Contextual Strength: the Whens and Hows of Context Effects   172
  Orna Peleg, Rachel Giora and Ofer Fein

9 Electrophysiology and Pragmatic Language Comprehension   187
  Seana Coulson

10 Speech Acts in Children: the Example of Promises   207
   Josie Bernicot and Virginie Laval

11 Reasoning and Pragmatics: the Case of *Even-If*   228
   Simon J. Handley and Aidan Feeney

Part III The Case of Scalar Implicatures

12 Implicature, Relevance and Default Pragmatic Inference 257
   Anne L. Bezuidenhout and Robin K. Morris

13 Semantic and Pragmatic Competence in Children's and Adults' Comprehension of *Or* 283
   Gennaro Chierchia, Maria Teresa Guasti, Andrea Gualmini, Luisa Meroni, Stephen Crain and Francesca Foppolo

14 Pragmatic Inferences Related to Logical Terms 301
   Ira A. Noveck

15 Conversational Implicatures: Nonce or Generalized? 322
   Anne Reboul

Name Index 334

Subject index 340

Acknowledgements

The authors first presented their work as part of a workshop that was sponsored by the European Science Foundation and held at the Institut des Sciences Cognitives in Lyon, France. We would like to thank other participants at the workshop whose comments kept the exchanges lively and no doubt improved the quality of the chapters: Robyn Carston, Dick Carter, Billy Clark, Vittorio Girotto, Stephen Levinson, Jacques Moeschler, François Recanati, Johan van der Auwera and Deirdre Wilson. We would also like to thank the ESF, specifically Professor Ekkehard Koenig, who joined in the proceedings, and Phillipa Rowe, who helped make the workshop run smoothly.

Contributors

Adrian Bangerter, University of Neuchâtel, Switzerland
Josie Bernicot, Université de Poitiers, France
Anne L. Bezuidenhout, University of South Carolina, USA
Gennaro Chierchia, Università di Milano-Bicocca, Italy
Herbert H. Clark, Stanford University, USA
Seana Coulson, University of California San Diego, USA
Stephen Crain, University of Maryland, USA
Aidan Feeney, University of Durham, UK
Ofer Fein, Academic College of Tel Aviv, Israel
Francesca Foppolo, Università di Milano-Bicocca, Italy
Raymond W. Gibbs, Jr., University of California, Santa Cruz, USA
Rachel Giora, Tel Aviv University, Israel
Sam Glucksberg, Princeton University, USA
Andrea Gualmini, Massachusetts Institute of Technology, USA
Maria Teresa Guasti, Università di Milano-Bicocca, Italy
Simon J. Handley, University of Plymouth, UK
Virginie Laval, Université de Poitiers, France
Luisa Meroni, University of Maryland, USA
Robin K. Morris, University of South Carolina, USA
Linda M. Moxey, University of Glasgow, UK
Ira A. Noveck, Institut des Sciences Cognitives, France
Orna Peleg, Tel Aviv University, Israel
Guy Politzer, CNRS-Université de Paris 8, France
Anne Reboul, Institut des Sciences Cognitives, France
A. J. Sanford, University of Glasgow, UK
Dan Sperber, Institut Jean Nicod, France
Jean-Baptiste Van der Henst, Institut des Sciences Cognitives, France

1
Introduction
Dan Sperber and Ira A. Noveck

How does our knowledge of language on the one hand, and of the context on the other permit us to understand what we are told, resolve ambiguities, grasp both explicit and implicit content, recognize the force of a speech act, appreciate metaphor and irony? These issues have been studied in two disciplines: pragmatics and psycholinguistics, with limited interactions between the two. Pragmatics is rooted in the philosophy of language and in linguistics and has spawned competing theories using as evidence a mixture of intuitions about interpretation and observations of behaviour.

Psycholinguistics has developed sophisticated experimental methods in the study of verbal communication, but has not used them to test systematic pragmatic theories. This volume lays down the bases for a new field, Experimental Pragmatics, that draws on pragmatics, psycholinguistics and also on the psychology of reasoning. Chapters in this volume either review pioneering work or present novel ways of articulating theories and experimental methods in the area. In this introduction we outline some core pragmatic issues and approaches and relate them to experimental work in psycholinguistics and in the psychology of reasoning. We then briefly present one by one the chapters of this collection.

1 Some core pragmatic issues and approaches

In a very broad sense, pragmatics is the study of language use. It encompasses loosely related research programmes ranging from formal studies of deictic expressions to sociological studies of ethnic verbal stereotypes. In a more focused sense, pragmatics is the study of how linguistic properties and contextual factors interact in the interpretation of utterances. We will be using 'pragmatics' only in this narrower sense. Here we briefly highlight a range of closely related, fairly central pragmatic issues and approaches that have been of interest to linguists and philosophers of language in the past thirty years or so, and that, in our opinion, may both benefit from, and contribute to, work in experimental psychology.

2 Experimental Pragmatics

A sentence of a language can be considered as an abstract object with phonological, syntactic and semantic properties assigned by the grammar of the language (the grammar itself being generally seen as a mental system). The study of these grammatical properties is at the core of linguistics. An utterance, by contrast, is a concrete object with a definite location in time and space. An utterance is a realization of a sentence (a realization that can be defective in various respects, for instance by being mispronounced). An utterance inherits the linguistic properties of the sentence it realizes and has further properties linked to its being uttered in a given situation by a speaker addressing an audience. In verbal communication, both linguistic and non-linguistic properties of utterances are involved. But what role exactly do these properties play and how do they interact? These are questions that pragmatic theories attempt to answer.

The pragmatic approaches we are concerned with here all accept as foundational two ideas that have been defended by the philosopher Paul Grice (Grice, 1989). The first idea is that, in verbal communication, the interlocutors share at least one goal: having the hearer recognize the *speaker's meaning*. The linguistic decoding of the sentence uttered provides the hearer with the *sentence meaning*, but this decoding is only a subpart of the process involved in arriving at a recognition of the speaker's meaning. This recognition does not involve any distinct awareness of the sentence meaning, that is, of the semantic properties assigned to the sentence by the grammar. Only linguists and philosophers of language have a clear and distinct notion of, and an interest in, sentence meaning proper. Unlike sentence meaning, which is an abstraction, a speaker's meaning is a mental state. More specifically, for a speaker to mean that P is for her to have the intention that the hearer should realize that, in producing her utterance, she intended him to think that P. A speaker's meaning is an overt intention that is fulfilled by being recognized by the intended audience. Consider, for instance, Mary's contribution to the following exchange:

(1) *Peter*: Do you like Fellini's films?
    Mary: Some of them.

In replying 'some of them', Mary intends Peter to realize that she intends him to think that she likes some of Fellini's films, but not all. The proposition *Mary likes some of Fellini's films but not all* is Mary's meaning. It is not the linguistic meaning of the sentence fragment 'some of them', which can be used in other situations to convey totally different contents. Mary's meaning goes well beyond the meaning of the linguistic expression she uttered.

Verbal comprehension is often seen in psycholinguistics as the study of linguistic decoding processes, drawing on grammar (with the possibility that grammar may extend above the level of the sentence to that of discourse) and using contextual factors in a limited way, to disambiguate ambiguous expressions and fix reference. The idea that successful communication consists in the recognition by the audience of the speaker's meaning suggests a different approach. Verbal comprehension should be seen as a special form of attribution of a mental state to the speaker. This attribution is dependent on linguistic decoding, but is essentially an inferential process using as input the result of this decoding and contextual information.

The second foundational idea defended by Paul Grice is that, in inferring the speaker's meaning on the basis of the decoding of her utterance and of contextual information, the hearer is guided by the expectation that the utterance should meet some specific standards. The standards Grice envisaged were based on the idea that a conversation is a cooperative activity. Interlocutors are expected to follow what he called a 'co-operative principle' requiring that they 'make [their] conversational contribution such as is required, at the stage at which it occurs, by the accepted purpose or direction of the talk exchange in which [they] are engaged'. This is achieved by obeying a number of 'maxims of conversation' which Grice expressed as follows:

Maxims of Quantity
1. Make your contribution as informative as is required (for the current purpose of the exchange).
2. Do not make your contribution more informative than is required.

Maxims of Quality

Supermaxim. Try to make your contribution one that is true.
1. Do not say what you believe to be false.
2. Do not say that for which you lack adequate evidence.

Maxim of Relation
Be relevant.

Maxims of Manner

Supermaxim. Be perspicuous.
1. Avoid obscurity of expression.
2. Avoid ambiguity.
3. Be brief (avoid unnecessary prolixity).
4. Be orderly.

In interpreting an utterance, the best hypothesis for the hearer to choose is the one that is the most consistent with the assumption that the speaker has indeed followed these maxims. For instance, in interpreting Mary's reply 'some of them' in the above dialogue, Peter is entitled to draw several inferences. He is entitled, in the first place, to treat this sentence fragment as elliptical for 'I like some of Fellini's films' since this is the interpretation most consistent with the assumption that Mary was following the maxims, and in particular the maxims 'be relevant' and 'be brief'. Peter is also entitled to understand Mary to mean that she does not like all of Fellini's films. If she did like all of them, she would be violating the maxim 'make your contribution as informative as is required' in talking only of 'some of them'.

Current pragmatic theories draw on Grice's idea that the existence of set expectations is what allows hearers to infer the speaker's meaning on the basis of the utterance and the context. These theories differ in their account of the precise expectations that drive the comprehension process. Neo-Griceans (Atlas, forthcoming; Gazdar, 1979; Horn, 1973, 1984, 1989, 1992; Levinson, 1983, 2000) stay relatively close to Grice's formulation. Levinson (2000), for instance, defines three basic principles linked to three of Grice's maxims (here in abridged form):

Q-Principle

Speaker's maxim. Do not provide a statement that is informationally weaker than your knowledge of the world allows.

Recipient corollary. Take it that the speaker made the strongest statement consistent with what he knows.

I-Principle

Speaker's maxim. Produce the minimal linguistic information sufficient to achieve your communicational ends.

Recipient corollary. Amplify the informational content of the speaker's utterance, by finding the most *specific* interpretation, up to what you judge to be the speaker's... point.

M-Principle

Speaker's maxim. Indicate an abnormal, non-stereotypical situation by using marked expressions that contrast with those you would use to describe the corresponding normal, stereotypical situations.

Recipient corollary. What is said in an abnormal way indicates an abnormal situation.

These principles provide heuristics for interpreting utterances. For instance, when Mary answers elliptically 'some of them', she can be seen by Peter as producing the minimal linguistic information sufficient to achieve her communicational ends (following the I-Principle), and this, together with the assumption that Mary obeyed the Gricean Maxim of relation, justifies his amplifying the content of her utterance up to what he judges to be her point (see Levinson, 2000, pp. 183–4). Moreover, the Q-Principle justifies

Peter in taking it that Mary made the strongest statement consistent with her knowledge, and that therefore it is not the case that she likes all of Fellini's films.

Relevance Theory (Bezuidenhout, 1997; Blakemore, 1987, 2002; Blass, 1990; Carston, 2002; Carston and Uchida, 1997; Gutt, 1991; Ifantidou, 2001; Matsui, 2000; Moeschler, 1989; Noh, 2000; Papafragou, 2000; Pilkington, 2000; Reboul, 1992; Rouchota and Jucker, 1998; Sperber and Wilson, 1986/1995; Yus, 1997), though still based on Grice's two foundational ideas, departs substantially from his account of the expectations that guide the comprehension process. For Griceans and neo-Griceans, these expectations derive from principles and maxims, that is, rules of behaviour that speakers are expected to obey but may, on occasion, violate. Such violations may be unavoidable because of a clash of maxims or of principles, or they may be committed on purpose in order to indicate to the hearer some implicit meaning. Indeed, in the Gricean scheme, the implicit content of an utterance is typically inferred by the hearer in his effort to find an interpretation which preserves the assumption that the speaker is obeying, if not all the maxims, at least the cooperative principle. For Relevance Theory, the very act of communicating raises in the intended audience precise and predictable expectations of relevance, which are enough on their own to guide the hearer towards the speaker's meaning. Speakers may fail to be relevant, but they may not, if they are communicating at all (rather than, say, rehearsing a speech), produce utterances that do not convey a presumption of their own relevance.

Whereas Grice invokes relevance (in his 'maxim of relation') without defining it at all, Relevance Theory starts from a detailed account of relevance and its role in cognition. Relevance is defined as a property of inputs to cognitive processes. These inputs include external stimuli, which can be perceived and attended to, and mental representations, which can be stored, recalled or used as premises in inference. An input is relevant to an individual when it connects with background knowledge to yield new cognitive effects, for instance by answering a question, confirming a hypothesis, or correcting a mistake. Slightly more technically, cognitive effects are changes in the individual's set of assumptions resulting from the processing of an input in a context of previously held assumptions. This processing may result in three types of cognitive effects: the derivation of new assumptions, the modification of the degree of strength of previously held assumptions, or the deletion of previously held assumptions. Relevance, that is, the possibility of achieving such a cognitive effect, is what makes an input worth processing. Everything else being equal, inputs which yield greater cognitive effects are more relevant and more worth processing. For instance, being told by the doctor 'you have the flu' is likely to carry more cognitive effects and therefore be more relevant than being told 'you are ill'. In processing an input, mental effort is expended. Everything else being equal, relevant inputs involving a smaller processing effort are more relevant and more worth processing. For instance, being told 'you have the flu' is likely to be more relevant than being told 'you have a disease spelled with the sixth, the twelfth and the twenty-first letter of the alphabet' because the first of these two statements would yield the same cognitive effects as the second for much less processing effort. Relevance is thus a matter of degree and varies with two factors; positively with cognitive effect, and inversely with processing effort.

Relevance Theory develops two general claims or 'principles' about the role of relevance in cognition and in communication:

Cognitive principle of relevance. Human cognition tends to be geared to the maximization of relevance.

Communicative principle of relevance. Every act of communication conveys a presumption of its own optimal relevance.

As we have already mentioned, these two principles of relevance are descriptive and not normative (unlike the principles and maxims of Gricean and neo-Gricean pragmaticists). The first, Cognitive Principle of Relevance, yields a variety of predictions regarding human cognitive processes. It predicts that our perceptual mechanisms tend spontaneously to pick out potentially relevant stimuli, our retrieval mechanisms tend spontaneously to activate potentially relevant assumptions, and our inferential mechanisms tend spontaneously to process them in the most productive way. This principle, moreover, has essential implications for human communication processes. In order to communicate, the communicator needs her audience's attention. If, as claimed by the Cognitive Principle of Relevance, attention tends automatically to go to what is most relevant at the time, then the success of communication depends on the audience taking the utterance to be relevant enough to be worthy of attention. Wanting her communication to succeed, the communicator, by the very act of communicating, indicates that she wants her utterance to be seen as relevant by the audience, and this is what the Communicative Principle of Relevance states.

According to Relevance Theory, the presumption of optimal relevance conveyed by every utterance is precise enough to ground a specific comprehension heuristic:

Presumption of optimal relevance
(a) The utterance is relevant enough to be worth processing.
(b) It is the most relevant one compatible with communicator's abilities and preferences.

Relevance-guided comprehension heuristic
(a) Follow a path of least effort in constructing an interpretation of the utterance (and in particular in resolving ambiguities and referential

indeterminacies, in going beyond linguistic meaning, in computing implicatures, etc.).
(b) Stop when your expectations of relevance are satisfied.

For instance, when Mary, in response to Peter's question 'Do you like Fellini's films?' utters 'some of them', she can be confident that, following a path of least effort, Peter will understand 'them' to refer to Fellini's films (since this is the plural referent most prominent in his mind) and the whole utterance to be elliptical for 'I like some of them' (since this is the resolution of the ellipsis closest to his expectations). The fact that there are films by Fellini that Mary likes is relevant enough to be worth Peter's attention (as he indicated it would be by asking the question). However, this does not yet fully satisfy Peter's expectations of relevance: Mary was presumably able, and not reluctant, to tell him whether she liked all of Fellini's films, and that too would be of relevance to Peter. Given that she did not say that she likes them all, Peter is entitled to understand her as meaning that she likes *only* some of them. Having so constructed the interpretation of Mary's utterance, Peter's expectations of relevance are now satisfied, and he does not develop the interpretation any further.

Grice's original theory, the Neo-Gricean theory and Relevance Theory are not the only theoretical approaches to pragmatics (even in the restricted sense of 'pragmatics' we adopt here). Important contributors to pragmatic theorizing with original points of view include Anscombre and Ducrot (1995); Bach (1987, 1994); Bach and Harnish (1979); Blutner and Zeevat (2003); Dascal (1981); Ducrot, (1984); Fauconnier (1975, 1985); Harnish (1976, 1994); Kasher (1976, 1984, 1998); Katz (1977); Lewis (1979); Neale (1990, 1992, forthcoming); Recanati (1979, 1988, 1993, 2000); Searle (1969, 1979); Stalnaker (1999); Sweetser (1990); Travis (1975); Van der Auwera, J. (1981, 1985, 1997); Vanderveken (1990-91); see also Davis (1991), Moeschler and Reboul (1994). However, the three approaches we have briefly outlined here are arguably the dominant ones, and the most relevant ones to the experimental research reported in this book.

2 What can pragmatic theories and experimental psycholinguistics offer each other?

Griceans, neo-Griceans, Relevance Theorists and other pragmaticists, all have ways to account for examples such as (1) above, and for pragmatic intuitions generally. It is hard to find in pragmatics crucial evidence that would clearly confirm one theory and disconfirm another. To experimental psychologists, it might be obvious that one should use experimental evidence in order to evaluate and compare pragmatic claims. Pragmatics, however, has been developed by philosophers of language and linguists who often have little familiarity with, or even interest in, experimental psychology. The only source of evidence most of them have ever used has been their own intuitions about how an invented utterance would be interpreted in a hypothetical situation. Provided that intuitions are systematic enough across subjects, there is nothing intrinsically wrong in using them as evidence, as the achievements of modern linguistics (which relies heavily on such intuitions) amply demonstrate. More sociologically oriented pragmaticists have insisted on the use of evidence from recordings of genuine verbal exchanges, or of genuine written texts, together with data about the speakers or authors and the situation. Even though the interpretation of these naturally occurring utterances is normally left to the pragmaticist's intuitive interpretive abilities, their use has been of great value in investigating a variety of pragmatic issues.

Pragmatic research is not to be censured, let alone discarded, on the grounds that it is mostly based on intuition and observational data and has hardly been pursued at all as an experimental discipline. However, this has meant that preference for one theory over another is justified not in terms of crucial empirical tests, but mostly on grounds of consistency, simplicity, explicitness, comprehensiveness, explanatory force and integration with neighbouring fields. For example, it has been argued that Grice's own formulation of his principle and maxims is too vague, and not explanatory enough: Gricean explanations are more like *ex post facto* rationalizations. Neo-Griceans are developing an approach to pragmatics in close continuity with linguistic semantics, and view this as an advantage. Relevance theorists feel that their approach is more explanatory, more parsimonious, and better integrated into the cognitive sciences. These considerations, however relevant to evaluating theories, can themselves be diversely evaluated.

Turning from pragmatics to experimental psycholinguistics – an older and more developed science – we find a rich and extensive domain of research dealing with diverse themes ranging from the child's first language acquisition to the mechanics of speech production. Among these themes is that of comprehension, which includes a variety of sub-themes from the perception and decoding of the acoustic (or visual) signal to the interpretation of discourse. In principle, the range of phenomena that pragmatics investigates is part of the much wider domain of psycholinguistics. However, with its own rich history, traditions and focus on experimental research, psycholinguistics has generally paid very little attention to the discipline of pragmatics, even when the phenomena studied have been standard pragmatic phenomena. Rather, it has developed its own theoretical approaches to pragmatic themes, in particular under the label of 'discourse processes'. To what extent, and on what specific points research on discourse processes might converge or conflict with specific pragmatic claims remains largely to be seen (for a comparison between the psycholinguistic notion of discourse coherence and the pragmatic notion of relevance see in particular Blakemore, 2001, 2002; Blass, 1990; Rouchota, 1998; Unger, 2000; Wilson, 1998; Wilson and Matsui, 2000).

It is reasonable to expect that two fields of research dealing in part with the same material at the same level of abstraction would gain by joining forces, or at least by interacting actively. For pragmatics the gain would be twofold. First, experimental evidence can be used, together with intuition and recordings, to confirm or disconfirm hypotheses. The high reliability and strong evidential value of experimental data puts a premium on this sort of data even though it is hard to collect and is generally more artificial than observational data (and therefore raises specific problems of interpretation). The three kinds of evidence – intuitions, observations and experiments – are each in their own way relevant to suggesting and testing pragmatic hypotheses, and they should be used singly or jointly whenever useful. Second, aiming at experimental testability puts valuable pressure on theorizing. Too often, armchair theories owe much of their appeal to their vagueness, which allows one to reinterpret them indefinitely so as to fit one's understanding of the data, but which also makes them untestable. Developing an experimental side to pragmatics involves requiring a higher degree of theoretical explicitness. Moreover, experimentally testing theories often leads one to revise and refine them in the light of new and precise evidence, and gives theoretical work an added momentum.

For experimental psycholinguistics, the gain from a greater involvement with pragmatics would be in taking advantage of the competencies, concepts and theories developed in this field, in order to better describe and explain a range of phenomena that are clearly of a psycholinguistic nature, and to develop new experimental paradigms. The experimental approach often results in unbalanced coverage of the domain of study. Topics for which an experimental paradigm has been developed get studied in great detail, whereas other topics of comparable empirical importance may remain largely untouched for lack of an *ad hoc* experimental tradition. There is, for instance, a wealth of psycholinguistic research on metaphor but very little on implicatures, when, from a pragmatic point of view, the two phenomena are of comparable importance. Typically, pragmatic theories have been more comprehensive and evenly detailed than psycholinguistic ones.

The small amount of existing Experimental Pragmatic work from psycholinguists and pragmaticists already shows what this collection is meant to demonstrate, namely that there is much to gain, both for pragmatics and for psycholinguistics, from systematically putting pragmatic hypotheses to the experimental test. Here we give a brief account of two examples: indirect speech acts and bridging.

An early illustration of the relevance of experiments to theoretical issues was provided by experimental work done in the 1970s on a topic of hot theoretical debate at the time: indirect speech acts (Searle, 1975). When a speaker says 'Could you stop fidgeting?', is the speech act a question or is it a request? The problem with categorizing this as a question is that, in ordinary circumstances, the proper response for the hearer is not to provide a verbal answer such as: 'yes, I could' or 'no, I couldn't' (as would be appropriate in response to a genuine question), but to actually stop fidgeting. The problem with categorizing it as a request is that the mood of the sentence is interrogative and not imperative. Sentential moods, it is generally assumed, indicate the kind of speech acts an utterance can be used to perform: declaratives serve to make assertions, interrogatives to ask questions, imperatives to make requests, and so forth. Indirect speech acts are called 'indirect' precisely because they don't seem to conform to the indication given by their mood: a declarative utterance may indirectly express a request (e.g., 'you could stop fidgeting') or a question (e.g., 'I would like to know where you have been'), an interrogative utterance may indirectly express a request (as in our example, 'could you stop fidgeting?') or an assertion (e.g., 'Who could remain indifferent in front of such injustice?'); an interrogative can also serve to ask an indirect wh-question, different from the yes-no question it would express directly (e.g., 'could you tell me the time?'), and so on. Indirect speech acts thus seem to threaten a basic assumption of much linguistic thinking. A possible way to go is to treat indirect speech acts as non-literal uses of language, comparable to metaphor and, like metaphor, explainable in pragmatic terms. Another way is to take indirect requests to be conventional or idiomatic. But are these descriptions really adequate? This is where experimental work comes in.

If an indirect speech act is like an idiom with a conventional meaning, then understanding it should not involve more processing than understanding a direct speech act. Reaction time studies, such as those by Clark and Lucy (1975), suggested that, in fact, indirect requests do take longer to comprehend than direct ones and therefore are not conventional (but see Gibbs, 1979, for a more complex picture). If indirect requests are like metaphor, then their literal interpretation should not be retained at all. After all, when a sentence is used metaphorically (e.g., 'John is a bulldozer'), the literal sense is not at all part of the speaker's meaning. Clark (1979) telephoned store owners with indirect questions such as 'Can you tell me what time you close?' and most answered with responses like 'Yes, we close at six'. *Yes*, in such an answer, seems to be an answer to the direct question ('Can you tell me?' 'yes, I can tell you') whereas the rest of the sentence ('we close at six') is an answer to the indirect question ('At what time do you close?'), suggesting that both the direct and the indirect questions were considered parts of the speaker's overall meaning (for further evidence and different analyses, see Munro, 1979; Gibbs, 1981). Not only did these experimental studies provide relevant evidence in the theoretical debate, they also suggested new and more specific hypotheses about indirect speech acts. However, this early dialogue between experimentalists and pragmaticists working on indirect speech acts largely ends here. The two groups failed to take as much advantage of each other's work as they could have.

Another example of interactions between psycholinguistics and pragmatics is provided by the case of *bridging*. A bridging inference, or bridging *implicature* (Clark, 1977), links a referring expression to an intended referent that is neither present in the environment nor mentioned in the ongoing discourse but that is nevertheless inferentially identifiable. For example, in the two sentences in (2) below:

(2) John walked into a room. The *window* was open.

The expression *the window* is a referring expression implicitly linked to the *room* mentioned in the preceding sentence. In order to establish the link, a bridging implicature such as *the room had a window* has to be retrieved. Bridging was the basis for one of the first innovative accounts of discourse from Clark and colleagues – the *Given–New contract* – which has inspired much valuable experimental work in psycholinguistics. This research has contributed to the development of innovative paradigms (e.g., using reading times and semantic probes), for the creation of typologies in texts (Sanders, Spooren and Noordman, 1992), and has fed theoretical debates (e.g., between the Constructionist vs Minimalist accounts of inference generation) in the psychological literature (Graesser, Singer and Trabasso, 1994; McKoon and Ratcliff, 1992).

Although Clark explicitly drew inspiration from Grice, and although bridging is obviously an important pragmatic topic, the exchanges between the pragmatic and psycholinguistic communities on the theme of bridging remained limited. A recent exception is provided by the work of Tomoko Matsui (Matsui, 2000), a pragmaticist who has become involved in experimental research. She makes a distinction between cases of bridging proper, like (2), where 'contextual assumptions [are] needed to *introduce* an intended referent which has not itself been explicitly mentioned' and cases where the intended referent is mentioned under a different description in a previous utterance, as in (3) and (4) below (both of which are bona fide bridging inferences according to most accounts):

(3) I met a man yesterday. The *nasty fellow* stole all my money.
(4) Peter took a cello from the case. The *instrument* was originally played by his grandfather.

Her definition allows for cases of bridging that are not normally considered by current theories, where the bridge is not to previous text but to salient background assumptions as in (5):

(5) [Peter and Mary are off to visit a flat]. *Mary*: I hope the bathroom is not too small.

Is Matsui right in assuming that the cognitive tasks involved in fixing reference in (2) and (5) have more in common than either does with the task involved in (3) and (4)? The issue is of obvious psycholinguistic relevance. Contrary to accounts that rely, for constructing the bridge, on the explicit linguistic information in a prior utterance (Clark, 1977) or on a situational model (Garrod and Sanford, 1982; Walker and Yekovich, 1987), Matsui predicts that 'in interpreting an utterance, the individual automatically aims at optimal relevance [which means] he will try to pick out, from whatever source, a context in which to process the utterance so that it gives at least adequate cognitive effects for no unjustifiable processing effort'. This prediction is supported by a series of investigations based on utterances (presented alone or in the context of a story), with two plausible intended referents. Consider (6):

(6) I prefer the restaurant on the corner to the student canteen. *The cappuccino* is less expensive.

Is it the cappuccino at the restaurant or at the canteen that is said to be less expensive? Eighty per cent of Matsui's participants indicate that one can generally get less expensive cappuccinos at student canteens. If such common knowledge were the determining factor, then participants should construct a bridge from cappuccino to student canteen. Similarly, if the determining factor were the shortness of the gap between the referring expression and a previous expression to which it could plausibly be bridged, then the canteen, the mention of which is the closest to that of cappucino, should provide the preferred bridge. Yet 100 per cent of participants respond *restaurant* when asked 'Where is the cappuccino less expensive?'. Unlike theories developed in psycholinguistics, Relevance Theory provides an explanation of these data. The sentence *The cappuccino is less expensive* achieves optimal relevance as an explanation of the speaker's preference for the restaurant over the canteen when the bridge is to the restaurant, and is of no obvious relevance if the bridge is to the canteen. This is why all participants understand the phrase 'the capuccino' to refer to the cappucino at the restaurant. Matsui's work provides striking examples of the mutual relevance of pragmatics and psycholinguistics (for further discussion, see Wilson and Matsui, 2000).

3 Pragmatics and the experimental psychology of reasoning

Fruitful interactions between pragmatics and experimental psychology are not limited to psycholinguistics. All experiments involving verbal communication with participants are affected by the way in which they understand what they are told. When an experimenter's expectations do not measure up with a participant's comprehension, this can have major consequences.

In the psychology of reasoning in particular, experiments typically involve not one but two levels of verbal communication from experimenter to participants: verbal instructions on how to perform the task and the task itself consist partly or wholly of verbal material. Experimenters (who are usually focused on rates of correct responses) often take it for granted that instructions and the verbal material are understood as intended, but this need not be the case. What happens if the instructions or text for a reasoning problem are not understood as intended? The performance of participants may fail to meet the experimenters' criteria of success because they have, in fact, performed a task different from the one intended. Their pragmatic comprehension processes may be functioning quite properly, and so may their reasoning processes, and yet their responses may seem mistaken to the experimenter. This is enough to give some plausibility to the claim that participants' apparent irrationality in reasoning tasks is linked to miscontruals or reconstruals of the task rather than to their reasoning incompetencies (Henle, 1962). Even apparently successful performance of a task may in some cases be due to an unforeseen interpretation that happens to yield the experimenter's normative response, not for logical, but for pragmatic reasons.

The role of pragmatic processes in reasoning experiments is generally acknowledged, but only in a vague sort of way. There has been no attempt to introduce systematic pragmatic considerations into experimental methodology. Nevertheless, there have been more and more studies investigating the role of pragmatic factors in standard paradigms in the psychology of reasoning, following the pioneering work of researchers such as Politzer (1986) and Mosconi (1990). A number of apparent irrationalities in people's performance have been shown to be explainable, at least in part, as resulting from these pragmatic factors. It is not an exaggeration to say that nearly every task in the reasoning literature has inspired a pragmatic analysis. Several illustrations can be found in the chapters by Politzer and by Van der Henst and Sperber. The relevance of this work to the study of reasoning is self-evident. Its relevance to the experimental study of pragmatics is also clear because, in each case, researchers have had to identify precise pragmatic factors at work and devise ways of testing their role experimentally.

4 The chapters

The book is divided into 3 parts devoted, respectively, to pioneering approaches (Chapters 2–6), to current issues in experimental pragmatics (Chapters 7–11), and to the special case of scalar implicatures (Chapters 12–15). Although this volume aims to develop and give a name to a budding field of inquiry, the chapters in Part I are devoted to researchers who have been working in this area all along.

4.1 Pioneering approaches

Chapter 2, by **Herb Clark** and **Adrian Bangerter**, provides both a historical and a contemporary perspective on reference, which is the ubiquitous activity involved in picking out an object for an addressee. Consider the utterance *Put the small coffee cup over there*. One would have to pick out the cup (presumably from among other candidate objects) and know where *over there* is (presumably from a gesture). Their chapter describes how reference was initially viewed as autonomous and addressee-blind before it came to be viewed as an activity that requires the coordination of both speaker and addressee. Among the features of referring highlighted are: (a) the multiple methods of directing an addressee's attention to individual objects; and (b) speaker-addressee pacts to arrive at a reference (i.e., to agree to certain provisional names). The coordination involved in referring is extensive, Clark and Bangerter argue, leading them to conclude that it is far from being an autonomous act. In fact, it requires more than mere coordination, it is an act that requires the full participation of both initiator and addressee. The chapter highlights how armchair reflection, field observations and careful experimentation have combined to lead to a more profound understanding of this fundamental communicative act. The chapter also provides an opportunity to appreciate Clark's well-known contributions to discourse analysis (the Given–New contract, common ground) in the context of pragmatic theory-making.

For more than 20 years, **Raymond W. Gibbs, Jr.**, has embodied the aim of this book, by specifically testing linguistic-pragmatic theories using experimental psychological methods. In Chapter 3, Gibbs describes how his experiments have constrained theories with respect to four areas that are at the heart of linguistic-pragmatics: making and understanding promises, understanding definite descriptions, making and interpreting indirect speech acts, and the distinction between what is said and what is meant. In each case, he has – like most accomplished experimentalists – come up with one or more clever designs that, in the end, either elucidate a given theory (e.g., the short-circuited nature of indirect requests) or force one to rethink a theory's claims (e.g., Searle's speech act theory with respect to promises). The aim of Gibbs's chapter is to convince experimentalists of the value of linguistic-pragmatic theories and to convince linguists of the value of experimentation.

Metaphor is a classic pragmatic form whose understanding has been greatly advanced by psycholinguistic investigations. As **Sam Glucksberg** shows in Chapter 4, metaphor comprehension in psycholinguistics was initially viewed through a Gricean lens, in which the literal interpretation of a metaphor is given priority. According to Grice (or Searle), a metaphor renders an utterance 'defective' and prompts one to look for another meaning. In his chapter, Glucksberg argues that this standard pragmatic model persisted in the literature because its literal-first hypothesis resonates with an approach that assumes that both semantics and syntax are primary while pragmatics is secondary, an assumption that is common in psycholinguistic circles. Through his and his colleagues' pioneering work on metaphor, Glucksberg demonstrates how metaphorical interpretations of sentences such as *Some jobs are jails* are carried out as automatically as other linguistic processes. He extends his analysis to other related phenomena (e.g., showing how novel features emerge in conceptual combinations like *peeled apples*) in order to show just how automatic pragmatic processes are in comprehension tasks. He concludes by suggesting that experimentation is needed to determine the correct division of labour between linguistic decoding and pragmatic inferencing, a central issue in current pragmatic theory. The pragmatic process, as shown by Sam Glucksberg, does not merit its 'stepchild' status; pragmatics is so automatic that it is arguably a module.

In Chapter 5, **Guy Politzer** – who was often a lone voice underlining the importance of linguistic-pragmatics to the field of reasoning – provides a pragmatic analysis of both classic and modern reasoning tasks along with experimental results that stress the importance of the way individual premises, conclusions and task information in general are interpreted. For a notable example, consider Piaget's famous class-inclusion problem, in which children are shown a picture of five daisies and three tulips and then asked, 'Are there more daisies or more flowers?' After presenting a 'microanalysis' of the way the task's demands are interpreted, Politzer shows that young children (5-year-olds) fail to answer correctly (to say *flowers*) because they interpret 'flowers' to mean *flowers-that-are-not-daisies*. He also shows how a short series of disambiguating questions prompts even the youngest children to demonstrate their class-inclusion skills. Such microanalyses can be applied equally to many of Kahneman and Tversky's tasks (e.g., the Linda problem and the Engineer-Lawyer problem), Wason's tasks (the 2-4-6 problem and the Selection Task), as well as to individual terms like conditionals and quantifiers. The implications for this approach are clear: one cannot do reasoning work without linguistic-pragmatics.

Chapter 6 by **Tony Sanford** and **Linda Moxey** reviews their previous work on the psychological processing of quantifier understanding and demonstrates how experimental approaches can inform linguistic-pragmatics. They begin by pointing out that not all quantifiers are alike. A large set of 'non-standard' quantifiers, such as *few, many* and *most*, convey much more than a rough notion of quantity or proportions; they have communicative functions as well. For example, polarity plays a determinative role in quantifier interpretation. A negative quantifier like *few* and a positive quantifier like *a few* have quite different effects on the interpretation of sentences. Compare *few*... versus *a few of the MPs attended the meeting*. *Few* is more likely than *A few* to place the focus on the complementary set, those MPs who did not show up. Their findings show that the interpretation of quantifiers goes well beyond the semantics of these terms.

4.2 Current issues in experimental pragmatics

The chapters in this section extend both the range of topics one can investigate in Experimental Pragmatics and the techniques one can use. The chapters here cover *inter alia* disambiguation, metaphor and joke comprehension, promise understanding, the import of saying *even-if*, and the telling of time. All these topics are addressed using various experimental paradigms from neuropsychology, developmental psychology, reasoning, psycholinguistics and anthropology.

In Chapter 7, **Jean-Baptiste Van der Henst** and **Dan Sperber** review experiments that test central tenets of Relevance Theory and in particular the cognitive principle of relevance ('human cognition is geared to the maximization of relevance'), and the communicative principle ('every utterance conveys a presumption of its own relevance'). Some of these experiments draw on two standard paradigms in the psychology of reasoning: relational reasoning and the Wason Selection Task. Others investigate the behaviour of people asked the time by a stranger in public places. All involve manipulating separately the two factors of relevance, effect and effort. These experiments illustrate how a pragmatic theory that is precise enough to have testable consequences can put previous experimental research in a novel perspective and can suggest new experimental paradigms.

Orna Peleg, **Rachel Giora** and **Ofer Fein** give an account of the role of the context in accessing the appropriate meaning of ambiguous terms in sentence comprehension in Chapter 8. They argue against: (a) a modular view which assumes that lexical access to all meanings of a word are automatic and encapsulated only to be refined by an independent non-modular system; and against (b) a direct access view which relies largely on just the context to arrive at a word's intended meaning. Rather, they propose the *graded salience hypothesis*, which assumes that: (a) more salient meanings are accessed faster from the start; and that (b) context also affects comprehension on-line. Their chapter presents four experiments whose results lend strong support to their claims.

In Chapter 9, **Seana Coulson** provides a review of the way Evoked Response Potentials' (ERP) methods can be applied to language comprehension, with a focus on what this technique has to offer pragmatics. The chapter is instructive in that it describes ERP's various dependent variables (P300, N400, P600 etc.) and the aspects of comprehension with which these measures are associated. Coulson cites studies of pragmatic import – for example, on joke comprehension and metaphor integration – including many that come from Coulson herself. She works from a model that predicts that processing difficulty is related to the extent to which comprehension requires the participant to align and integrate conceptual structure across domains. She goes on to suggest ways in which ERP experiments could be exploited to investigate other linguistic-pragmatic issues, such as prosody and the distinction between explicatures and implicatures. Overall, her chapter shows very clearly how imaging can be exploited and indicates what one should expect from this technique in the future.

In Chapter 10, **Josie Bernicot** and **Virginie Laval** focus on children between the ages of 3 and 10 and their developing understanding of promises, based on the theoretical framework of Speech Act Theory (Austin 1962; Searle, 1969, 1979; Searle and Vanderveken, 1985). The authors summarize a programme of research that has been investigating promise comprehension among children from the point of view that language is a communication system and that language competence is the acquisition and use of that system. What counts as a promise? Here the authors present two experiments investigating the extent to which interlocutors' intentions (listener's wishes about the accomplishment of an action) and textual characteristics of utterances (verb tense) play a role in understanding that a promise was made.

In Chapter 11, **Simon J. Handley** and **Aidan Feeney** develop a psychological account of the way in which people reason with *even-if*, working in a mental models' framework (Johnson-Laird, 1983). According to the mental model approach, many errors of reasoning arise because people represent only one or a few of all the models of a given set of premises and leave the other models implicit. They then draw their conclusions on the sole basis of the explicitly represented models. Handley and Feeney compare two possible ways in which this partial representation of problems might arise. In one, all models are represented before being pared down by extra-logical, namely pragmatic, factors; in the other, which the authors advocate, initial representations are limited to one model while pragmatic considerations add new models. They present two experiments based on inference making from *even-if* premises that lend support to their account. They discuss the implications of their work for experimental pragmatics in general.

4.3 The case of scalar implicatures

The chapters in the third section of the book focus on one pragmatic phenomenon, *scalar implicature*, which is at the heart of ongoing debates in pragmatic theory. As described earlier, there are two main accounts of these inferences. One assumes that such implicatures are automatically associated with the use of a weak term (as exemplified by Levinson, 2000) and the other assumes that the implicature is drawn out effortfully (as exemplified by Relevance Theory). In these chapters, four authors (or groups of authors) present experimental findings that lend support either to Relevance Theory or to some form of the default view.

In Chapter 12, **Anne Bezuidenhout** and **Robin Morris** first describe how they operationalized the two theoretical accounts into testable pragmatic-processing models. This is less obvious than it might seem because it is hard to do justice to the rich and detailed accounts that have been offered by

18 Experimental Pragmatics

these rival theories on the topic of scalar implicatures. They then report on two eye-movement experiments that test predictions generated from the models as participants read a series of sentence-pairs such as *Some books had colour pictures. In fact all of them did, which is why the teachers liked them*. One can determine whether *Some* in the first sentence readily prompts *Not all* by investigating potential slowdowns and look-backs when processing the second sentence. They argue that the weight of the evidence favours the Underspecification Account (which is the one inspired by Relevance Theory); however, they argue that their Default Model (the one inspired by a neo-Gricean account) could be modified to accommodate their results.

In Chapter 13, **Gennaro Chierchia, Maria Theresa Guasti, Andrea Gualmini, Luisa Meroni, Stephen Crain** and **Francesca Foppolo** present a novel account of implicatures based on the Semantic Core Model, which challenges a way of interpreting Grice's proposal that has become dominant in the field. According to the dominant view, one first retrieves the semantics of a whole root sentence and then processes the implicatures associated with it (in a strictly modular way). The Semantic Core Model proposes, instead, that semantic and pragmatic processing take place in tandem. Implicatures are factored in recursively, in parallel with truth conditions. They go on to present experimental evidence from adults and children that support this new model. One of the novel findings from this work demonstrates how particular grammatical contexts predict the non-existence of scalar implicatures.

In Chapter 14, **Ira A. Noveck** reviews the two rival accounts and the processing predictions they engender, before summarizing his laboratory's findings from experiments investigating those logical terms (i.e., *might, some, or* and *and*) that could be interpreted either minimally (i.e., with just their linguistically encoded meanings) or as pragmatically enriched. His developmental studies show how children are less likely than adults to pursue pragmatic inferences, leading to a robust experimental effect in which children actually appear more logical than adults. Follow-ups show how task-demands, and not just age, can affect the production of pragmatic inference making, pointing to the important role of context in these paradigms. The adult studies, which include an ERP investigation, primarily explore the time-course of scalar inferences. Whereas participants' pragmatic treatments of underinformative statements (e.g., the time taken to respond *False* to *Some cows are mammals*) are very time consuming, *True* responses are not. Furthermore, time pressure encourages *True* responses. Noveck presents his findings as support for Relevance Theory.

In Chapter 15, **Anne Reboul** presents a novel task, which she calls Koenig's puzzle, as promising ground for testing between the two rival theories. Imagine that after being handed a glass of wine, a speaker says *Better red wine than no white wine*. The puzzle consists in determining the speaker's wine preference and inferring what she was actually given. While referring to the two sides of the debate as localists and globalists (for the Default and

Relevance accounts, respectively), Reboul describes Koenig's puzzle in detail and proposes a solution to it. Reboul then explains why such sentences may be used to test between the two accounts. Finally, her paper reports two experiments whose results show how implicatures are actually involved in the puzzle. Her results are presented as support for global over local theories for this specific pragmatic phenomenon.

5 How to approach the book

The chapters are representative of what we are calling Experimental Pragmatics. Each summarizes previous experimental work or presents original experiments that address topics central to pragmatic theory – metaphor, quantifier interpretation, scalar inference, disambiguation, reference and promise understanding, to name a few. Many of the chapters share common themes, especially the last four, but each can be read and appreciated separately. Our intention has been to illustrate how Experimental Pragmatics may contribute to linguistics and psychology, and to the cognitive sciences in general.

References

Anscombre, J.-C., and Ducrot, O. (1995). *L'Argumentation dans la langue*. Paris: Madarga.
Atlas, J. (forthcoming) *Logic, Meaning, and Conversation: Semantical Underdeterminacy, Implicature, and Their Interface*. Oxford: Oxford University Press.
Austin, J. (1962). *How to do Things with Words*. New York: Oxford University Press.
Bach, K. (1987). *Thought and Reference*. Oxford: Clarendon Press.
Bach, K. (1994). Conversational implicature. *Mind and Language* 9: 124–62.
Bach, K., and Harnish, R. M. (1979). R. *Linguistic Communication and Speech Acts*. Cambridge, MA: MIT Press.
Bezuidenhout, A. (1997). Pragmatics, semantic underdetermination and the referential – attributive distinction. *Mind* 106: 375–409.
Blakemore, D. (1987). *Semantic Constraints on Relevance*. Oxford: Blackwell.
Blakemore, D. (2001). Discourse and relevance theory. In D. Schiffrin, D. Tannen and H. Hamilton (eds), *Handbook of Discourse Analysis*: 100–18. Oxford: Blackwell.
Blakemore, D. (2002). *Linguistic Meaning and Relevance: The Semantics and Pragmatics of Discourse Markers*. Cambridge: Cambridge University Press.
Blass, R. (1990). *Relevance Relations in Discourse: A Study with Special Reference to Sissala*. Cambridge: Cambridge University Press.
Blutner, R., and Zeevat, H. (2003). *Optimality Theory and Pragmatics*. Basingstoke: Palgrave Macmillan.
Carston, R., and Uchida, S. (eds). (1997) *Relevance Theory: Applications and Implications*. John Benjamins, Amsterdam.
Carston, R. (2002). *Thoughts and Utterances: The Pragmatics of Explicit Communication*. Oxford: Blackwell.
Clark, H. H. (1977). Bridging. In P. N. Johnson-Laird and P. C. Wason (eds), *Thinking*: 411–20. Cambridge: Cambridge University Press.
Clark, H. H. (1979). Responding to indirect speech acts. *Cognitive Psychology* 11: 430–77.

Clark, H. and Lucy, P. (1975). Inferring what was meant from what was said. *Journal of Verbal Learning and Verbal Behavior* 14: 56–72.
Clark, H., and Schunk, D. (1980). Polite responses to polite requests. *Cognition* 8: 111–43.
Dascal, M. (1981). Contextualism. In H. Parret et al. (dir.), *Possibilities and Limitations of Pragmatics*. Amsterdam: Benjamins.
Davis, S. (ed.) (1991). *Pragmatics: A Reader*. Oxford: Oxford University Press.
Ducrot, O. (1984). *Le Dire et le Dit*. Paris: Minuit.
Fauconnier, G. (1975). Pragmatic scales and logical structure. *Linguistic Inquiry* 6: 353–75.
Fauconnier, G. (1985). *Mental Spaces: Aspects of Meaning Construction in Natural Language*, Cambridge, MA: MIT Press/Bradford Books.
Garrod, S. C., and Sanford, A. J. (1982) Bridging inferences in the extended domain of reference. In A. Baddeley and J. Long (eds), *Attention and Performance IX* (vols 331–46). Hillsdale, NJ: Earlbaum.
Gazdar, G. (1979). *Pragmatics: Implicature, Presupposition, and Logical Form*. New York: Academic Press.
Gibbs, R. (1979). Contextual effects in understanding indirect requests. *Discourse Processes* 2: 1–10.
Gibbs, R. (1981). Your wish is my command: convention and context in interpreting indirect requests. *Journal of Verbal Learning and Verbal Behavior* 20: 431–44.
Graesser, A. C., Singer, M., and Trabasso, T. (1994). Constructing inferences during narrative text comprehension. *Psychological Review* 101(3): 371–95. URLJ: http://www apa org/journals/rev html.
Grice, H. P. (1989). *Studies in the Way of Words*. Cambridge, MA: Harvard University Press.
Gutt, E.-A. (1991). *Translation and Relevance: Cognition and Context*. Oxford: Blackwell.
Harnish, R. M. (1976). Logical form and implicature. In T. Bever, J. Katz and D. T. Langendoen (eds). *An Integrated Theory of Linguistic Ability*: 313–91. New York: Crowell. Reprinted in S. Davis (1991): 316–64.
Harnish, R. M. (1994). Mood, meaning and speech acts. In S. Tsohatzidis (ed.), *Foundations of Speech-Act Theory: Philosophical and Linguistic Perspectives*: 407–59. London: Routledge.
Henle, M. (1962). On the relation between logic and thinking. *Psychological Review*.
Horn, L. R. (1973). Greek Grice: A brief survey of proto-conversational rules in the history of logic. Paper presented at the *Proceedings of the Ninth Regional Meeting of the Chicago Linguistic Society*, Chicago.
Horn, L. R. (1984). Towards a new taxonomy for pragmatic inference: Q- and R-based implicature. In D. Schiffrin (ed.), *Meaning, Form, and Use in Context*: 11–42. Washington DC: Georgetown University Press.
Horn, L. R. (1989). *A Natural History of Negation*. Chicago: University of Chicago Press.
Horn, L. R. (1992). The said and the unsaid. *SALT II: Proceedings of the Second Conference on Semantics and Linguistic Theory*: 163–202. Ohio State University Linguistics Department, Columbus, Ohio.
Ifantidou, E. (2001). *Evidentials and Relevance*. Amsterdam: John Benjamins.
Johnson-Laird, P. N. (1983) *Mental Models*. Cambridge: Cambridge University Press.
Kasher, A. (1976) Conversational maxims and rationality. In A. Kasher (ed.), *Language in Focus: Foundations, Methods and Systems*: 197–211. Dordrecht: Reidel. Reprinted in A. Kasher (ed.) (1998), vol. IV.
Kasher, A. (1984). Pragmatics and the modularity of mind. *Journal of Pragmatics* 8: 539–57. Revised version reprinted in S. Davis (ed.) (1991): 567–82.
Kasher, A. (ed.) (1998). *Pragmatics: Critical Concepts*, vols I–V. Routledge, London.

Katz, J. J. (1977). *Propositional Structure and Illocutionary Force*. New York: Crowell.
Kempson, R. (1996). Semantics, pragmatics and deduction. In S. Lappin (ed.), *Handbook of Contemporary Semantic Theory*: 561–98. Oxford: Blackwell.
Levinson, S. (1983). *Pragmatics*. Cambridge: Cambridge University Press.
Levinson, S. (2000). *Presumptive Meanings: The Theory of Generalized Conversational Implicature*. Cambridge, MA: MIT Press.
Lewis, D. (1979). Scorekeeping in a Language Game, in his *Philosophical Papers*, vol. 1: 233–49. New York: Oxford University Press (1983).
Matsui, T. (2000). *Bridging and Relevance*. Amsterdam: John Benjamins.
McKoon, G., and Ratcliff, R. (1992). Inference during reading. *Psychological Review* 99(3): 440–66.
Moeschler, J. (1989). *Modélisation du dialogue*. Paris: Hermès.
Moeschler, J., and Reboul, A. (1994). *Dictionnaire Encyclopédique de Pragmatique*. Paris: Seuil.
Mosconi, G. (1990). *Discorso e pensiero*. Bologna: Il Mulino.
Munro, A. (1979). Indirect speech acts are not strictly conventional, *Linguistic Inquiry* 10, 353–6.
Neale, S. (1990). *Descriptions*. Cambridge, MA: MIT Press.
Neale, S. (1992). Paul Grice and the philosophy of language. *Linguistics and Philosophy* 15: 509–59.
Neale, S. (forthcoming). *Linguistic Pragmatism*. Oxford: Oxford University Press.
Noh, E.-J. (2000). *Metarepresentation: A Relevance-Theory Approach*. Amsterdam: John Benjamins.
Papafragou, A. (2000). *Modality: Issues in the Semantics–Pragmatics Interface*. Amsterdam: Elsevier Science.
Pilkington, A. (2000). *Poetic Effects: A Relevance Theory Perspective*. Amsterdam: John Benjamins.
Politzer, G. (1986). Laws of language use and formal logic. *Journal of Psycholinguistic Research* 15(1): 47–92.
Reboul, A. (1992). *Rhétorique et stylistique de la fiction*. Nancy: Presses Universitaires de Nancy.
Recanati, F. (1979). *La Transparence et l'Énonciation*. Paris: Seuil.
Recanati, F. (1988). *Meaning and Force*. Cambridge: Cambridge University Press.
Recanati, F. (1993). *Direct Reference: From Language to Thought*. Oxford: Basil Blackwell.
Recanati, F. (2000). *Oratio Obliqua, Oratio Recta*. Cambridge, MA: MIT Press/Bradford Books.
Rouchota, V., and Jucker, A. (eds) (1998). *Current Issues in Relevance Theory*. Amsterdam: John Benjamins.
Rouchota, V. (1998). Connectives, coherence and relevance. In V. Rouchota and A. Jucker (eds), *Current Issues in Relevance Theory*: 11–57. Amsterdam: John Benjamins.
Sanders, T. J., Spooren, W. P., and Noordman, L. G. (1992). Toward a taxonomy of coherence relations. *Discourse Processes* 15(1): 1–35.
Searle, J. (1969). *Speech Acts*. Cambridge: Cambridge University Press.
Searle, J. (1975). Indirect Speech Acts. In *Syntax and Semantics: vol. 3 Speech Acts*: 59–82. Ed. P. Cole and J. L. Morgan. New York: Academic Press.
Searle, J. (1979). *Expression and Meaning*. Cambridge: Cambridge University Press.
Searle, J., and Vanderveken, D. (1985). *Foundations of Illocutionary Logic*. Cambridge: Cambridge University Press.
Sperber, D., and Wilson, D. (1986/1995). *Relevance: Communication and Cognition*. Oxford: Basil Blackwell.

Stalnaker, R. (1999). *Context and Content*. Oxford: Clarendon Press.
Sweetser, E. (1990). *From Etymology to Pragmatics: Metaphorical and Cultural Aspects of Semantic Structure*. Cambridge: Cambridge University Press.
Travis, C. (1975). *Saying and Understanding*. Oxford: Basil Blackwell.
Unger, C. (2000). On the Cognitive Role of 'Genre': A Relevance-Theoretic Perspective. University of London Ph.D. dissertation.
Van der Auwera, J. (1981). *What Do We Talk About When We Talk? Speculative Grammar and the Semantics and Pragmatics of Focus*. Amsterdam: John Benjamins.
Van der Auwera, J. (1985). *Language and Logic. A Speculative and Condition-Theoretic Study*. Amsterdam: Benjamins.
Van der Auwera, J. (1997). Conditional perfection. In A. Athanasiadou and R. Dirven (eds.), *On Conditionals Again*: 169–190 Amsterdam: Benjamins.
Vanderveken, D. (1990–1). *Meaning and Speech Acts* (2 vols). Cambridge: Cambridge University Press.
Walker, C. H., and Yekovich, F. R. (1987). Activation and use of script-based antecedents in anaphoric reference. *Journal of Memory and Language*, 26(6): 673–91.
Wilson, D., and Matsui, T. (2000). Recent approaches to bridging: Truth, coherence and relevance. In J. de Bustos Tovar, P. Charaudeau, J. Alconchel, S. Iglesias Recuero and C. Lopez Alonso (eds), *Lengua, Discurso, Texto*, vol. 1: 103–32. Visor Libros: Madrid. (Also published in *UCL Working Papers in Linguistics* 10: 173–200 (1998).)
Wilson, D. (1998). Discourse, coherence and relevance: A reply to Rachel Giora. *Journal of Pragmatics* 29: 57–74.
Yus, F. (1997) *Cooperación y relevancia. Dos aproximaciones pragmáticas a la interpretación*. Alicante: Universidad de Alicante, Servicio de Publicaciones.

Part I
Pioneering Approaches

2
Changing Ideas about Reference

Herbert H. Clark and Adrian Bangerter

1 Introduction

How do people refer to things? At first, the answer seems simple: they produce the right expression in the right situation. According to John Searle (1969), for example, to refer to a dog, speakers must produce a referring expression (such as *the dog she had with her*) with the intention that it pick out or identify the dog for their addressees. But is the answer really this simple? Accounts of how people refer have changed again and again since about 1960, often dramatically. But how have they changed, and why? In this chapter, we offer a selective, largely personal history of these changes as they have played out in the experimental study of reference. Our goal is not a complete history – an impossible ambition – but a better understanding of what reference really is.

Language use isn't easy to study. It has been investigated largely by three methods – intuition, experiment and observation. With intuitions, you imagine examples of language used in this or that situation and ask yourself whether they are grammatical or ungrammatical, natural or unnatural, appropriate or inappropriate. This was Searle's method. With experiments, you invite people into the laboratory, induce them to produce, comprehend or judge samples of language, and measure their reactions. With observations, you note what people say or write as they go about their daily business. We will name these methods by their characteristic locations: *armchair*, *laboratory* and *field*.

Each of these methods has its plusses and minuses. Almost every analysis of language use begins in the armchair. There you imagine a wide range of utterances and situations and draw your conclusions. You are limited only by what you can imagine, but that turns out to be quite a limitation. It is impossible to imagine the hidden processes behind planning and word retrieval, and it is difficult mentally to simulate the opportunistic back and forth processes of social interaction. And armchair judgments are known to suffer from bias, unreliability, and narrowness (Schütze, 1996). The laboratory, in contrast, is especially useful for isolating and measuring hidden processes, as inferred from reaction times, eye movements or brain activation. But when you bring language into the laboratory, you are forced to strip it of its everyday features – often in unknowable ways. The field is the best place to see how ordinary people, unencumbered by theoretical preconceptions and laboratory wiring, actually use language, but it, too, has its dangers. There you are forced to choose what quarry to track, where to track it and what to record, and these lead to their own biases. And in the field it is hard to infer what causes what.

To return to the original question, how, then, do people refer to things? Since 1960, ideas have changed not merely about the *process* by which people refer, but about the very *conception* of what reference is. Both changes, we suggest, came about when scientists got out of their armchairs and went into the laboratory and the field. Language use, we argue, cannot be studied without all three methods. You cannot even begin without armchair observations. You cannot easily draw causal claims outside the laboratory. And yet you cannot really know what language use is, in all its richness, without venturing out into the field.

2 Referring as a cooperative process

Early on, definite reference was viewed as a rather simple act: the uttering of a referring expression adequate to pick out the intended referent uniquely. In an influential paper by David Olson in 1970, the idea was this: 'Language is merely the specification of an intended referent relative to a set of alternatives' (p. 272). So speakers consider a *set of alternatives* – an array of things that they could potentially refer to – and try to find an expression that will uniquely specify the intended referent in that array. To do this, they select a description that is just sufficient for the purpose. Suppose the set of alternatives is this (in an unknown arrangement):

And you want to refer to the figure on the left. You cannot use 'the figure' or 'the circle' or 'the small figure', because these don't specify the left-most figure uniquely. You have to use 'the small circle'. For Olson, reference is the relation between a referring expression and an element of an array.

Olson's account was built on a number of unstated assumptions. Among them were these:

- Referring is an *autonomous act*. It consists of planning and producing a referring expression, which speakers do on their own.
- Referring is a *one-step process*. It consists of the planning and uttering of a referring expression and nothing more.

- Referring is *addressee-blind*. It depends on the context – the set of alternatives in the situation – but doesn't otherwise depend on beliefs about the addressees.
- Referring is *ahistorical*. It doesn't take account of past relations between speakers and their addressees.
- The referent belongs to a *specifiable set of alternatives*.

Olson's account was hardly unique. A variety of these assumptions were common to most accounts of reference in 1970, including Searle's, and some are still taken for granted. The problem is that they are each suspect – indeed, we will argue, wrong. Some have been challenged by field observations, others in the laboratory, and still others in the armchair. We begin with challenges to Olson's assumptions about addressees.

2.1 Reference and cooperation

Are speakers blind to their addressees? Unbeknownst to Olson, Paul Grice had already argued in 1967 that speakers and addressees cooperate in their use of language (Grice, 1975, 1978; see also Sperber and Wilson, 1986). In his account, they adhere to a *cooperative principle* and therefore try to follow four maxims – to be informative, but not too informative; to have evidence for what they say; to be relevant; and to be orderly in how they speak. Indeed, according to Grice's maxim of quantity ('Make your contribution as informative as is required; do not make your contribution more informative than is required'), speakers should design definite descriptions much as Olson predicted. But in Grice's proposal speakers do that because they are trying to cooperate with addressees.

By Grice's maxim of manner, speakers should also be orderly in their use of language. One way to be orderly, according to Clark and Haviland (1974, 1977), is to follow conventional practices in the design of utterances. One of these practices specifies how to express *given* and *new information*. According to Halliday (1967) originally, speakers obligatorily mark utterances for what he called *information focus*. They distinguish between '(1) information the speaker considers given – information he believes the listener already knows and accepts as true; and (2) information the speaker considers new – information he believes the listener does not yet know' (Clark and Haviland, 1977).[1] When June tells David, 'It was George who bought Julia's car', she takes it as given that someone bought Julia's car, and she takes it as new that the buyer was George.

---

[1] The Given–New distinction here is closely related to 'old' and 'new' information as introduced by Chafe (1970), and to 'presupposition' and 'focus' as used by Akmajian (1973), Chomsky (1971) and Jackendoff (1972). See also Prince (1981).

Speakers and listeners should therefore adhere to the tacit contract expressed here as a directive to the speaker (Clark and Haviland, 1977, p. 9):

Given–New contract. Try to construct the given and the new information of each utterance in context: (a) so that the listener is able to compute from memory the unique antecedent that was intended for the given information; and (b) so that he will not already have the new information attached to that antecedent.

June, in saying 'It was George who bought Julia's car', must have assumed that David was able to compute, from memory, the unique antecedent to the given information, the event of someone buying Julia's car, and that he didn't already believe, or was unable to compute, that the buyer was George. The contract in turn enables listeners to rely on a *Given–New strategy*. By that strategy, David would divide June's utterance into 'X bought Julia's car' (given information) and 'X was George' (new information), search memory for a unique event of X buying Julia's car, and replace X with an index to George.

The Given–New strategy leads to a class of Gricean implicatures called *bridging inferences* (Clark and Haviland, 1974, 1977; Clark, 1975, 1977; Haviland and Clark, 1974; see Matsui, 2000). Suppose June describes a scene for David with one of these two sequences:

(1) I went for a walk this afternoon. The walk was invigorating.
(2) I went for a walk this afternoon. The park was beautiful.

When she says 'The walk was invigorating', David can find a unique antecedent for 'the walk', namely the walk mentioned in the first sentence. But when she says 'The park was beautiful', he has to do more. He must assume that she intended him to draw an implicature, or *bridging inference*, namely:

(2′) *Bridging inference*: June's walk took her through a park.

According to this proposal, a great many definite descriptions require bridging inferences.

Bridging inferences should take extra time or effort. In the original test of this prediction (Haviland and Clark, 1974), people were brought into a laboratory and given sequences of two sentences, to read one sentence at a time:

(3) Horace got some beer out of the car. The beer was warm.
(4) Horace got some picnic supplies out of the car. The beer was warm.

Inferring the referent for 'the beer' in 'The beer was warm' requires a bridging inference in (4), namely 'There was beer among the picnic supplies'. No such bridging inference is required in (3). Indeed, it took people about 200 msec longer to read and understand 'The beer was warm' in (4) than in (3). In a later study (Garrod and Sanford, 1982), people were given sequences like these:

(5) Keith took his car to London. The car kept overheating.
(6) Keith drove to London. The car kept overheating.

To identify the referent of 'the car' in 'The car kept overheating' requires a bridging inference in (6) ('What Keith drove was a car'), but not in (5); however, it took no longer to read and understand 'The car kept overheating' in (6) than in (5). Apparently, the bridging inference took little time to create, or the proposition was already part of the discourse model (see, e.g., Brewer and Treyens, 1981), so the bridging itself added no *measurable* time to the reading process. Replace *car* in (5) and (6) with *motorcycle*, and (6) would have taken more time. Over the years it has been found that the more readily inferable the bridging inference is in context, the less extra time is needed to infer it (see Matsui, 2000).

What is remarkable about the work on bridging is how little of it has come from the laboratory. The original model was built on field observations – from books, newspapers and magazines (Clark and Haviland, 1977), and from spontaneous speech (Halliday, 1967). And the predictions of that model – we have mentioned only some of them – were sharpened by armchair judgments about a vast range of examples analogous to (1) through (6) (e.g., Clark, 1977; Clark and Haviland, 1977), and that tradition has continued (Matsui, 2000). In short, the model was refined with evidence from the armchair, laboratory *and* field. No one method would have been enough.

2.2 Reference and common ground

Given information was originally characterized as 'information the speaker believes the listener already knows and accepts as true'. But this cannot be right, as bridging inferences themselves show. When June referred to 'the park' in (2), she did *not* need to believe David already knew she had walked in a park. All she needed to believe was that David could *infer* she had walked in a park – from what he already knew plus her definite reference. But can speakers work merely from what they believe their addressees know? Not in general. Speakers and addressees must rely on *shared* information – but shared in a particular way. The needed concept is what has come to be called *common ground*.[2]

Suppose June points at a cup of coffee on a nearby table and tells David, 'The cup of coffee is for you'. How can she refer to that as 'the cup of

---
[2] For early formulations of common ground, see Karttunen (1977), Karttunen and Peters (1975), but especially Stalnaker (1978).

coffee'? According to armchair examples in Clark and Marshall (1981), she needs to find a basis for the following belief: She believes that, with this reference, (*i*) she and David will come to believe, or assume, both that there is a cup of coffee nearby and that she and David believe, or assume, *i*. This, technically, is a *reflexive* belief – her and David's belief *i* *includes* that very belief *i*.[3] In general, for A to have information that a proposition *p* is common ground for A and B, A needs information that:

(i) A and B have information that *p* and that *i*.

'Having information' may include knowing, believing, assuming, seeing, hearing, even feeling, so common ground can range from mutual knowledge to mutual supposition, and its basis can range from abstract inferences to immediate perception.[4] The idea is that to make a definite reference is to presuppose that the referent can be readily and uniquely inferred from the current common ground of speaker and addressees. To make an indefinite reference is to presuppose that the referent *cannot* be so inferred.

Common ground, according to Clark and Marshall (see also Clark, 1996), divides into two broad types, *communal* and *personal* common ground:

Communal common ground. Suppose June and David are strangers, but in talking to each other at a party, they discover they are both on the faculty at the University of Illinois. This way they mutually establish that they are both members of at least these communities – United States, Illinois, Champaign, American academics, University of Illinois employees. The two of them can therefore take as *communal* common ground all the information they believe is common to members of these communities. On the basis of residence, June can refer, for example, to 'South Prospect', 'the psychology department', 'the stadium', 'the road to Chicago', confident that David will be able to identify the referents uniquely.

Personal common ground. Once June and David begin interacting, June may infer that certain items are *co-present* to the two of them. That includes items

---

[3] Reflexive beliefs, as formulated here, are analogous to the reflexive intentions on which Grice formulated the notion of speaker's meaning (Grice, 1957, 1968). See Harman (1977).
[4] As an aside, we note that common ground was characterized as reflexive by Lewis (1969), but as an *infinite regress* of beliefs by Schiffer (1972). The point of Clark and Marshall's paper was to argue for Lewis's characterization and against Schiffer's – at least, as a psychological model of common ground. All too many investigators have cavilled about common ground based on Schiffer's, not Lewis's characterization (see Clark, 1996, ch. 4).

they have talked about – say, her walk that afternoon. It also includes items that are perceptually co-present – say, the cup of coffee on the table. On the basis of these joint personal experiences, she can take the walk and cup to be part of their *personal* common ground and refer to them as *the walk* and *the cup of coffee*.

Do people rely on common ground in referring? Early examples from the armchair suggested yes (see Clark and Marshall, 1981; Hawkins,1978). But it soon became clear that the question is too simple. Common ground plays a role in every aspect of language use, from word meanings to politeness formulae. Here we consider its role in convention and coordination; later, we see how it forced a change in the conception of reference itself.

2.3 Reference and coordination

Even more fundamental than Grice's notion of cooperation is the notion of *coordination*: For June and David to cooperate on a task, they have to coordinate their individual actions (see Clark, 1996, ch. 2). In 1960, Thomas Schelling had already laid out several principles of coordination, and in 1969, David Lewis showed how these applied to language use. Lewis argued, in particular, that principles of coordination are needed to account for the linguistic notion of convention. Take the word *dog*. Speakers and their addressees have a recurrent problem of coordination: how are the speakers to refer to a domesticated canine such that their addressees will understand them as intended? They need what Lewis called a *coordination device*. The solution that evolved historically is the convention that the word-form /dôg/ mean 'domesticated canine.' What makes it a convention, according to Lewis, is that it is a regularity in behaviour, partly arbitrary, that is common ground in the community of English speakers as a coordination device for a recurrent coordination problem.

Conventions, of course, are essential to most forms of reference. When people refer with proper names – *Washington, Bonaparte, van Gogh* – they count on addressees picking out the right referents by means of conventions associated with the names. And when English speakers refer to walks, parks, beer, cars, stadiums and cups of coffee, they do so in part by means of the conventions associated with the nouns *walk, park, beer, car, stadium, cup and coffee*. Coordinating with words requires conventions, and deploying conventions requires common ground.

Coordination *à la* Lewis and Schelling, however, goes beyond convention, for there are other coordination devices as well. What these devices have in common is that people coordinate with them by means of *salience against current common ground*. It is this formulation of coordination that is needed for demonstrative reference.

The point was made clear in a series of experiments reported by Clark, Schreuder and Buttrick (1983). In one experiment, a student named Sam approached people on the Stanford University campus, handed them a photograph of a garden with four types of flowers in it, and asked:

(7) How would you describe the colour of this flower?

According to a model like Olson's, and even Grice's, the referent of *this flower* is underdetermined because the description *flower* is not informative enough to distinguish among the four flowers. If so, people should find the question impossible to answer. But by a Lewis-style model of coordination, listeners expect speakers to coordinate with them so that they can readily pick out the referent by considering the referring expression (here 'this flower') against their current common ground. By the coordination model, people should find Sam's question acceptable if they can pick out such a referent.

People in this experiment had been handed one of two pictures. In the first, one flower was a bit more salient perceptually than the other three; in the second, no flower was particularly salient. For the first picture, people immediately described the salient flower 55 per cent of the time, requesting clarification the rest of the time ('Which one?' or 'This one?'). But for the picture with no salient flower, people described a flower only 15 per cent of the time, requesting clarification the rest of the time. People took the perceptual salience of a flower against common ground as the means by which they were supposed to coordinate on identifying the referent.

In the coordination model, people don't coordinate by perceptual salience *per se*, but by salience against common ground more broadly. In a second experiment (Clark et al., 1983), Sam handed people on campus a photograph of Ronald Reagan and David Stockman equally salient standing side by side. At that time, Reagan was president of the United States, and Stockman was the director of the Office of Management and Budget. Sam then asked one of two questions:

(8) You know who this man is, don't you?
(9) Do you have any idea at all who this man is?

By many models, people should either: (a) pick out the same man for 'this man' in both questions; or (b) refuse to pick out either man. By the coordination model, people should take Sam's presuppositions into account and find one man more salient than the other against their current common ground. With (8), Sam presupposed 'this man' was familiar, but with (9), he presupposed 'this man' was *not* familiar.

The results were clear. With (8), people answered 'Reagan' 80 per cent of the time, requesting clarification the rest of the time. They never answered 'Stockman'. But with (9), people answered 'Stockman' 20 per cent of the time, requesting clarification the rest of the time. They never answered 'Reagan'. Afterwards, all of the participants were able to identify both men in the photograph, and all rated Reagan as the more familiar for Stanford students.

For these addressees, then, the current common ground included not only the photograph, but the public familiarity of the two men, and Sam's presuppositions. They selected the referent by judging the most salient possibility against that common ground.

The picture so far is this. Speakers ordinarily perform their acts of referring for particular addressees. They design referring expressions in the expectation that their addressees will be able to pick out the referents as the most salient possibility against their current common ground. Listeners, in turn, rely on the same logic to identify the referents. This is no more, really, than an elaboration of cooperation and coordination as characterized by Grice and Lewis.[5] Once again, it took evidence from armchair, laboratory and field to create this picture. Indeed, we see with Sam's questions on the Stanford University campus the beginning of a combined method – the field experiment, or laboratory in the field.

3 Referring as an interactive process

Coordination and cooperation are profoundly social. June and David try to keep close track of what each other is doing, thinking, looking at, and when they act, they try do so by anticipating what the other will do. Yet in the models reviewed so far, referring is treated not merely as an *individual* act – something speakers do themselves – but as an *autonomous* act – something speakers do *by* themselves. Referring is a one-shot process that is complete once the speaker has uttered 'the beer', 'Washington' or 'this man'. As Searle (1969) put it: 'The unit of linguistic communication is not. . . the symbol, word, or sentence, but rather the production or issuance of the symbol, word, or sentence, in the performance of a speech act' (p. 16). For him, linguistic communication is like writing a postcard and dropping it in the mailbox. It doesn't matter whether the addressee receives, reads or understands it. But this feels wrong. How can the concepts of coordination and cooperation be so social, and the act of referring be so non-social?

3.1 Language and joint action

The answer was already present, though not entirely explicit, in Lewis's 1969 characterization of coordination. Suppose that Michael, an accomplished pianist, plays Bach's 'Two-part Invention Number 3' *as a solo*. Next, June and David, also pianists, play the same two-part invention, but *as a duet*, June playing the right-hand, and David the left.[6] We make recordings of both

---

[5] Note that speakers don't design their references for just any hearer. They can prevent bystanders and overhearers from understanding, as when June tells David knowingly, 'You know who has just arrived' (Clark and Carlson, 1982; Clark and Schaefer, 1987, 1992). This, too, is evidence that speakers aren't blind to their audience.
[6] This example is based on a childhood experience of Clark, who would team up with a girl named Jane to play Bach fugues that he couldn't play by himself.

performances. If June and David are successful, we shouldn't be able to tell the duet from the solo. But how did June and David manage, with their two individual actions, to sound just like Michael, with his single individual action? Duets are performed by two people acting in concert. Let us represent what June and David are doing mid-performance this way:

(a) The pair June-and-David are playing a two-part invention that consists of b and c.
(b) The individual June is playing the right-hand part *as a part of (a)*.
(c) The individual David is playing the left-hand part *as part of (a)*.

The duet is represented in (a), and June's and David's individual acts in (b) and (c). The schema shows how June cannot intend to play her part *as part of the duet* without believing that David is trying to play his part *as part of the same duet*, and vice versa. June's and David's acts are individual acts, but ones that cannot be carried out without the other. These are called *participatory acts* (Clark and Schaefer, 1989). Yes, June is playing the left-hand part herself, which makes it an individual act, but she is doing it as part of the duet, which makes it a participatory act. The duet itself is a *joint action*. Although Michael's performance is also an individual act, it is an autonomous one.

Face-to-face conversation, the primary form of language use, is also a *joint activity*. For June and David to talk to each other, they have to coordinate their individual actions at at least these four levels (Clark, 1996):

Level 1 June produces vocalizations and gestures for David; David tries to attend to them.
Level 2 June presents words and phrases for David; David tries to identify them.
Level 3 June means something for David; David tries to understand what June means.
Level 4 June proposes something for David; David tries to consider her proposals.

June and David coordinate not only on the *content* of these four levels of action, but on the *timing*: They synchronize what they are doing. In face-to-face conversation, speaking and listening are participatory acts that require as-much coordination as playing the two parts of the Bach two-part invention.

If language is a joint activity, then we need to treat referring as part of a joint activity. An obvious solution is to treat it as a participatory act by the speaker – as one half of a joint act by the speaker and addressees. Let us review evidence for such a change.

3.2 Referring and interaction

Referring had been shown to be a social act long before Olson's proposals. In 1964 and 1966, Robert Krauss and Sidney Weinheimer described a referential communication task in which participants were given an ordered array of four figures and asked to describe them so that a partner, hearing those descriptions, could put the figures into the same order. The figures were abstract difficult-to-describe drawings; one, for example, pictured a figure that was first described by one participant as 'an upside-down martini glass in a wire stand'. The participants repeated this task with the same few figures used many times. Olson's account would predict that the martini figure should be referred to each time as 'the upside-down martini glass in a wire stand', because that was the description originally needed to distinguish it from its alternatives. In fact, the descriptions got shorter on successive references, going from 'the upside-down martini glass in a wire stand' to 'the inverted martini glass', to 'the martini glass', and finally to 'the martini'. In a model like Olson's, referring shouldn't reflect previous references, except in pronouns. But as Krauss and Weinheimer demonstrated, it does. Referring is a historical process.

Even more remarkable was Krauss and Weinheimer's (1966) finding on feedback. In the same task, some participants were paired with active partners who provided on-going feedback such as 'uh-huh', 'which one?' and 'got it'. Others were paired with a tape-recorder into which they spoke to future partners. When there was feedback, referring expressions became shorter over successive references, as in the martini example; the average length dropped from ten to two words. But when there was no feedback, there was no shortening.[7] A model like Olson's has no explanation for such a phenomenon.

Krauss and Weinheimer's experiments raised two important issues. What role does feedback play in acts of referring? And how do speakers appeal to previous references in the current act of referring?

3.3 Referring and collaboration

Common ground isn't a homogeneous body of well-established propositions. When June and David are talking, some elements are firmly established at the moment one of them speaks, but others are in doubt. Most propositions haven't even been assessed for whether or not they are in common ground. Also, June and David must reassess their common ground with each new utterance – indeed, each new bit of utterance – and with each new joint perceptual experience. So people in conversation are faced with an ongoing practical problem: how are they to assess and establish what is common ground as they race along in their conversation?

Establishing common ground is a central problem in acts of referring, and people have evolved strategies for doing that. In one documented example

---

[7] Later studies (e.g., Bavelas, Coates and Johnson, 2000, 2002; Clark and Krych, 2004; Fox Tree 1999) showed that references designed without feedback are also inferior in quality; some are even impossible to grasp.

(from Sacks and Schegloff, 1979), Ann wanted to refer to a couple named Ford. She thought her friend Betty might recognize the couple by that name, but she wasn't sure. How should she design her reference? Ann's solution was ingenious:

(10) Ann: ...well I was the only one other than than the uhm tch *Fords?*, uh Mrs Holmes Ford? You know uh
⌈the the cellist?
Betty: ⌊Oh yes. She's she's the cellist.
Ann: Yes. Well she and her husband were there.

Ann produced *the Fords* with a so-called *try marker*, a rising intonation (indicated by the question mark) followed by a slight pause (Sacks and Schegloff, 1979). In doing that, she implied that Betty might *not* recognize the couple by that name, and she was requesting Betty to say yes – 'uh-huh' or a nod – if she *did*. When Betty didn't say yes immediately, Ann tried *Mrs Holmes Ford*? And when Betty didn't say yes to that, she tried *the cellist*? This time Betty said 'Oh yes', even interrupting Ann to keep her from trying yet another expression, and confirmed her understanding with 'She's she's the cellist'. Only then did Ann go on.

Most models of definite reference assume that speakers refer as if they were writing to distant readers. Searle's and Olson's accounts are two examples. These *literary models*, as Clark and Wilkes-Gibbs (1986) called them, make several tacit assumptions: (1) References are expressed with standard noun phrases (e.g., proper names, definite descriptions, pronouns). (2) Speakers use these noun phrases intending their addressees to be able to identify the referents uniquely against their common ground. (3) Speakers discharge their intentions simply by the issuing of such noun phrases (see Searle, 1969). And (4) the course of the process is controlled by the speakers alone.

Ann's reference to the Fords is at odds with all four assumptions: (1) The noun phrase *the Fords?* with its added try marker is not standard. (2) In using it, Ann did not necessarily expect Betty to be able to identify the referent uniquely; that is why she added the try marker. (3) Ann did not discharge her intention to get Betty to identify the referent simply by producing the noun phrase. On the contrary, (4) she used *the Fords?* to initiate a process that required Betty's collaboration. She recognized that the process was controlled not just by her but by Betty too. If Betty had nodded immediately after *the Fords?*, Ann would have stopped there. As it was, she had to try three noun phrases (*the Fords?, Mrs Holmes Ford?* and *the cellist?*) before going on. Ann's act of referring was not an autonomous act, but a *participatory* act. It required Betty's coordinated participation in a duet-like joint action. Indeed, her act was opportunistic in a way that she could not have done alone.

3.4 Referring and grounding

The collaborative nature of referring was documented in an experiment by Clark and Wilkes-Gibbs (1986). The experiment was a referential communication task, but with several changes from Krauss and Weinheimer's method. Instead of one naïve participant and a confederate of the experimenter, there were two naïve participants – a *director* and a *matcher*. Instead of arrays of four abstract line drawings, there were 12 hard-to-describe Tangram figures – abstract geometric shapes that vaguely depicted people. And there were six trials on which the director got the matcher to arrange his 12 figures in the same order as the director's.

Although the new experiment confirmed many features of the original task, it raised new issues. As in the original task, directors needed fewer words per figure as they referred repeatedly to each figure – from 42 words on the first trial to ten on the sixth. But along with the drop in words was a drop in the number of *turns* per figure – from five on the first trial to two on the sixth. Why? Because of the way directors and matchers collaborated with each other on each reference.

The most striking finding was how often the partners made references with techniques not described in grammars of English – techniques that required contributions from both participants. Speakers initiated the process of referring with a variety of standard and non-standard noun phrases. Most of the following types are from the Tangram task, although a few are from other conversations:

1. **Elementary noun phrases**, such as *the guy leaning against the tree*.
2. **Self-repaired noun phrases**, such as *the guy reading with, holding his book to the left*.
3. **Other-repaired noun phrases**, as when B repairs A's 'Monday' in this example (Schegloff, Jefferson, and Sacks, 1977, p. 369):
   B: How long y'gonna be here?
   A: Uh- not too long. Uh just til uh Monday.
   B: Til- oh yih mean like a week f'm to*mo*rrow.
   A: Yah.
4. **Episodic noun phrases**. These are noun phrases produced in two or more intonation units, such as *the person ice skating, with two arms*.
5. **Instalment noun phrases**. In these, the speaker produces a first instalment of a noun phrase, waits for an acknowledgement, and then produces the next instalment, and so on, as here:
   Director: And the next one is *the one with the tail to the right*,
   Matcher: Okay.
   Director: *With the square connected to it*.
6. **Expanded noun phrases**. These are noun phrases expanded at the instigation of the addressee, as in the italicized expansion instigated by the matcher's 'uhhh':
   Director: Okay, the next one is *the rabbit*.

Matcher: Uhhh –
Director: That's asleep, you know. It looks like it's got ears and a head pointing down?
Matcher: Okay.

7. **Trial noun phrases**, or noun phrases with try markers, such as *the person ice skating that has two arms?* (or *the Fords?*). With these, speakers request a yes or no answer, which then leads to two different courses of action.

8. **Holder noun phrases**. These are conventional expressions used as stand-ins until the speaker or addressee can come up with a better noun phrase, as in 'it may take a hell of a long time to come, . if he puts it into the diplomatic bag, as u:m – *what's his name,*. Mickey Cohn did', (1.1.83, Svartvik and Quirk, 1980). Other examples: *the whatchamacallit, what's-its-name,* and *you-know-who* (see Enfield, 2003).

9. **Invited noun phrases**. With these, speakers invite a completion from their addressees (see Lerner, 1987; Wilkes-Gibbs, 1986), as here:
Director: And number twelve is, uh,
Matcher: Chair
Director: With a chair, right.

Examples like these challenge the notion of referring as an autonomous act. In every case, speakers require coordinated actions from their addressees. In cases (3), (5), (6), (7), (8) and (9), there is an explicit request for help. In the rest, there is an implicit one.

What do all these noun phrases have in common? It is clearly not that they constitute complete references in and of themselves. When speakers produce *the rabbit* in (6) or *the whatchamacallit* in (8) or *uh* in (9), they recognize that these are hardly adequate to complete the references. Rather, they use these noun phrases to *initiate* a process with two intertwined goals:

Identification. Speakers are trying to get their addressees to identify or pick out a particular figure under a particular description.

Grounding. Speakers and their addressees are trying to establish the mutual belief that the addressees have identified the referent well enough for current purposes.

Identification is the traditional goal of referring – it is central to Olson's and Searle's models. What is new here is *grounding*.

Grounding is a very general notion. *To ground a thing is to establish it as part of common ground well enough for current purposes* (Clark, 1996; Clark and Brennan, 1991; Clark and Wilkes-Gibbs, 1986). In conversation, speakers and addressees ordinarily try to ground everything that gets said (Clark and Schaefer, 1989), and that includes the identification of referents (Clark and

Wilkes-Gibbs, 1986; Schober and Clark, 1989). The grounding process is neatly illustrated in Ann's reference to the Fords. It consisted roughly of these steps:

Step 1	Ann says *the Fords*?	Betty doesn't respond, implying failure of identification
Step 2	Ann says *Mrs. Holmes Ford*?	Betty still doesn't respond, implying failure of identification
Step 3	Ann says *the cellist*?	Betty says *oh yes*, confirming the identification with *she's the cellist*
Step 4	Ann accepts Betty's evidence by going on	Betty accepts Ann's judgment

Each type of noun phrase we listed initiates a different grounding process, one with different presuppositions and different prospects.

With examples like these, it is no longer possible to maintain that referring in conversation is an autonomous, one-shot act by the speaker. Rather, it is a participatory act that requires coordinated actions from the addressee. We would never have discovered this in an armchair. We needed conversation from the field – or at least from field-like conditions in the laboratory.

3.5 Reference and conceptual pacts

In history books, the same individual might be referred to as *Napoleon, the loser at Waterloo*, and *the man who instituted modern French civil law*. Although these pick out the same individual, they do so under different descriptions – *qua* person named Napoleon, *qua* loser at Waterloo, and *qua* man who instituted modern French civil law. Each description reflects a different conceptualization, a different perspective. Indeed, every reference (so far as we can tell) is intended to pick out: (i) an individual; (ii) under a description. Descriptions play an important role in referring (a role absent from Searle's account, implicit in Olson's, and undeveloped in others). Let us see how.

One of the Tangram figures was referred to by various people as 'the rice bag', 'the whale', 'the complacent one', and 'the stretched-out stop sign' (Schober and Clark, 1989), and another one as 'the ice skater' and 'the ballerina' (Clark and Wilkes-Gibbs, 1986). Reaching these perspectives took coordination. It wasn't enough for a director to see a figure as, say, an ice skater. He or she had to get the matcher to see it that way too. Consider this first reference to a Tangram figure:

(11) *Director*: Okay, the next one looks, is the one with the person standing on one leg with the tail.
 Matcher: Okay.

Director: Looks like an ice skater.
Matcher: Okay.

First, the director got the matcher to identify the figure as 'one with the person standing on one leg with the tail'. But then he offered a more memorable perspective ('ice skater'), which the matcher agreed to ('okay').[8] What the director and matcher established in this exchange was a *conceptual pact*, a temporary agreement about how they were to conceptualize the referent (Brennan and Clark, 1996). Participants establish conceptual pacts each time they achieve a reference. Typically, speakers offer a conceptualization that is tentative or provisional, using hedges such as *sort of, kind of* and *looks like*. Once their addressees agree to the conceptualization, speakers drop the hedges and treat it as a conceptual pact that they can appeal to later. The director in (11), for example, referred to the figure from then on as *the ice skater*. To ignore the pact and call it *the ballerina* would implicate that the first perspective needed to be replaced by the second. Once two people have established a conceptual pact, it is cooperative to continue using it – unless there are reasons to change.

Conceptual pacts lead to several predictions about referring (Brennan and Clark, 1996; Van Der Wege, 2000). In one referential communication task, a first group of participants were given a set of 12 figures that included three shoes, whereas a second group were given a set that included just one of the three shoes. As expected, the shoe common to both sets was called, for example, *the dress shoe* by the first group, but *the shoe* by the second group. Speakers were no more informative than they needed to be – conforming to Grice's Maxim of Quantity. After a few trials with the same sets, the first group was given a set of 12 figures that had only the one shoe in it. On Olson's or a Gricean account, they should immediately switch to calling it simply *the shoe*. But they didn't. Most continued calling it *the dress shoe*, even though that was more informative than necessary. Why? Once the two partners had established the conceptual pact 'dress shoe', it was cooperative to continue using that pact. Indeed, the more firmly they had established that pact in the initial trials, the longer they continued using *the dress shoe* before simplifying it to *the shoe*. More than that, speakers tended *not* to retain *dress shoe* for the lone shoe when they got a new partner (see also Wilkes-Gibbs and Clark, 1992).

The picture of referring has changed once again, and this time into a truly social act. Speakers perform acts of referring *in collaboration with* their addressees. Referring is a participatory act: like playing in a duet, it requires the co-participation of the addressee. In conversation, therefore, speakers go

---

[8] One pair was unable to reach a single perspective, so the director referred to the figure thereafter as 'your monk and my machine gun'.

beyond standard noun phrases. They use both standard and non-standard forms to initiate the grounding of that reference – establishing the mutual belief that their addressees have understood it as intended. Speakers and addressees come to agree not only on which individual is being referred to, but on how that individual is to be conceptualized. Once again, these conclusions come from the judicious combination of evidence from armchair, laboratory and field.

4 Referring with language and gestures

At first, referring was treated as an autonomous act based mostly on literary or armchair examples. But once investigators began studying conversations on audiotape that conception changed, and referring was treated as a participatory act – one half of a joint act by speakers and addressees coordinating with each together. Even in this conception, referring was treated as a *linguistic* act, one achieved exclusively through language. But once investigators began studying conversations on *video*tape, that conception changed yet again. Let us use the term signal as an act by which speakers mean something for their addressees *à la* Grice (1957). When people were videotaped in conversation, they were found to exploit a range of signals that were not linguistic at all. Referring was seen to be a multi-method process, one that normally requires more than one *method* of signalling.

Work from the field shows that speakers routinely *anchor* their references to the material world – to actual people, artifacts, rooms, buildings, landscapes, events, processes (Clark, 2003; Goodwin, 2003; Hindmarsh and Heath, 2000). When June refers to a nearby building, 405 North Mathews Avenue, as 'that building', it isn't enough for David to understand that she is referring to a particular building in their current common ground. He is to understand that she is referring to that huge building *over there* – the one she is now pointing at – and she expects him to look at it to confirm. Recall that the act of referring establishes two things: (i) an individual as the referent; (ii) a conceptualization or perspective on that individual. Schematically,

referring = indicating + describing

Describing may be doable through language alone, but indicating requires something more – the locating of an individual in relation to the speaker and addressee's here and now.

4.1 Methods of indicating

Indicating is a method of signaling built on C. S. Peirce's notion of *index* (see Buchler, 1940). Pointing is a prototypical example. When June points at 405 North Mathews Avenue, she intends David to construe her pointing as an index to that building. Unlike symbols (such as the words *building* and *work*), indexes work by means of *intrinsic connections* between them and their objects. To quote Peirce, an index designates its object 'because it is in dynamical (including spatial) connection both with the individual object, on the one hand, and with the senses or memory of the person for whom it serves as a sign, on the other hand' (Buchler, 1940, p. 109). There is an intrinsic connection between June's pointing gesture (the timing and direction of her finger and hand motions) and 405 North Mathews Avenue. Indexes often work by directing our attention. According to Peirce, 'A rap at the door is an index. Anything which focuses the attention is an index' (Buchler, 1940, p. 108).

In indicating, then, speakers get their addressees to *focus attention* on individual objects. In discourse there is always a focus of attention, and it is often exploited, for example, in referring with pronouns (Brennan, 1995; Grosz and Sidner, 1986). Indicating is analogous but focuses *visual* or *auditory* attention on *physical* objects, events and states. Speakers can indicate in two main ways (Clark, 2003). One is by directing the addressees' attention to an object. This is what June does in pointing at 405 North Mathews Avenue for David. This technique is called *directing-to*. Another way is by placing an object within the addressees' current or future locus of attention. June, at a bookstore, places a book on the checkout counter to indicate that book to the clerk as one she wants to buy. This technique is called *placing-for*. Directing-to and placing-for are exploited in acts of referring both when required and when circumstances permit.

Focusing attention creates even more difficulties for Olson's model. By focusing addressees' attention on a subset of potential referents, speakers don't have to specify the referents in relation to *all* alternatives; they can get away with an otherwise ambiguous description (Beun and Cremers, 1998; Clark, Schreuder and Buttrick, 1983). But then referring is not a one-step process anymore: indicating (focusing attention) and describing are at least two components. We will return to this question later.

4.2 Indicating within acts of referring

Deictic references, which contain words like *this, that, here, there, I* and *you*, are often incomplete without visible or audible acts of indicating. Such acts are common in field observations, as in this example (Schegloff, 1984, p. 280):

(12) *Frank*: why:nchu put that t the end uh the ta:ble there [pointing]

Although Frank refers to the dish 'that' in front of Marge without a gesture, he refers to the end of the table 'there' by pointing at it. Without the gesture, he couldn't get Marge to recognize where precisely on the table he meant. So in referring to that location, he used a *composite signal* – a combination of describing (the use of *there* to mean 'in that place') and indicating (his pointing gesture).

The problem is that many indicative gestures are performed *without* affiliated noun phrases. Here is an example from the field (Schegloff, 1984, p. 284):

(13) *Linda*: en I'm getting a sun tan [pointing at her two cheeks in turn]

As Linda says this, she points first at her left cheek and then at her right, asserting that she is getting a tan on her cheeks. She might be paraphrased as saying 'I'm getting a sun tan **on my face**', but she doesn't say 'on my face'. Examples like this – and they are legion – bring out the radical change required for multi-method communication. Linda and her addressee are to combine Linda's linguistic signal ('en I'm getting a sun tan') with her gestural signal (her pointing to her cheeks) to get an interpretation something like 'I'm getting a sun tan on my face'. We don't know of formal systems of pragmatics in which this is possible.

Indicative gestures can also be performed by addressees. In one experiment (Clark and Krych, 2004), one participant (called the director) was required to tell another participant (called the builder) how to put Lego blocks together to form a small, seven-block abstract sculpture. When the directors could watch their partners work, the builders would often ground the directors' references with forms of directing-to and placing-for. The following is an exchange from the middle of a sculpture (with overlapping actions marked):

(12) *Director*: and now get
⌈[.75 sec] a-uh eight piece green, [1.50 sec]
Builder: ⌊[begins reaching into a collection of blocks, picks one out, then **exhibits** it]
Director: and join the two so that it's all
⌈symmetric,
Builder: ⌊[**poises** block above a location]
Director: yeah right in the center.

In (12), once the builder finds what he thinks is the right block, he holds it out for the director to inspect, and the director confirms its correctness by going on. That is, the builder *exhibits* the block to the director, a form of directing-to. Once the builder believes he knows where the block goes, he then *poises* it just above the location, to which the director says 'yeah'. Poising is also a form of directing-to.

As examples like these show once again, referring is a participatory act: the course of an utterance – its phrasing and timing – is determined not by the speaker alone, but by the speaker in collaboration with the addressees. In (12), the director delays going on for fully 1.5 seconds while the builder retrieves and exhibits the right block. In another example, the director says,

'Put it at the end of the red that's o- the other end'. She alters the course for her utterance (at 'o- the other end') when the builder points at the wrong location. In another example, a director says, 'And put it on the right hand half of the yes [0.30 sec] of the green rectangle'. She interrupts her reference to say 'yes' to confirm the location where the builder has poised the block. In the building of Lego models, examples like this were common.

Referring, indeed, is far more accurate and efficient when it is bilateral and multimethod. In the Lego experiment (Clark and Krych, 2004), half the directors were able to see the builders' workspace – their hands and blocks – and the other half were not. Partners took less than half the time to build their models when the builders' workspace was visible than when it was not. Language alone was no substitute for language and gesture together. Still other directors were asked to tape-record their instructions for future builders, so they were forced to design references unilaterally. The result was disastrous. Builders made over ten times as many errors when they worked from the tape-recorded instructions as when they collaborated with the directors live.

4.3 Indicating and describing in acts of referring

Indicating works, fundamentally, by *locating* referents for addressees. Recall Sam handing people a photograph of Reagan and Stockman and asking 'You know who this man is, don't you?' To refer to Reagan, Sam needed to get his addressees to locate the right *indicatum* – the image of Reagan in the photograph. That image, of course, was merely an index to the actual referent of 'this man' – to Reagan himself. Most demonstrative references work by means of two indexes: (i) the pointing locates a proximal indicatum (e.g., Reagan's image); and (ii) the indicatum is an index to the intended referent (e.g., Reagan himself).

But what does it take to locate an indicatum? In Clark and Marshall (1981), it was argued that some methods of locating things should be preferred to others. It requires fewer assumptions to point at a book and say 'that book' than to describe its location 'the book third from the left on the top shelf'. Indeed, if David asks June, 'In which building do you work?', it would be obtuse of her to answer '405 North Mathews Avenue' when she could simply point and say 'that building'. June's aim is not simply to name the building's location, but to bring the building into her and David's *joint focus of attention*.

Pointing itself can be more or less precise in establishing a joint focus of attention. Let us distinguish *close pointing* from *distant pointing* (Wilkins, 1999). With close pointing, the things pointed at are more or less within arms' reach; they can often be touched. With distant pointing, the things pointed at are out of arms' reach. *A priori*, close pointing should be more precise than distant pointing. If so, speakers should need to supplement distant pointing with other locative descriptions. They shouldn't need to do that with close pointing.

Pointing and locative descriptions do indeed trade off in referring at different distances, as shown in an experiment by Bangerter (2004). In that experiment, two participants sat next to each other facing a large board with 20 pictures of people placed at random locations on it. The director had a sheet with the people's names, and his or her job was to get the matcher to write down the right name for the right picture on an answer sheet. The two of them could talk as much as they liked. The board of pictures was placed at five distances away from the participants, from arm's length to a metre away.

Several things happened as the targets got further away. The directors pointed less often, from 82 per cent to 23 per cent of the time. They also used fewer deictic expressions (*this, that, here* or *there*), from 57 per cent to nearly none (4 per cent). Remarkably, as they used fewer deictic expressions, they included more descriptions of the pictures' locations (e.g., 'she's on the very right hand side' or 'the person right below the person laughing'). Indeed, the locative descriptions compensated precisely for the absence of deictic expressions, as they rose from 43 per cent to nearly all (97 per cent). In a control condition, partners could see the board but not each other, and thus could not point. They identified almost every picture by describing its location and features. But the location descriptions preceded the feature descriptions 91 per cent of the time, suggesting that speakers were trying to focus their addressees' attention before describing the pictures.

In short, the two partners started the referring process by bringing the location of the picture into their joint focus of attention and only then described its features. They used pointing to do this whenever possible. When pointing was ambiguous or impossible, they compensated by describing the location.

5 Conclusions

Since about 1960, theoretical conceptions of reference have gone through remarkable changes. It was originally treated as an autonomous, one-shot act by the speaker to enable a listener to identify the intended referent. It was viewed as addressee-blind or at least addressee-myopic. But with the arguments of Grice and Lewis, it came to be viewed as a cooperative, or coordinated, act that required speakers to consider their addressees. Furthermore, if face-to-face conversation is the primary setting for language use – it was the only setting for most of the history of both humans and their languages – then referring cannot be an autonomous act. It must be a participatory act – one half of a joint act by speakers and addressees working together. According to this view, speakers may initiate the process of referring, but they count on the active participation of their addressees. The act of referring was no longer viewed as unilateral, but as bilateral, its course determined by the actions of both speakers and addressees. Still later, referring came to be seen as multi-method. It required indicating as well as describing, and indicating ordinarily requires locating referents in a joint focus of attention.

It took all three methods of language analysis – armchair, laboratory and field – to change these theoretical conceptions of reference. No one method did it alone. There are lessons to be learned, therefore, from the history we have presented. One is that the final arbiter of a theory or model of referring must be whether or not it can account for the acts that arise in everyday language. One can imagine many of these uses from an armchair, but many other uses can only be discovered in the field. And one can examine these acts under the microscope of laboratory experiments, in which people are constrained to behave in certain ways. But the evidence about what people can do with constraints applied can only be interpreted against what people can do, and do do, in the field. Language has evolved as a natural phenomenon. Just as we can never know the true behaviour of bears, penguins or dolphins by studying them in a zoo, we can never know the true nature of language without studying it in the wild.

References

Akmajian, A. (1973). The role of focus in the interpretation of anaphoric expressions. In S. R. Anderson and P. Kiparsky (eds), *A Festschrift for Morris Halle*: 215–26. New York: Holt, Rinehart and Winston.

Bangerter, A. (2004). Using pointing and describing to achieve joint focus of attention in dialogue. *Psychological Science* 15(6): 415–19.

Bavelas, J. B., Coates, L., and Johnson, T. (2000). Listeners as co-narrators. *Journal of Personality and Social Psychology* 79(6): 941–52.

Bavelas, J. B., Coates, L., and Johnson, T. (2002). Listener responses as a collaborative process: The role of gaze. *Journal of Communication* 52(3): 566–80.

Beun, R.-J., and Cremers, A. H. M. (1998). Object reference in a shared domain of conversation. *Pragmatics and Cognition* 6(1–2): 121–52.

Brennan, S. E. (1995). Centering attention in discourse. *Language and Cognitive Processes* 10(2): 137–67.

Brennan, S. E., and Clark, H. H. (1996). Conceptual pacts and lexical choice in conversation. *Journal of Experimental Psychology: Learning, Memory, and Cognition* 22(6): 1482–93.

Brewer, W. F., and Treyens, J. C. (1981). Role of schemata in memory for places. *Cognitive Psychology* 13(2): 207–30.

Buchler, J. (ed.) (1940). *Philosophical Writings of Peirce*. London: Routledge and Keegan Paul.

Chafe, W. L. (1970). *Meaning and the Structure of Language*. Chicago: University of Chicago Press.

Chomsky, N. (1971). Deep structure, surface structure, and semantic interpretation. In L. A. Jakobovits and D. D. Steinberg (eds), *Semantics: An Interdisciplinary Reader in Philosophy, Psychology, Linguistics, and Anthropology*. Cambridge: Cambridge University Press.

Clark, H. H. (1975). Bridging. In R. C. Schank and B. L. Nash-Webber (eds), *Theoretical Issues in Natural Language Processing*. New York: Association for Computing Machinery.

Clark, H. H. (1977). Bridging. In P. N. Johnson-Laird and P. C. Wason (eds), *Thinking: Readings in Cognitive Science*: 411–20. Cambridge: Cambridge University Press.
Clark, H. H. (1996). *Using Language*. Cambridge: Cambridge University Press.
Clark, H. H. (2003). Pointing and placing. In S. Kita (ed.), *Pointing: Where Language, Culture, and Cognition meet*. Hillsdale, NJ: Lawrence Erlbaum.
Clark, H. H., and Brennan, S. A. (1991). Grounding in communication. In L. B. Resnick, J. M. Levine and S. D. Teasley (eds), *Perspective on Socially Shared Cognition*: 127–49. Washington, DC: APA Books.
Clark, H. H., and Carlson, T. B. (1982). Hearers and speech acts. *Language* 58: 332–73.
Clark, H. H., and Haviland, S. E. (1974). Psychological processes in linguistic explanation. In D. Cohen (ed.), *Explaining Linguistic Phenomena*: 91–124. Washington: Hemisphere Publication Corporation.
Clark, H. H., and Haviland, S. E. (1977). Comprehension and the Given–New contract. In R. O. Freedle (ed.), *Discourse Production and Comprehension*: 1–40. Hillsdale, NJ: Erlbaum.
Clark, H. H., and Krych, M. A. (2004). Speaking while monitoring addressees for understanding. *Journal of Memory and Language* 50(1): 62–81.
Clark, H. H., and Marshall, C. R. (1981). Definite reference and mutual knowledge. In A. K. Joshi, B. L. Webber and I. A. Sag (eds), *Elements of Discourse Understanding*: 10–63. Cambridge: Cambridge University Press.
Clark, H. H., and Schaefer, E. F. (1987). Concealing one's meaning from overhearers. *Journal of Memory and Language* 26: 209–25.
Clark, H. H., and Schaefer, E. R. (1989). Contributing to discourse. *Cognitive Science* 13: 259–94.
Clark, H. H., and Schaefer, E. F. (1992). Dealing with overhearers. In H. H. Clark (ed.), *Arenas of Language Use*: 248–97. Chicago: University of Chicago Press.
Clark, H. H., Schreuder, R., and Buttrick, S. (1983). Common ground and the understanding of demonstrative reference. *Journal of Verbal Learning and Verbal Behavior* 22: 1–39.
Clark, H. H., and Wilkes-Gibbs, D. (1986). Referring as a collaborative process. *Cognition* 22: 1–39.
Enfield, N. J. (2003). The definition of WHAT-d'you-call-it: semantics and pragmatics of recognitional deixis. *Journal of Pragmatics* 35(1): 101–17.
Fox Tree, J. E. (1999). Listening in on monologues and dialogues. *Discourse Processes* 27(1): 35–53.
Garrod, S. C., and Sanford, A. J. (1982). Bridging inferences in the extended domain of reference. In A. Baddeley and J. Long (eds), *Attention and Performance*, Vol. IX: 331–46). Hillsdale, NJ: Erlbaum.
Goodwin, C. (2003). Pointing as situated practice. In S. Kita (ed.), *Pointing: Where Language, Culture, and Cognition Meet*. Hillsdale, NJ: Lawrence Erlbaum.
Grice, H. P. (1957). Meaning. *Philosophical Review* 66: 377–88.
Grice, H. P. (1968). Utterer's meaning, sentence-meaning, and word-meaning. *Foundations of Language* 4. 225–42.
Grice, H. P. (1975). Logic and conversation. In P. Cole and J. L. Morgan (eds), *Syntax and Semantics, Vol. 3: Speech Acts*: 113–28. New York: Seminar Press.
Grice, H. P. (1978). Some further notes on logic and conversation. In P. Cole (ed.), *Syntax and Semantics 9: Pragmatics*: 113–27. New York: Academic Press.
Grosz, B., and Sidner, C. (1986). Attention, intentions, and the structure of discourse. *Computational Linguistics* 12: 175–204.
Halliday, M. A. K. (1967). Notes on transitivity and theme in English. Part 2. *Journal of Linguistics* 3: 199–244.

Harman, G. (1977). Review of 'Linguistic behavior' by Jonathan Bennett. *Language* 53: 417–24.
Haviland, S. E., and Clark, H. H. (1974). What's new? Acquiring new information as a process in comprehension. *Journal of Verbal Learning and Verbal Behavior* 13: 512–21.
Hawkins, J. A. (1978). *Definiteness and Indefiniteness: A Study of Reference and Grammaticality Prediction*. London: Croom Helm.
Hindmarsh, J., and Heath, C. (2000). Embodied reference: A study of deixis in workplace interaction. *Journal of Pragmatics* 32(12): 1855–78.
Jackendoff, R. S. (1972). *Semantic Interpretation in Generative Grammar*. Cambridge, MA., MIT Press.
Karttunen, L. (1977). Presupposition and linguistic context. In A. Rogers, B. Wall and J. P. Murphy (eds), *Proceedings of the Texas Conference on Performatives, Presuppositions, and Implicatures*: 149–60. Arlington, VA: Center for Applied Linguistics.
Karttunen, L., and Peters, S. (1975). Conventinal implicature in Montague grammar. *Berkeley Linguistics Society* 1: 266–78.
Krauss, R. M., and Weinheimer, S. (1964). Changes in reference phrases as a function of frequency of usage in social interaction: A preliminary study. *Psychonomic Science* 1: 113–14.
Krauss, R. M., and Weinheimer, S. (1966). Concurrent feedback, confirmation, and the encoding of referents in verbal communication. *Journal of Personality and Social Psychology* 4: 343–6.
Lerner, G. H. (1987). *Collaborative Turn Sequences: Sentence Construction and Social Action*. Unpublished Ph.D. dissertation, University of California, Irvine.
Lewis, D. K. (1969). *Convention: A Philosophical Study*. Cambridge, MA: Harvard University Press.
Matsui, T. (2000). *Bridging and Relevance*. Amsterdam: J. Benjamins.
Olson, D. R. (1970). Language and thought: Aspects of a cognitive theory of semantics. *Psychological Review* 77(4): 257–73.
Prince, E. F. (1981). Toward a taxonomy of Given–New information. In P. Cole (ed.), *Radical Pragmatics*: 223–55. New York: Academic Press.
Sacks, H., and Schegloff, E. (1979). Two preferences in the organization of reference to persons in conversation and their interaction. In G. Psathas (ed.), *Everyday Language: Studies in Ethnomethodology*: 15–21. New York: Irvington Publishers.
Schegloff, E. A. (1984). On some gestures' relation to talk. In J. M. Atkinson and J. Heritage (eds), *Structures of Social Action: Studies in Conversation Analysis*: 262–96. Cambridge: Cambridge University Press.
Schegloff, E. A., Jefferson, G., and Sacks, H. (1977). The preference for self-correction in the organization of repair in conversation. *Language* 53: 361–82.
Schelling, T. C. (1960). *The Strategy of Conflict*. Cambridge, MA: Harvard University Press.
Schiffer, S. R. (1972). *Meaning*. Oxford: Oxford University Press.
Schober, M. F., and Clark, H. H. (1989). Understanding by addressees and overhearers. *Cognitive Psychology* 21: 211–32.
Schütze, C. T. (1996). *The Empirical Base of Linguistics: Grammaticality Judgments and Linguistic Methodology*. Chicago: University of Chicago Press.
Searle, J. R. (1969). *Speech Acts: An Essay in the Philosophy of Language*. Cambridge: Cambridge University Press.
Sperber, D., and Wilson, D. (1986). *Relevance: Communication and Cognition*. Oxford: Basil Blackwell.

Stalnaker, R. C. (1978). Assertion. In P. Cole (ed.), *Syntax and Semantics 9: Pragmatics*: 315–32. New York: Academic Press.

Svartvik, J., and Quirk, R. (eds). (1980). *A Corpus of English Conversation*. Lund: Gleerup.

Van Der Wege, M. M. (2000). *Agreeing on Reference*. Unpublished Ph.D. dissertation, Stanford University.

Wilkes-Gibbs, D. (1986). *Collaborative Processes of Language Use in Conversation*. Unpublished Ph.D. dissertation, Stanford University.

Wilkes-Gibbs, D., and Clark, H. H. (1992). Coordinating beliefs in conversation. *Journal of Memory and Language* 31(2): 183–94.

Wilkins, D. P. (ed.). (1999). *Manual for the 1999 Field Season. Language and Cognition Group*. Nijmegen: Max Planck Institute for Psycholinguistics.

3
Psycholinguistic Experiments and Linguistic-Pragmatics
Raymond W. Gibbs, Jr.

1 Introduction

The field of psychology has always had a curious relationship with the study of linguistic-pragmatics. Linguists, philosophers, anthropologists and sociologists have over the past 40 years offered important analytic insights into the ways people employ pragmatic knowledge in using and understanding language. Some psychologists, most notably psycholinguists and social psychologists, have exploited the findings from scholars working in linguistic-pragmatics to conduct psychological experiments. Social psychologists, for instance, examine the ways language helps structure social interactions. Cognitive psychologists, on the other hand, focus on the underlying mental processes involved when people acquire, produce and comprehend language in real-life social settings. In both cases, ideas from linguistic-pragmatics are critical sources of hypotheses for various experimental investigations.

Yet there remains in psychology a persistent scepticism about experimental studies in linguistic-pragmatics. Psycholinguists are typically less concerned with the pragmatics of language use than they are with the architecture of the language processor (or production system) where the emphasis is on a single person comprehending words, sentences or texts apart from real-life communicative situations. Psychologists too often feel that the complexities of realistic language use make the topics of linguistic-pragmatics too difficult to study scientifically.

Furthermore, scholars working on linguistic-pragmatics outside of psychology sometimes question the utility of experimental studies on pragmatics. These scholars question some of the methods employed by psychologists to test pragmatic theories, and even, mostly privately, wonder whether studies employing ordinary speakers (e.g., college students) necessarily help distinguish between competing pragmatic accounts of different language phenomena. Ordinary speakers presumably lack the training needed to draw significant pragmatic distinctions and, therefore, may not be the best adjudicators of competing pragmatic theories. Only expert linguists and philosophers presumably possess the necessary skills to recognize subtle nuances of how different pragmatic knowledge shapes language use. Scholars embracing these sceptical beliefs about psychological experiments adopt the traditional view that the best theories focus on 'idealized speakers/hearers' and need not be concerned with the regularities and variations among real speakers.

My aim in this chapter is to address the concerns of both psychologists sceptical of studying realistic language use in a scientific manner, and linguists or philosophers critical of the value of experimental studies employing ordinary speakers/listeners. I describe four case studies of experimental research on linguistic-pragmatics that are part of my own empirical work over the past 15 years. These include studies on the pragmatics of making and understanding promises, understanding attributive and referential definite descriptions, making and understanding indirect speech acts, and recent work on inferring what speakers pragmatically say and implicate. Each project was originally motivated by theoretical debates in linguistic-pragmatics. I demonstrate how different hypotheses may be experimentally investigated using appropriate research methods. These experimental tasks involve systematic examination of ordinary people's linguistic intuitions, ratings of the contextual relevance of linguistic expressions, utterance production, reading-time experiments and priming studies. My purpose is to show how ideas from linguistic-pragmatics can profitably be used in conducting experimental studies in psycholinguistics.

2 Making and understanding promises

We are all accustomed to making promises to other people in our daily conversations. A speaker might say 'I promise to meet you for lunch at noon' or 'I'll meet you for lunch at noon' to indicate that he or she will be at some place at noon to eat lunch. When a speaker utters any of these statements, is he or she obligated to actually show up at the scheduled time? If so, where does the obligation to show up on time actually come from: is it from the utterance or does the obligation lie somewhere else?

There has been considerable discussion in the philosophy of language about the rules for the use of promises, the most explicit of which is in Searle's (1965, 1969) theory of speech acts. Searle's analysis of the illocutionary act of promising suggests that certain conditions, known as felicity conditions (Austin, 1962), must hold true for a promise to be made. These include the following:

1. The speaker's utterance (as he or she intends it) counts as the undertaking of an obligation to do what he or she has promised because it is conventionally recognized as a promise given the use of the words 'I promise'. This may be called the felicity of 'Obligation'.

2. The speaker believes that the hearer would prefer him to do what he has promised rather than not to do it. This rule distinguishes promises from threats. This may be called the 'Hearer Preference' felicity condition.
3. It is not obvious to either the speaker or the hearer that the speaker would do what he has promised in the ordinary course of events. This may be called the 'Nonevident' felicity condition.

Each of these conditions is seen as necessary for the performance of a promise, and taken collectively the set of conditions will be sufficient for a promise to have been made. In this sense, Searle suggests that his set of felicity conditions are the 'constitutive' rules (Rawls, 1955), which enable people to create an obligation just by saying something.

Many philosophers have disagreed over whether the felicity conditions Searle proposed are necessary ingredients in the practice of promising. For instance, speakers' utterances may not place them under any obligation to do the action mentioned in the promise. Instead, speakers' utterances only reaffirm a previously existing obligation and, as such, do not by themselves create obligations (Hare, 1964). Furthermore, promises can be made about actions that are performed in the normal course of events. Thus, a speaker who has carried out some action repeatedly in the past can promise to do it even if it is obvious to all concerned that he or she will continue to do it (Atiyah, 1981). Finally, hearers may under some circumstances prefer the speaker to not do the action mentioned, and yet the hearer views the utterance as a promise and not simply as a threat (Peetz, 1977).

Gibbs and Delaney (1987) experimentally examined the pragmatic factors that determine how people actually make and understand promises. We investigated whether people have tacit knowledge, similar to the felicity conditions proposed by Searle, which affect how promises are made and understood. Participants were presented with stories that depicted a person about to say something concerning a future event. These stories were either consistent with the three felicity conditions (i.e., Obligation, Hearer Preference and Nonevident) or violated one of them. Participants read each story and produced an utterance that they would say in the situation. Afterwards, participants went back and rated their utterances on the extent to which each one represented a promise.

We hypothesized that if people have tacit knowledge governing how they make promises, then violating any of these should affect participants' ratings of the utterances they produce as promises. For example, suppose that you have been mowing the lawn once a week for the past three summers and that everyone in your family expects you to do so. One day you say to a member of your family, 'I'll mow the lawn this afternoon'. According to Searle, this utterance should not count as a promise because it is obvious to both you and the hearer that you would have mowed the lawn in the normal course of events. This violates the Nonevident felicity condition in that there is no point in making a promise if the action to be performed would have been done anyway.

The participants in this study produced a range of utterance types, including most frequently statements of future acts (e.g., 'I'll take out the garbage'), reassurance and future act (e.g., 'I'll take out the garbage today, for sure'), statements of fact (e.g., 'I don't mind taking out the garbage when it's my turn'), and reassurance alone (e.g., 'Don't worry about the garbage, really'). These utterances generally fit into Searle's scheme that in making promises a speaker predicates a future action. Yet only 2 per cent of all the utterances produced contained the explicit performative 'I promise', a finding that contradicts Searle's claim that promises cannot be made without the explicit use of the words 'I promise'. (See Chapter 10 for a similar result among children, p. 215.)

Analysis of the promise ratings showed that the utterances produced in the normal and different violation conditions were not equally promise-like. Thus, people gave higher promise ratings to utterances generated in the normal condition than in each of the Obligation, Hearer Preference, and Nonevident violation conditions. But people gave higher ratings to utterances produced in the Nonevident condition than in the other two violation conditions, indicating that promises can be made in situations where the speaker would have done the action in the normal course of events. A second study found similar results when a different group of participants rated the utterances produced by people from the first experiment. Overall, the rating data are consistent with the idea that people believe that certain conditions should hold for promises to be felicitous, and this seems especially true for the Obligation condition, showing that promises normally obligate the speaker to fulfil the action described in the utterance.

A third study asked participants what they understood when they encountered promises in different social situations. Participants read a series of normal scenarios, each one ending with a speaker making an utterance that could be viewed as a promise. The participants read each story and then answered four questions about the speaker's final statement. These questions explicitly addressed: (a) whether the last statement was a promise; (b) whether the speaker was obligated to fulfill the promise stated; (c) whether the addressee in each story actually preferred the speaker doing the action mentioned to his not doing it; and (d) whether the addressee in each story actually preferred the speaker would have done the action mentioned in the normal course of events. Each of these questions, therefore, directly queried people about one of Searle's main felicity conditions for making promises.

The results of this study showed that participants gave lower ratings to questions regarding the validity of the Nonevident condition than they did to questions regarding either the Obligation or Hearer Preference conditions. This suggests that the Nonevident condition is least important in making and understanding promises. This finding is consistent with the view that it is not necessarily the speaker's utterance *per se* that marks his intention to make a promise. Instead, it is some pre-existing obligation, tacitly assumed in most cases, but reaffirmed by the speaker's utterance, that makes one's utterance count as a promise. Therefore, the speech act of making promises does not conventionally create an expectation on the part of the addressee, but it reaffirms an obligation that may already exist.

Finally, a fourth experiment had participants read stories depicting situations that varied in their degree of obligation (Obligation condition) and in terms of whether the obligation already existed or was brought about solely by the speaker's utterance (Nonevident condition). The participants' task was to read each story and then rate the degree to which the last utterance in each story constituted a promise.

The data showed that participants gave high promise ratings to utterances stated in contexts in which the speaker was expected to perform some action in the normal course of events. Thus, people view utterances as being most promise-like in situations where there is some mutually held belief between the speaker and the addressee about some future action on the part of the speaker. The obligation derives from the fact that the speaker is normally expected to do the action, and not just from what the speaker says to reaffirm the existence of this obligation. Moreover, promises depend crucially on the speaker's belief that the action to be performed is desired by the addressee.

In summary, there is good empirical support for Searle's (1965, 1969) set of felicity conditions as being important pragmatic factors which people abide by in making and understanding promises. The main exception to this, however, concerns the Nonevident condition. Searle (1964) originally proposed that to utter the words 'I promise to pay you, Smith, five dollars' is a promise, and as such, places the speaker under an obligation to give Smith five dollars. But surely one cannot go up to a stranger and say this and expect anyone to believe that the speaker is truly obligated to give Smith the money, even if Smith prefers the speaker to do so. A promise is only binding if in making a promise a speaker is acting under the rules and practices of some institution. The obligation, then, to fulfil a promise may be anchored in previous verbal agreements or in different social rules and conventions (Lewis, 1969). The main function of verbal promises is to remind the addressee of the existence of some prior obligation and to specify when an action is to be performed.

3 Understanding definite descriptions

There are a variety of linguistic devices available to talk about things in the world. Perhaps the most common way of talking about a particular object is when speakers use a 'definite description' to designate the object or person they wish to refer to. For instance, speakers can refer to people by uttering definite descriptions, such as 'Smith's murderer is insane' or 'The woman by the bus stop had bright red hair'. Definite descriptions, such as 'Smith's murderer is insane', can be used to refer in two distinct ways (Donnellan, 1966). The 'referential' use of the noun phrase refers to a particular person whom the speaker chooses to describe in a specific way, even though many other ways of referring to the same individual, including a proper name, may be equally appropriate. In the referential case, then, the speaker refers to a specific individual, for example, 'Bob Jones', and intends that the listener understand the utterance as referring to that person. The 'attributive' use of the noun phrase 'Smith's murderer is insane' refers to 'whomever fits the description'. When a police detective finds the brutally murdered body of Smith and comments attributively that 'Smith's murderer is insane', the reference is to whomever murdered Smith without specifying one particular, known individual. Definite descriptions, then, represent a type of pragmatic ambiguity because the same referring statement can have either an attributive or referential role depending on the context and the speaker's intentions.

There has been a great deal of interest in the problem of interpreting attributive and referential definite descriptions within philosophy and linguistics. Most of this concern focuses on the truth-conditional basis of definite descriptions and whether the attributive–referential distinction is best understood as a matter of semantics or pragmatics (Cole, 1978; Donnellan, 1966, 1968, 1978: Fauconnier, 1985; Kripke, 1972; Over, 1985; Quine, 1956; Searle, 1979; Strawson, 1950).

But psychological empirical studies have examined the relative difficulty in processing the attributive and referential descriptions too to see how comprehension of these referring descriptions makes use of previously mentioned antecedents in discourse. Mueller-Lust and Gibbs (1991) examined comprehension of definite descriptions that do not have clear antecedents, namely attributive referring phrases. What, if anything, is activated when attributive descriptions are understood? There are several hypotheses worth considering.

Searle (1979) suggested that there are two aspects under which reference is made. The primary aspect is the act of stating which object or person is the referent, and must be true for the sentence to be true. Compare 'Smith's murderer is insane', in the context of a bloody murder scene, to a courtroom scene. At the scene of the crime, 'Smith's murderer' states who is being referred to, namely whoever is the killer. This Searle called the primary aspect of the reference, because if the death occurred by accident and there is no killer, the statement cannot be true. Reference in attributive cases are made under the primary aspect, because it is the only way possible to indicate what is the referent.

In the courtroom context, 'Smith's murderer is insane' states that the man being referred to, namely the defendant, is insane. Again, this is called the primary aspect of reference because if the man is not insane, the statement cannot be true. However, in addition to the primary aspect, there is a secondary aspect that elaborates on who is being referred to, which Searle suggested is not intended as part of the truth conditions for this utterance. The elaboration in this case is that the man being referred to is the defendant. It may be that the man did not really murder Smith, a case of where the secondary aspect is false. Nonetheless, the description of 'Smith's murderer' uniquely identifies and secures who is being referred to. This use is an instance of a referential definite description.

According to Searle's view, the first step in understanding both attributive and referential descriptions is to uncover the primary aspect that some object or person is being referred to. At this point, the intended meaning of an attributive description is uncovered. But understanding referential uses requires a second step to secure the actual identity of the object or person being referred to. It is unclear whether Searle believed that both aspects must be uncovered in referential cases, because the primary aspect indicates that there is some referent. Nonetheless, his analysis suggested that attributive definite description should be understood faster than referential definite description because uncovering two aspects takes a greater number of inferential steps than uncovering only one aspect. We called this possibility the 'secure referent' hypothesis.

An alternative theory of how attributive and referential descriptions are understood stems from earlier work which revealed that reference comprehension is slowed down when there is no explicit antecedent (Haviland and Clark, 1974). Given this previous finding, then, in the referential case, a single inference is made between the description and the explicit antecedent. In the attributive example, even though there is an assumed antecedent, the fact that there is no explicit antecedent may cause problems in understanding the description's meaning. Understanding attributive definite descriptions, on the other hand, demands that people first try to uncover the explicit antecedent. When they are unable to do this, they then may make the inference that the description stands for whoever, or whatever, fits that description. This additional inferential step requires more processing effort and should result in longer reading-times to understand attributive reference than to understand referential reference. This theory may be dubbed the 'indirect reference' hypothesis.

Although attributive descriptions refer to something, it is not clear what types of inferences are used to determine this. One way to deal with this issue is to propose that a speaker and hearer share the belief that there is a hypothetical entity to which the attributive description refers. If a speaker says 'The first person to sail to America was an Icelander', the speaker and hearer mutually create the mental representation (i.e., a mental model) of this 'first person'. Even though no known specific person fills this role (for these particular conversants), the mutual knowledge is secured that this representation exists. Thus, when a reference is made, a token is introduced into the discourse model. This token may have a specific referent connected to it, as in referential cases. In general, attributive definite descriptions may not be more difficult to understand than referential definite descriptions, because in both instances, mutual beliefs allow resolution of the reference being made by the token 'The first person to sail to America was an Icelander'. This view is called the 'token' hypothesis'.

Several experiments tested the predictions of these three views of attributive and referential description comprehension (Mueller-Lust and Gibbs, 1991). A first study asked participants to choose the contextually appropriate paraphrase of statements such as 'The piano player is very talented'. The data showed that people readily distinguish between attributive and referential definite descriptions. Participants viewed attributive descriptions as referring to a general class of objects or persons and referential descriptions as referring to specific tokens of that group. These findings lend credence to Donnellen's (1966) original distinction between two types of definite descriptions and generally show that contextual information can eliminate the pragmatic ambiguity of statements such as 'I would like to meet the mayor of New York City'.

A second study found that people took longer to understand attributive descriptions than they did referential uses, a finding that suggests additional processing effort is needed to recognize that attributive expressions describe something or someone but do not specify particular entities. This reading-time difference between attributive and referential descriptions is most consistent with the indirect inference model. The extra time needed to process attributive expressions that describe newly introduced topics suggests that additional processing capacity is required when an attributive expression describes something that is left unspecified.

Even though the results of the second study imply that people understand attributive phrases as describing something, it is not clear that such phrases actually refer. Consider the definite description at the end of the following attributive context:

Kristin and Todd were studying in the music building.
They could hear piano music coming from another room.
A musician was practising but they didn't know who it was.
'That's a pretty piece', observed Kristin.
'It's a Chopin prelude', said Todd.
'Those are very complicated pieces', said Kristin adding,
'The piano player is very talented.'

The first experiment indicated that people most often chose 'Whoever is playing is a good musician' as an appropriate paraphrase for 'The piano player is very talented'. Although 'whoever' is not a specific referent, it does signify that someone captures the implied meaning. A second study employed a priming methodology to examine how attributive and referential descriptions are linked to their antecedents. Participants were presented with stories on a computer screen, one line at a time. The last sentence of each story was a definite description that was presented in one of two reinstatement contexts. One context primed a referential reading of the target sentence and the other context primed an attributive reading. The participants' task was to read each story and decide whether a probe word was presented in the preceding passage. The probe words referred back to antecedent nouns that may be activated when attributive and referential sentences were read.

Two measures were used to explore what processes are employed to understand attributive and referential definite descriptions. The first was the time to read the final sentence in each story. The second index of comprehension measured people's times to respond to the probe words. Experiment 2 showed that attributive definite descriptions primed recognition of a group of objects or persons relative to when no reference was made. This priming effect is somewhat weaker than the priming that was found for referential descriptions that specify an actual object or person, but the fact that any priming occurred suggests that the notion of reference need not rely upon actual securing of a single entity (e.g., the indirect inference view). Attributive descriptions refer and retain their sense or meaning because they refer to an unspecified member of a class of objects.

Comprehension of attributive and referential descriptions is different when the descriptions refer to famous persons. A third experiment revealed that people take longer to understand referential uses of definite descriptions than attributive uses of definite description when they refer to proper names. The fact that this effect only occurred when the referents were proper names is most likely due to the accessibility of information in the text. Thus, when people encounter a proper name in a referential text, they tag it as important information for the continuing conversation. When a referential description is then encountered, a backward inference is drawn to establish a connection with the proper name that is highly accessible in the discourse or mental model. This search cannot be avoided because the proper name itself serves to prime the search. This is consistent with the notion that a token exists in the discourse model (the proper name) that is linked to the description. In an attributive situation there is no proper name tagged as a salient token. When an attributive description is then encountered, although a token is introduced into the mental model, no backward search need be initiated because there is no proper name in the text to cause the search to occur and the description itself adequately specifies who is being described. The data from the third study are in accord with the view that people can short-circuit the search and quickly understand the attributive reference when the description is so closely associated with the person. This short-circuited process automatically occurs because the definite description is of a known person, not a referring phrase.

The Mueller-Lust and Gibbs (1991) studies do not bear directly on philosophical and linguistic questions about the truth-conditional semantics of attributive and referential referring phrases. But they do bear on psychological, pragmatic theories of reference and anaphora resolution. The fact that people can easily interpret attributive definite descriptions which do not have clear antecedents, as referring to an unspecified member of a general class of objects or persons, highlights a need for further expansion of existing psychological models of reference. We suggested that theoretical notions of 'situation models', 'discourse models' and 'mental models' can all be appropriately modified to handle instances of reference denoting classes of individuals not explicitly mentioned in some previous discourse, as well as cases when specific, known individuals have been previously introduced. At the same time, our results provided another example of how inferential strategies used during the recovery of referents vary as a function of the particular pragmatic intentions (e.g., attributive vs referential) behind a speaker's use of a definite description.

4 Making and interpreting indirect speech acts

Whenever a speaker requests something from someone, it costs the addressee some effort to supply what is desired. This could, and does in many situations, threaten the addressee's face value (Brown and Levinson, 1987). Face is defined as consisting of the freedom to act unimpeded (negative face) and the satisfaction of having one's values approved of (positive face) (Brown and Levinson, 1987). People usually act to maintain or gain face and to avoid losing face. A speaker's request often imposes on addressees and can potentially threaten the hearer's face. People are polite to the extent that they enhance or lessen the threat to another's face (Brown and Levinson, 1987; Clark and Schunk, 1980). To eliminate any threat to the addressee's face caused by a request, speakers usually formulate their requests indirectly, as in 'Could you lend me ten dollars?'. Making indirect speech acts provides addressees with options which enable them to either comply with requests, or give some good reason why they cannot or will not respectfully do so without losing face (Lakoff, 1973).

Indirect speech acts can be made in several ways. Each linguistic form specifies some part of the transaction of goods between speaker and listener, in which the listener's task is to infer the entire sequence of actions that the speaker wishes the listener to engage in to comply with the request. Requesting that someone shut the door, for example, can be done by questioning the ability of the listener to perform the action ('Can you shut the door?'), questioning the listener's willingness to shut the door ('Will you shut the door?'), uttering a sentence concerning the speaker's wish or need ('I would like the door shut'), questioning whether the act of shutting the door would impose on the listener ('Would you mind shutting the door?'), making a statement about some relevant fact in the world ('It's cold in here'), or simply asking about what the listener thinks about shutting the door ('How about shutting the door?').

Different kinds of indirect speech acts, however, may not be equally appropriate for a given social situation. Ordering a Big Mac at 'McDonald's' by saying 'I'll have a Big Mac' appears to be more appropriate than is the request 'Do you have a Big Mac?'. Traditional theories of indirect speech acts are unable to specify why speakers view some indirect requests as appropriate in some situations and not others. Most theories simply stipulate that the decision to use one kind of indirect request as opposed to another is an arbitrary phenomenon (or a matter of convention).

Formulating the right request in a situation depends on designing a transaction which takes into account a good deal of different information. A transaction requires the exchange of 'goods', such as tangible objects, commitments or obligations, between people. The speaker must then find a way of inserting his/her plan for the addressee's contribution (i.e., to respond to the request by providing the information) into what the addressee is doing or planning to do at the moment. In many situations, the speaker interrupts the addressee's activities, or projected plans, to impose his or her own goals. These 'detour' situations are more difficult to plan for, but in each case, the speaker finally designs his or her request as a turn in the transaction. To do this, the speaker must first assess what reasons there may be for the addressee not giving the desired information. The speaker will then formulate an utterance to deal with the greatest potential obstacle. By doing so, the speaker thereby implicates that the addressee will tell what the speaker wants to know.

The possibility that speakers formulate their requests to deal with the main obstacles to compliance is called the 'obstacle' hypothesis (Francik and Clark, 1985; Gibbs, 1986). This idea is interesting because it suggests that the apparent conventionality of an indirect request depends largely on the extent to which an utterance specifies the projected obstacles for an addressee in complying with the speaker's request (Gibbs, 1986). Thus, 'Do you have the time?' may be conventional to use in requesting the time of a passerby on the street because the greatest obstacle to the listener in providing the information may be that he or she simply doesn't know it and has no access to a timepiece. Since the speaker cannot rule out this most limiting case, he or she must design the request around it. However, saying 'Do you know what time you close?' as a request to a store owner to find out what time the store closes is inappropriate because the owner presumably knows what time his or her business closes. The most likely obstacle in this situation is the store owner's willingness to provide the desired information.

A good deal of experimental evidence supports the obstacle hypothesis. One set of studies had participants read various scenarios depicting a protagonist about to make a request (Gibbs, 1986). In some situations, the obstacles were general or even unknown. An example of this kind of scenario is presented below:

> Tracy and Sara were tired of eating dinner at their college's dining hall. So they went downtown to find something exciting to eat. They decided to go to Tampico's. Sara wanted an enchilada, so when the waitress came to take their order Sara said to her '...'

For other situations, the potential obstacle for the addressee in fulfilling the request was specific, as shown in the following scene:

> Tracy and Sara were tired of eating dinner at their college's dining hall. So they went downtown to find something exciting to eat. They decided to go to Tampico's. Sara wanted an enchilada, but was unsure whether the restaurant had them or not. The waitress came up to take their order and Sara said to her '...'

The main obstacle in this situation for the addressee (i.e., the waitress) in complying with Sara's request was whether the restaurant actually served enchiladas.

The participants' task in a first experiment was to read each scenario and simply write down what they would say in such a situation. Across all the different scenarios, the participants employed a variety of surface forms in making their requests (e.g., 'May I...? I would like..., Can you...?, Would you mind...? Do you have...?'). Although these forms of indirect requests were used most often, each request form was not equally appropriate for a particular situation. This was seen in how different types of requests were generated in different obstacle contexts. For instance, the participants generated Possession utterances, like 'Do you have...?' 68 per cent of the time when they read stories where the main obstacle concerned the addressee's possession of the object desired by the speaker. But people produced Possession requests only 8 per cent of the time in contexts where the obstacle concerned the addressee's ability to fulfil the request and participants never generated Possession utterances in situations with State-of-World obstacles.

A second study provided a better assessment of how speakers make requests in more realistic situations. Participants were brought to six locations on a university campus, each of which was carefully designed to highlight a different potential obstacle. For example, an experimenter and a participant went inside the university library and walked over to a table where a student was busily working on a paper assignment. The participant was told to imagine sitting near the student and also working on a paper when his or her pen suddenly ran out of ink. Participants were then asked to state what they would say to the nearby student in order to get that addressee to lend them a pen. Overall, participants produced appropriate requests 74 per cent of the time. Thus, when people were asked to make requests in situations which closely approximated the real world, they had an even stronger tendency to produce utterances that specified the obstacles present for the addressees.

Specifying the potential obstacles for addressees in making indirect requests makes it easier for listeners to comprehend these speech acts. The results of a reading-time experiment indicated that people process indirect requests that adequately specified the reasons for an addressee not complying with a request faster than they understand indirect requests that did not specify such obstacles (Gibbs, 1986). For instance, people read a statement like 'Can you possibly lend me your blue sweater?' faster in a context where the main obstacle concerned the addressee's ability to fulfil the request than in a situation where the obstacle focused on the addressee's possession of the desired object. On the other hand, people were faster to read 'Do you have a sweater to lend me?' than to comprehend 'Can you possibly lend me your blue sweater?' in a context where possession was the main obstacle. Most generally, people learn to associate specific obstacles for hearers in different social situations and know which sentence forms best fit these circumstances. A separate control study showed that, without context, participants found both types of indirect requests equally difficult to process, suggesting that the conventionality of an indirect speech act is not a property of an utterance itself, but is due to some relationship between an utterance and a particular social context.

Although scholars have claimed that many indirect requests are understood via some sort of short-circuited process (Bach and Harnish, 1979; Clark and Schunk, 1980; Gibbs, 1979, 1981; Morgan, 1978; Searle, 1975), no one has specified what it is about some requests that make them different. The results of the Gibbs (1986) reading-time study established that people take less time to process indirect requests that specify the projected obstacles for the addressee in complying with the request. Seeing indirect speech acts specified in this way makes it easier for listeners to determine speakers' intended meanings. What makes some indirect speech acts apparently 'conventional' is the appropriateness of the sentence forms in matching the obstacles present for addressees in a social context.

5 What speakers say and literal meaning

Understanding what speakers communicate in conversation often appears to first depend on recognizing what they actually say. Consider the following exchange between two college students:

Jim: Is your dormitory noisy?
Bill: I usually sleep wearing earplugs.

Bill's statement about his use of earplugs only conveys part of the meaning he wishes to communicate (i.e., that his dormitory is indeed noisy). In this way, what speakers say often vastly underestimates what they imply, or implicate, in conversation.

H. Paul Grice's theory (1975, 1978) suggests that any linguistic act conveys two levels of communicated propositional content: (a) the level of 'what is said', which is the proposition explicitly expressed, closely relevant to its linguistic, semantic content and usually equated with the truth-conditional, literal content of the utterance; and (b) the level of 'what is implicated', or the further propositions intended by the speaker which depend on pragmatics for their recovery. Although even Grice acknowledged that some contextual information must play a role in determining what speakers say, such as that needed to resolve ambiguity and fix indexical reference, what speakers say is essentially a minimally pragmatic meaning.

This section reports empirical research suggesting that Grice and others were incorrect in assuming that what a speaker says is equivalent to an utterance's context-free, semantic, literal or truth-conditional meaning. My claim is that significant aspects of what speakers say, and not just what they totally communicate, are fundamentally dependent upon enriched pragmatic knowledge (Recanati, 1989; Sperber and Wilson, 1986). In fact, recent experimental work suggests that people may analyse aspects of what speakers pragmatically say as part of understanding what speakers conversationally implicate. Under this revised theory, some aspects of pragmatic knowledge shape listeners' understanding of what speakers say, while other pragmatic information enables listeners to construct reasonable interpretations of what speakers imply in context. This new theory casts a very different, and more complete, role for pragmatics in a psychological theory of linguistic understanding than has previously been envisioned.

Several linguists and philosophers have recently argued that the traditional, Gricean view of implied meaning ignores the fact that essentially the same sorts of inferential processes used to determine conversational implicatures also enter into determining what it is that speakers say (Carston, 1993; Recanati, 1989, 1993; Sperber and Wilson, 1986). Gibbs and Moise (1997) demonstrated in several experimental studies that pragmatics indeed plays a major role in people's intuitions of what speakers say (for further discussion of these findings see Nicolle and Clark, 1999; and Gibbs, 1999). Consider the expression 'Jane has three children'. According to the Gricean view, the interpretation that 'Jane has exactly three children' comes from applying specific pragmatic information to the minimally pragmatic proposition of what is said (e.g., 'Jane has at least three children'), a process that results in what Grice referred to as a 'generalized conversational implicature' (i.e., implicatures that are normally drawn regardless of the context). But we showed in a series of experiments looking at students' intuitions about what speakers say that people do not equate the minimal meaning with what a speaker says. A first study showed that participants chose significantly more enriched pragmatic paraphrases of what speakers say (e.g., 'Jane has exactly three children') than they did paraphrases that were minimally pragmatic (e.g., 'Jane has at least three children and may have more than three'). A second study revealed that even when alerted to the Gricean position (i.e., what is said is equivalent to the minimal proposition expressed), people still reply that enriched pragmatics is part of their interpretation of what a speaker says and not just what the speaker implicates in context.

The fact that people prefer enriched pragmatic paraphrases for what speakers say doesn't mean that they are unable to distinguish between what speakers say and what they implicate. The findings of another study reported in Gibbs and Moise (1997) demonstrated that people recognize a distinction between what speakers say, or what is said, and what speakers implicate in particular contexts. For instance, consider the following story:

Bill wanted to date his co-worker Jane.
Being rather shy and not knowing Jane very well,
Bill asked his friend, Steve, about Jane.
Bill didn't even know if Jane was married or not.
When Bill asked Steve about this, Steve replied:
'Jane has three children.'

What does Steve say and what does he implicate by his utterance? Steve implicates by his statement 'Jane has three children' in this context that 'Jane is already married'. To the extent that people can understand what Steve says, but not implicates, by 'Jane has three children', they should be able to distinguish between the enriched and implicated paraphrases of the final expressions.

The results of one study showed this to be true. When participants were asked to choose the best paraphrase of what a speaker says in a context like the one above, they chose one that reflected the enriched pragmatic meaning (i.e., 'Jane has exactly three children') and not implicature paraphrases (i.e., 'Jane is married'). These findings show that pragmatics strongly influences people's understanding of both what speakers say and communicate. It appears that Grice's examples of generalized conversational implicatures are not implicatures at all but understood as part of what speakers say. More generally, the Gibbs and Moise (1997) findings suggest that the distinction between saying and implicating is orthogonal to the division between semantics and pragmatics.

One possibility is that comprehending what speakers pragmatically say serves as the foundation, in part, for further contextual elaborations to infer what speakers pragmatically imply. There may be two kinds of pragmatic information or knowledge, primary and secondary, that become activated during normal language understanding (Gibbs and Moise, 1997; Recanati, 1993). Primary pragmatic knowledge applies deep, default background knowledge to provide an interpretation of what speakers say. Under this view, primary pragmatic knowledge relates to deeply held, perhaps non-representational (Searle, 1983) knowledge that is so widely shared as to seem invisible. To take a classic example (Searle, 1978), our interpretation of the expression 'The cat is on the mat' presupposes an enumerable set of assumptions, such as that the cat chose for some reason to sit on the mat, and that the cat and mat are on the ground operating under the constraints of physical laws like gravity and are not floating in space in such a way that the cat is on the mat by virtue of touching the underneath part of the mat as in 'The fly is on the ceiling'. Our ability to infer what speakers say when uttering any word or expression rests, in large part, with deeply held background knowledge that is very much part of our pragmatic understanding of the world.

Secondary pragmatic knowledge, on the other hand, refers to information from context to provide an interpretation of what speakers implicate in discourse. For instance, a speaker who utters 'The cat is on the mat' might implicate that the addressee should get up and let the cat outside. Listeners draw the appropriate inferences about what speakers intend by recognizing specific features of the local context based on the common ground between themselves and speakers (i.e., their mutual beliefs, attitudes, knowledge). Thus, a speaker and listener may have as part of their common ground that the cat usually desires to go outside when it sits on the mat by the front door. Overall, though, listeners' stereotypical background knowledge dominates the application of secondary pragmatic information to reveal what is said by a speaker's utterance as distinct from what the speaker implicates (see Recanati, 1993).

Three recent studies by Hamblin and Gibbs (2002) examined the speed with which people understand expressions in which speakers' communicative intentions were either identical to what they pragmatically said or varied in some way, thus requiring listeners/readers to derive a conversational implicature. Consider the following stories, each of which ended with the same sentence:

Said/implied identical
Ted and Michele ran into each other at the mall.
Ted asked Michele what she had been doing lately.
Michele said that she had been busy car shopping.
Looking for ideas, Michele decided to consult Ted.
Michele asked Ted about his own car.
Ted mentioned:
'I drive a sports utility vehicle.' (enriched pragmatic meaning)

Said/implied different
Ted and Michele are planning a trip to Lake Tahoe.
Michele had heard that there was a terrible storm there.
She wondered if it was going to be safe for them to go.
Michele was concerned about the vehicle they would drive.
She asked Ted if he thought they would be okay.
Ted replied:
'I drive a sports utility vehicle.' (implicature)

In the first context, what the speaker pragmatically says by 'I drive a sports utility vehicle' is identical to what he implies in that there is no further pragmatic meaning he wishes for listeners to infer beyond that he drives a particular kind of car. But in the second context, the speaker not only says one thing (i.e., about the kind of car he drives), but also implies something beyond that meaning, namely that his particular car is safe to drive in a storm.

If people access primary pragmatic information sooner than they do secondary pragmatic knowledge, readers should take less time to comprehend utterances in which what speakers mean is identical to what they pragmatically say than to understand messages in which what speakers say underdetermines what they mean. This is exactly what we found. Drawing conversational implicatures increased processing effort over that needed to understand what speakers say. The data are consistent with the idea that people analyse what speakers say as part of their determination of what speakers imply.

Follow-up studies showed that people do not view conversational implicatures as ambiguous, but recognize that more than one meaning is specifically intended for them to understand. This is consistent with the view that inferring implicatures requires processing of both what speakers pragmatically say and pragmatically implicate. Moreover, participants in another separate study suggested that they only understood enriched pragmatic meaning in the Said/Implied Identical condition, but inferred both enriched pragmatic and pragmatically implied meanings in the Said/Implied Different condition. Although these data only reflect people's intuitions about the meanings of what they read, the findings are clearly consistent with the idea that people analysed what speakers pragmatically said as part of their understanding of what speakers conversationally implicated.

A second main experiment investigated processing of what speakers say and imply in a different way. Consider the following story, and two different final expressions:

Bill is a new tenant in an apartment building.
His neighbour Jack has lived there for four years.

Bill was concerned that the building might be too loud.
Bill decided to ask a neighbour about it.
Bill asked Jack since he was the only neighbour Bill had met.
Jack replied:
1. 'This is a very noisy building.' (said/implied identical)
2. 'I usually sleep with earplugs.' (said/implied different)

Understanding 'I usually sleep with earplugs' demands that listeners draw a pragmatic inference beyond that needed to understand what this same expression pragmatically says. However, understanding 'This is a very noisy building' in this context only requires listeners/readers to comprehend what the speaker pragmatically said. For this reason, participants should take less time to read 'This is a very noisy building' than 'I usually sleep with earplugs' in this context.

The results showed that people took significantly more time to read sentences necessitating the implicatures than they did the sentences requiring only enriched pragmatic said meanings. Once again, it appears that people more easily understand speakers' messages when these are identical to what they pragmatically say than when what is said underdetermines what the speakers intend to communicate (i.e., conversational implicatures).

The third study in this series specifically examined processing of the five types of indicative utterances studied by Gibbs and Moise (1997). These sentences were placed at the end of contexts designed to convey meanings where what speakers say was identical to what they implied (i.e., direct assertions), or where what speakers implied differed from what they pragmatically said (e.g., conversational implicatures). Consider the following example ending in a cardinal target sentence:

Bill wanted to date his co-worker Jane.
But Bill really didn't know much about her.
Being a bit shy, he first talked to another person, Fred.
Fred knew Jane fairly well.
Bill wondered is Jane was single.
Fred replied:
1. 'Jane is already married.' (said/implied identical)
2. 'Jane has three children.' (said/implied different)

Sentence 1 conveyed what the speaker implied directly. Yet Sentence 2 conveyed a conversational implicature in which what the speaker pragmatically said underdetermines what he implied in context (i.e., a conversational implicature). We examined the time it took people to interpret these two kinds of final statements. The results revealed that people took less time overall to read the final sentences when these conveyed the implied messages directly than when the sentences conveyed conversational implicatures.

Moreover, separate studies showed that participants did not see conversational implicatures as being ambiguous and that both enriched pragmatic and pragmatically implied meanings were understood when reading statements conveying conversational implicatures. These findings provide further evidence that people take longer to draw conversational implicatures than to understand assertions that only convey what speakers pragmatically say. Understanding indicative expressions such as 'Jane has three children' to imply in context that Jane is married requires additional time over that needed to interpret the same expression when it directly conveys what the speaker pragmatically says (i.e., Jane has exactly three children). Once more, people's complex pragmatic knowledge appears to be applied differently when understanding what speakers pragmatically say and when interpreting what speakers implicate in context. One possibility is that people more easily access primary pragmatic information to enable them to analyse what speakers pragmatically say and then more slowly access secondary pragmatics to infer what speakers implicate.

Hamblin and Gibbs concluded from their reading-time experiments that pragmatics is not simply used in understanding speakers' implicated meaning, but plays a role in utterance interpretation from the earliest stages of linguistic processing. In this sense, Grice was right in suggesting that people may analyse what speakers say as part of their inferring what is implicated. But Grice and others are incorrect in assuming that understanding what speakers say refers to minimally-pragmatic meaning and that enriched pragmatics only has a role in deriving conversational implicatures.

6 Conclusions

The psychological experiments described in this chapter were motivated by discussions originally provided by linguists and philosophers. My interest in different aspects of speech acts (e.g., promises, indirect requests, implicatures) led me to closely read the work of scholars outside of my own field of cognitive psychology. With the exception of a very few psychologists, most notably Herb Clark, and several psycholinguists interested in figurative language (e.g., Sam Glucksberg), there has been comparatively little experimental work on pragmatics within psychology (especially in contrast to the huge volume of work on other topics in psycholinguistics). My own efforts over the years have focused on showing how the work in linguistic-pragmatics can be profitably used to conduct well-designed psycholinguistic experiments. I take seriously the possibility that linguistic-pragmatics has important implications for psychological theory. Not every philosopher or linguist agrees that their respective views of pragmatic phenomena are necessarily intended as psychological hypotheses. Yet there is an increasing, encouraging trend within both linguistics and philosophy to formulate pragmatic theories that are constrained by psychological principles and data (e.g., Relevance

Theory). My research has been conducted as a contribution to this interdisciplinary effort on mind and meaning.

The empirical research on making and understanding promises, comprehending definite descriptions, making and interpreting indirect requests, and inferring what speakers pragmatically say and implicate has several benefits for linguistic-pragmatics. First, my work demonstrates the possibility of making good psychology out of linguistic and philosophical proposals. Each research project reviewed above formulated explicit competing hypotheses that were systematically investigated under controlled experimental conditions. In this way, psycholinguistic experiments aim to study linguistic-pragmatics within a falsification framework. I strongly embrace the belief that the best ideas in linguistic-pragmatics are those that can be experimentally examined and potentially falsified (where failing to falsify allows one to claim scientific evidence in support of a hypothesis). Similarly, making good psychology out of linguistic-pragmatics demands that each proposal be contrasted with some theoretical alternative. Although the falsification framework often leads to a 'winner takes all' view of theory testing, sophisticated experiments often enable researchers to tease apart several parts of any theory, showing which aspects may be psychologically plausible and which part may be invalid (e.g., the work on promising illustrated how only some parts of Searle's original intuitions were correct).

A second implication of my empirical research is that different research methods are required to experimentally investigate various pragmatic phenomena and theories. Although individual scholars' trained intuitions are key sources of evidence, they may not accurately reflect the ways ordinary speakers find meaning in everyday language. Systematically studying a wide range of people who are naïve to the hypotheses under consideration is essential to proving the psychological validity of any pragmatic theory. At the same time, many aspects of language production and comprehension occur so quickly (i.e., in mere milliseconds) that it is impossible to capture through conscious introspection anything about underlying cognitive processes in speaking and listening. Experimental psycholinguistics has devised a variety of indirect measures of online linguistic performance that can adequately explore the dynamics of many hypotheses from linguistic-pragmatics. Some of my experimental work (e.g., reading-time and priming tasks) allows me to distinguish between the psychological reality of varying hypotheses on pragmatic language understanding. I strongly claim that these online experiments are the only true adjudicators of competing ideas about mostly unconscious, rapid comprehension pragmatic processes.

Finally, psycholinguistic experiments are terribly important for convincing other psychologists of the necessity of incorporating ideas from linguistic-pragmatics into psychological models of language production and comprehension. For better or worse, few psychologists take the time to read work outside their discipline. The only way to demonstrate that pragmatics is an essential component in psychological theories is to conduct relevant experiments employing many of the experimental methods used by psycholinguists studying other aspects of linguistic performance. Thus, the experimental tasks themselves, while really being means to ends, may be persuasive tools to convince psychologists that pragmatics can be scientifically studied.

References

Atiyah, P. (1981). *Promises, Morals, and Law*. Oxford: Clarendon.
Austin, J. (1962). *How to do Things with Words*. New York: Oxford University Press.
Bach, K., and Harnish, R. (1979). *Linguistic Communication and Speech Acts*. Cambridge, MA: MIT Press.
Brown, P., and Levinson, S. (1987). *Politeness: Some Universals in Language Usage*. Cambridge: Cambridge University Press.
Carston, R., (1993). Conjunction, explanation, and relevance. *Lingua* 90: 27–48.
Clark, H., and Schunk, D. (1980). Polite responses to polite requests. *Cognition* 8: 111–43.
Cole, P. (1978). On the origins of referential opacity. In P. Cole (ed.), *Syntax and Semantics: Vol. 9, Pragmatics*: 1–22. New York: Academic Press.
Donnellan, K. (1966). Reference and definite description. *Philosophical Review* 75: 281–304.
Donnellan, K. (1968). Putting Humpty Dumpty back together again. *Philosophical Review* 77: 203–15.
Donnellan, K. (1978). Speakers reference, description, and anaphora. In P. Cole (ed.), *Syntax and Semantics: Vol. 9, Pragmatics*: 131–62. New York: Academic Press.
Fauconnier, G. (1985). *Mental Spaces*. Cambridge, MA: MIT Press.
Francik, E., and Clark, H. (1985). How to make requests that overcome obstacles to compliance. *Journal of Memory and Language* 24: 560–568.
Gibbs, R. (1979). Contextual effects in understanding indirect requests. *Discourse Processes* 2: 1–10.
Gibbs, R. (1981). Your wish is my command: Convention and context in interpreting indirect requests. *Journal of Verbal Learning and Verbal Behavior* 20: 431–44.
Gibbs, R. (1986). What makes some indirect speech acts conventional? *Journal of Memory and Language* 25: 181–96.
Gibbs, R. (1999). Speakers' intuitions and pragmatic theory. *Cognition* 69: 355–9.
Gibbs, R., and Delaney, S. (1987). Pragmatic factors in making and understanding promises. *Discourse Processes* 10: 107–26.
Gibbs, R., and Moise, J. (1997). Pragmatics in understanding what is said. *Cognition* 62: 51–74.
Grice, H. (1975). Logic and conversation. In Peter Cole and James Morgan (eds), *Syntax and Semantics: Vol. 3, Speech Acts*: 41–58. New York: Academic Press.
Grice, H. (1978). Further notes on logic and conversation. In P. Cole (ed.), *Syntax and Semantics: Vol. 9, Pragmatics*: 113–27. New York: Academic Press.
Hamblin, J., and Gibbs, R. (2002). Processing the meanings of what speakers say and implicate. Manuscript submitted for publication.
Hare, R. (1964). The promising game. *Revue Internationale de Philosophie* 18: 389–404.
Haviland, S., and Clark, H. (1974). What's new? Acquiring new information as a process in comprehension. *Journal of Verbal Learning and Verbal Behavior* 13: 512–21.

Kripke, S. (1972). Naming and necessity. In D. Davidson and G. Harmon (eds), *Semantics of Natural Language*: 253–355. Dordrecht: Reidel.

Lakoff, R. (1973). The logic of politeness: Or minding your p's and q's. In *Papers from the Ninth Regional Meeting*, Chicago Linguistic Society: 292–305.

Lewis, D. (1969). *Convention*. Cambridge, MA: Harvard University Press.

Morgan, J. (1978). Two types of convention in indirect speech acts. In P. Cole and J. Morgan (eds), *Syntax and Semantics: Vol. 3, Speech Acts*: 45–61. New York: Academic Press.

Mueller-Lust, R., and Gibbs, R. (1991). Inferring the interpretation of attributive and referential definite descriptions. *Discourse Processes* 14: 107–31.

Nicolle, S., and Clark, B. (1999). Experimental pragmatics and what is said: A response to Gibbs and Moise. *Cognition* 69: 337–54.

Over, D. (1985). Constructivity and the referential/attributive distinction. *Linguistics and Philosophy* 8: 415–30.

Peetz, V. (1977). Promises and threats. *Mind* 86: 578–81.

Quine, W. (1956). Quantifiers and propositional attitudes. *Journal of Philosophy* 53: 177–86.

Rawls, J. (1955). Two concepts of rules. *Philosophical Review* 64: 3–52.

Recanati, F. (1989). The pragmatics of what is said. *Mind and Language* 4: 295–329.

Recanati, F. (1993). *Direct Reference: From Language to Thought*. Cambridge: Blackwell.

Searle, J. (1964). How to derive Ought from Is. *Philosophical Review* 64: 43–58.

Searle, J. (1965). What is a speech act? In M. Black (ed.), *Philosophy in America*: 221–39. Ithaca, NY: Cornell University Press.

Searle, J. (1969). *Speech Acts*. Cambridge: Cambridge University Press.

Searle, J. (1975). Indirect speech acts. In P. Cole and J. Morgan (eds), *Syntax and Semantics: Vol. 3, Speech Acts*: 59–82. New York: Academic Press.

Searle, J. (1978) Literal meaning. *Erkenntnis* 13: 207–24.

Searle, J. (1979). *Expression and Meaning: Studies in the Theory of Speech Acts*. Cambridge: Cambridge University Press.

Searle, J. (1983). *Intentionality*. Cambridge: Cambridge University Press.

Sperber, D., and Wilson, D. (1986). *Relevance: Communication and Cognition*. Oxford: Blackwell.

Strawson, P. (1950). On referring. *Mind* 49: 320–4.

4
On the Automaticity of Pragmatic Processes: a Modular Proposal
Sam Glucksberg

1 Introduction

In the US legal system, experts are often called upon to testify on behalf of one or the other side of a legal dispute. On one particular occasion, I testified that employees in a financial benefit plan had been led to buy company stock in various implicit but persuasive ways. For example, a company brochure provided step-by-step instructions on how employees could allocate retirement funds to various investment options. Employees were instructed, 'first, you decide how much of your investment should be in Company stock'. I argued that the pragmatic concept of presupposition applied to this statement. The instruction carries the presupposition that at least some of the investment would be in that stock, and so people who would read it would be implicitly led to accept that presupposition. The opposing lawyer asked, 'wouldn't you agree, Doctor, that pragmatics is the fuzziest and least precise field in linguistics?'. He went on to ask if I also agreed that pragmatics was essentially a grab-bag for everything not covered by syntax and semantics, and hence not to be taken seriously. I disagreed with both of his attempted assertions.

The disregard, indeed the disrespect, for pragmatics and for those of us who work in the field can have serious consequences. One such consequence is the assumption that syntax and semantics are primary for models of language processing. This assumption leads to the choice of the sentence as the unit of analysis, and this has had far-reaching consequences for how the field of psycholinguistics has developed over the last 30 years. I will argue that this assumption has also led to process models of language comprehension that, because of their inappropriately exclusive focus on sentence meaning, turned out to be of limited scientific utility (i.e., they were incomplete if not wrong).

2 The primacy of the literal in process models of comprehension

How we talk about pragmatics can mislead people to assume that it is secondary to syntax and semantics, both in terms of importance and in terms of process. For example, Sperber and Wilson (2003) recently wrote that 'the goal of pragmatics is to explain the gap between sentence meaning and speaker's meaning'. Although Sperber and Wilson emphatically do not assign strict temporal priority to syntax and semantics over pragmatics, they could easily be misinterpreted. Indeed, in their original formulation of Relevance Theory (Sperber and Wilson, 1986), utterance interpretation was explicitly characterized as a two-stage process: a decoding stage followed by an inferential stage. Their current approach provides for a far more subtle, sophisticated interactive process, in which all levels of interpretation are active as an utterance unfolds (Sperber, personal communication). Their approach also affords a more privileged status to the putative inferential stage, along lines suggested by Carston (1988). For example, aspects of utterance meaning, such as temporal and causal connotations, need not be treated as extra-linguistic inferences, but instead as computationally accessible aspects of the propositions that are expressed. Within a truth-conditional view of meaning, Sperber and Wilson (1993) suggest that 'the primary bearers of truth conditions are not utterances but conceptual representations; to the extent that utterances have truth conditions, we see these as inherited from the propositions that those utterances express' (p. 24). These kinds of arguments call into question the traditional sharp distinction between linguistic decoding, which presumably generates a literal meaning, and inferential procedures that operate upon the literal meaning in a strictly sequential fashion.

In addition to temporal priority, literal meanings also enjoyed unconditional automaticity, at least with respect to single assertions. According to modularity theory, the language comprehender is a dedicated input module that is stimulus driven. Any linguistic input, irrespective of the context of utterance, automatically triggers not only phonological but also syntactic and semantic analyses that result in sentence comprehension (Fodor, 1983). When Miller and Johnson-Laird proposed that language comprehension 'occurs automatically without conscious control by the listener' (1976, p. 166), they were referring to literal comprehension, not inferential pragmatics. Grice (1975) and Searle (1979) were explicit on this issue. Non-literal interpretations, be they indirect requests or metaphors, are generated only when an utterance is 'defective': 'Where the utterance is defective if taken literally, look for an utterance meaning that differs from sentence meaning' (Searle, 1979, p. 114). Utterances are defective if, when taken literally, they seem to violate rules of conversation. One such rule is that utterances should be truthful, and so literally false assertions, (as when someone says 'what a beautiful day' in the midst of a rain storm) will trigger a conversational implicature (Grice, 1975). Utterances are also defective when they are literally true but make no sense in context, as when someone says 'dogs are animals' during a conversation about how to keep the streets of Lyon clean. In all such cases, literal meanings are generated automatically, only to be followed by additional inferences to arrive at an interpretable non-literal meaning, that is, speaker's meaning.

3 Theories of metaphor comprehension

How does this general model play out for theories of metaphor comprehension? The most influential instantiation of the general model is the standard pragmatic theory of metaphor comprehension (see Ortony, 1979). Metaphors are understood via a three-stage process that begins with the automatic generation of literal meanings:

1. Derive the literal meaning of the utterance.
2. Assess that meaning against the context of the utterance.
3. If the literal meaning does not make sense in context, seek an alternative meaning that does.

For nominal metaphors such as *alcohol is a crutch*, people recognize that the literal meaning is false. Following the Gricean Maxim of Truth, people reject that meaning as the speaker's meaning and seek an alternative non-literal meaning by implicitly converting the false categorical assertion into a true comparison, *alcohol is like a crutch*. All comparisons are, of course, true, because any two things can be alike in any number of ways (Goodman, 1972). The trick is to discover just how the two terms of the metaphor (the topic, *alcohol*, and the vehicle, *crutch*) are alike. How are alcohol and crutch alike? They are both English words, can be bought in stores, can be put in containers that are smaller than a house, but surely none of these would be what a speaker might intend. The Gricean principle of relevance now comes into play. Only those properties that might be relevant would be considered as plausible interpretations. How are candidate interpretations generated? The comparison theory of metaphor holds that the properties of the metaphor topic and vehicle are extracted, and then matched against one another. Those properties that are matched but irrelevant are discarded, while those that match and are considered relevant are selected as the ground for the comparison, and hence for the metaphor (Gentner and Wolff, 1997; Miller, 1979; Ortony, 1979).

We now have two hypotheses that had their roots in the primacy of the literal assumption. The first hypothesis is that literal meaning is automatically generated, while figurative meaning is optional. This is consistent with a core assumption of modularity theory, that sentence meaning processes are stimulus driven, but 'post-comprehension' inferences are not (Fodor, 1983). The second hypothesis, specific to metaphor comprehension, is that metaphors are understood via a comparison process that operates on true similes rather than on the original, putatively false, metaphors. This view has at least three testable implication. First, metaphor comprehension involves more processing steps than literal comprehension, and so it should take longer. This straightforward implication can be rejected. People can understand metaphors as rapidly and easily as comparable literal expressions (Blasko and Connine, 1993; Glucksberg, 2001; Ortony, Schallert, Reynolds and Antos, 1978). We now turn to the other two implications of the standard pragmatic theory.

3.1 Is metaphor understanding stimulus driven?

The second implication, that metaphor understanding is optional, can also be rejected. People cannot ignore simple transparent metaphors, even when literal meanings make perfect sense in context. Instead, the pragmatic operations of making sense of a putatively false literal categorical assertion such as *alcohol is a crutch* may be as automatic, that is, as stimulus driven, as the syntactic and semantic operations that produce sentence meanings. The evidence for this conclusion comes from a series of experiments in which people would have performed optimally if they did not seek non-literal meanings. In these experiments, we asked people to attend exclusively to literal meanings (Gildea and Glucksberg, 1983; Glucksberg, Gildea and Bookin, 1982; Keysar, 1989). These experiments were modelled after Stroop's (1935) classic demonstration that people cannot inhibit processing word meanings even when asked to ignore them. Stroop presented words printed in various colours and asked people to name the colour of the ink, *not* to read the words themselves. When colour words such as *red* were printed in any colour other than red, for example, in green, then people had difficulty saying 'green', indicating response competition from the involuntary reading of the word itself, 'red'. This colour-word interference effect was taken to mean that people could not inhibit reading words that are attended to, even when such inhibition would improve task performance. Word meanings cannot be ignored, just as the modularity hypothesis predicts.

We applied this logic to literally false but metaphorically true sentences such as 'some roads are snakes' and 'some offices are icebergs' (Glucksberg et al., 1982). Our experimental participants were shown sentences one at a time and were instructed to decide whether each sentence was literally true or false. We used four different kinds of sentences: literally true (e.g., *some birds are robins*); literally false (e.g., *some birds are apples*); metaphors (e.g., *some jobs are jails, some flutes are birds*); and scrambled metaphors (e.g., *some jobs are birds, some flutes are jails*). The metaphors were literally false category-membership assertions, but they were readily interpretable if taken non-literally. The scrambled metaphors were also literally false, but not readily interpretable.

If people could ignore metaphorical meanings, then the metaphors should take no longer to be judged literally false than the scrambled metaphors. If, on the other hand, people automatically register any metaphorical meanings that are available, then the metaphor sentences should take longer to judge as false than their scrambled counterparts. This would be caused by the response competition between the metaphors' 'true' non-literal meanings and their 'false' literal ones. Our results were clear-cut. People had difficulty in judging that metaphors were literally false. The mean response time to reject metaphor sentences (1239 msec) was reliably longer than the time to reject either literally false sentences (1185 msec) or scrambled metaphors (1162 msec). Furthermore, this effect is not due to mere associations between metaphor topics and vehicles, but to an appreciation of metaphorical meaning itself. If an association between topic and vehicle is sufficient for the interference effect, then the quantifier, *all* versus *some*, should make no difference, but it does. Metaphors that are judged to be good in the *some* form but poor in the *all* form behave differentially. For example, people tend to agree that *some surgeons are butchers*, but don't agree that *all surgeons are butchers*. Not surprisingly, *some surgeons are butchers*, which is literally false but metaphorically true, produces the metaphor interference effect. In contrast, *all surgeons are butchers*, which is both metaphorically and literally false, does not. We interpreted this metaphor interference effect in the same way that Stroop interpreted his colour-word interference effect: people could not inhibit their understanding of metaphorical meanings, even when literal meanings were acceptable in the context of our experiment.

Metaphorical meanings, then, like literal meanings, cannot be ignored. It seems as if the pragmatic operations that generate metaphorical meanings are, like syntactic and semantic operations, stimulus driven. If we have semantic and syntactic input modules, then we might have to add a pragmatic module as well. Indeed, Sperber and Wilson (2003) have proposed a dedicated comprehension module that computes inferences about speakers' intentions. But before considering what such a module might look like, let us examine the third hypothesis of standard pragmatic theory, that metaphor comprehension is accomplished via a comparison process.

3.2 Are metaphors understood via comparison?

According to the comparison view, metaphors of the form X is a Y are understood by converting them into simile form, X is *like* a Y. The simile is then understood by comparing the properties of X and Y. This view has been challenged on both theoretical and empirical grounds. One finding is particularly telling. Metaphors in class-inclusion form, such as *my lawyer is a shark*, take less time to understand than when in simile form, such as *my lawyer is like a shark* (Johnson, 1996). That metaphors can be understood more easily than similes argues that metaphors are exactly what they seem to be, namely class-inclusion assertions (Glucksberg, 2001; Glucksberg and Keysar, 1990). In such assertions, a metaphor vehicle such as *shark* is used to refer to the category of predatory creatures in general, not to the marine creature that is also named 'shark'. This dual reference function of metaphor vehicles is explicit in statements such as *Cambodia was Vietnam's Vietnam*.

Here, the first mention of *Vietnam* refers to the nation of Vietnam. In contrast, the second mention of *Vietnam* does not refer to that nation, but instead to the American involvement in Vietnam which has come to epitomize the category of disastrous military interventions. That intervention has become a metaphor for such disasters, and so the word 'Vietnam' can be used as a metaphor vehicle to characterize other ill-fated military actions, such as Vietnam's invasion of Cambodia. More generally, metaphor vehicles such as *Vietnam* can be used as names for categories that have no names of their own (Brown, 1958).

Metaphors are thus attributive assertions, not comparisons. To say that someone's job is a jail is to attribute salient properties of the category JAIL to a particular job (Ortony, 1979). That particular job is now included in the general, abstract category of JAIL, and as a consequence of that categorization is now similar in relevant respects to literal jails (Glucksberg, McGlone and Manfredi, 1997). Predicative metaphors, in which verbs are used figuratively, function similarly (Torreano, 1997). The verb TO FLY literally entails movement in the air. Flying through the air epitomizes speed, and so expressions such as *he hopped on his bike and flew home* are readily understood via the same strategies that nominal metaphors, such as *his bike was an arrow*, are understood. Arrows are prototypical members of the category of speeding things; flying is a prototypical member of the category of fast travel. For both nominal and predicative metaphors, prototypical members of categories can be used as metaphor vehicles to attribute properties to topics of interest.

The categorization view also explains why metaphorical comparisons can be paraphrased as category statements and vice versa, while literal comparisons cannot. We can convert similes such as *my lawyer was like a shark* into the categorical form *my lawyer was a shark* and not appreciably change the interpretation (but see below). However, we cannot convert literal comparisons such as *barracudas are like sharks* into the categorical form *barracudas are sharks* and still make sense. Metaphoric comparisons can be expressed as categorical statements because the simile is an implicit category statement (Glucksberg and Keysar, 1990). The metaphor vehicle, *shark*, refers to the

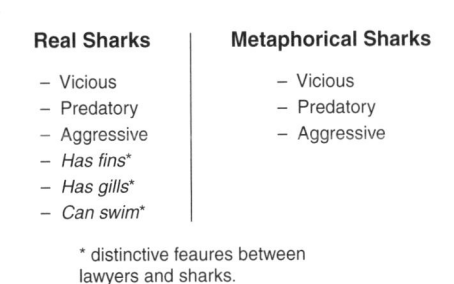

Real Sharks	**Metaphorical Sharks**
– Vicious	– Vicious
– Predatory	– Predatory
– Aggressive	– Aggressive
– *Has fins**	
– *Has gills**	
– *Can swim**	

* distinctive feaures between lawyers and sharks.

Figure 4.1 Features of 'real' and metaphorical sharks

superordinate category of predatory creatures exemplified by the literal shark, not to the subordinate-level, the fish that we call a shark (see Figure 4.1).

3.3 Metaphors as implicit similes revisited

We have argued that metaphors are not implicit similes. Instead, metaphors are explicit categorical assertions and are understood as class inclusion statements. Metaphor vehicles such as 'shark' can be used to refer either at the subordinate level, that is, to the literal concept, or at the superordinate level, to the metaphorical category. Via dual reference, people can refer either to the literal or to the metaphorical, or both. In metaphors, the 'IS A' usage is a cue to interpret the metaphor vehicle at the superordinate level – as referring to the metaphorical category that the literal concept exemplifies. In similes, 'LIKE' leads people to interpret the vehicle at the subordinate literal level, at least initially (Gernsbacher, Keysar, Robertson and Werner, 2001; see also Glucksberg, Newsome and Goldvarg, 2001).

These considerations suggest that even though metaphors can be paraphrased as similes and vice versa, they should not yield identical interpretations. Indeed, Relevance Theory suggests that whenever there is a choice between alternative ways to say something, that choice should have consequences. Given that speakers have a choice and that listeners know that speakers have a choice, there should be a consequential difference between the two alternatives. How should similes and metaphors differ in their interpretation? Recall that in metaphors, the vehicle should be taken to refer at a superordinate level, while in similes it should be at the subordinate level. Two predictions follow from this hypothesis. The first is that the degree of similarity between a metaphor topic and vehicle should differ between metaphors and similes. Specifically, the perceived similarity between topic and vehicle will be greater for metaphors than for similes. The second prediction is that people's interpretations of metaphors and similes will also be different: There should be fewer literal-related and more metaphoric properties in metaphor interpretations than in simile interpretations.

Tversky's (1977) contrast model of similarity provides a basis for the similarity prediction. According to Tversky, the perceived similarity between two items will be a positive function of properties in common to the two items and a negative function of properties that are not in common. Consider the expressions 'My lawyer is a shark' versus 'My lawyer is like a shark'. If the former use of 'shark' is at the superordinate level, then subordinate-level literal properties such as having fins, leathery skin, gills and so forth would not be part of the conceptual representation of the utterance (see Figure 4.1). The only properties of shark that would be included are the superordinate relevant ones, for example, vicious, predatory, tenacious and so forth (Gernsbacher et al., 2001). Given that the metaphor is understood, these properties are now in common to both the metaphor vehicle, 'shark', and the metaphor topic 'lawyer' (Camac and Glucksberg, 1984). Now consider the simile, in which 'shark' is initially interpreted at the subordinate-level of abstraction. In addition to metaphor-relevant properties such as 'vicious', literal properties such as having fins and leathery skin would also be included in the conceptual representation of the word 'shark'. Because these properties are distinctive (included in 'shark' but not in 'my lawyer'), the perceived similarity between 'shark' and 'my lawyer' should be attenuated.

To test these hypotheses, we gave people statements in either metaphor or simile form, and asked them to rate the similarity of topic to vehicle for each one. For example, for the trope *some ideas are/are like diamonds*, one group of people answered the question, 'how similar are ideas to diamonds?' on a scale of zero (not at all similar) to six (extremely similar). The people who had read the metaphor form provided a mean topic-vehicle rating of 2.94, which was reliably higher than the mean rating after similes, 2.49. Metaphors, as predicted from the dual-reference hypothesis and Tversky's contrast model, induce a higher similarity between topic and vehicle than do similes (Hasson, Estes and Glucksberg, 2001).

The second issue is whether or not metaphors produce more metaphorical properties to be attributed to the topic than do similes. We can identify two types of properties. One type is common to both the subordinate-level literal meaning and the superordinate, metaphorical meaning of the vehicle. The property of *predatory* for the term *shark* in the lawyer-shark metaphor is applicable to both the literal shark and the metaphorical one, albeit differently instantiated. Such properties are context-independent, belonging to sharks unless specifically negated. The second property type is context-dependent, applying to the metaphor vehicle only in the context of the metaphor topic.[1] The property of *unethical* for the combination *lawyer-shark* is an example of such a property. It is not a property of the literal shark, but could be a property of metaphorical sharks in the context of the metaphor topic, lawyer. Thus, loan sharks and pool sharks could also be unethical, but hammerhead sharks could not be. These two types of properties should be differentially accessible when people interpret metaphors versus similes. Specifically, we should expect people to include relatively more context-dependent metaphorical properties of metaphor vehicles when interpreting metaphors than when interpreting similes. This is because the vehicle term in a metaphor refers directly at the superordinate metaphorical level, while in a simile it refers directly at the subordinate, literal level of abstraction.

We gave participants statements in either metaphor or simile form, and then asked them to paraphrase them, one at a time. We then coded the

---

[1] The classification of properties as context-independent (literal) and context-dependent (metaphorical) is analogous to Barsalou's (1983) distinction between context-independent and dependent properties in his discussion of taxonomic vs *ad hoc* functional categories. For example, round is a context-independent property of *ball*; *oval* is a context-dependent property when the ball is a *football* (American style).

paraphrases in terms of whether each property mentioned was either a literal property of the vehicle, or a property that, in isolation, would not belong to the vehicle but was nonetheless attributed to the topic. For example, for the statement *some ideas are/are like diamonds*, properties such as rare, desirable, shiny, glitter and very valuable would be coded as literal properties, applicable to the vehicle (diamonds) in isolation, as well as to the topic in the context of the metaphor. In contrast, properties such as incredible, insightful, fantastic, creative and unique would be coded as non-literal, applicable to the superordinate (metaphorical) sense of diamonds rather than the subordinate-level literal sense. The coding was, of course, done blind. The coder did not know under which condition the paraphrases came from.

As shown in Figure 4.2, there were significantly more superordinate-level (metaphorical) properties than subordinate-level (literal) properties in the paraphrases of metaphors (M = 0.89 vs 0.57). In contrast, the mean number of superordinate-and subordinate-level properties did not differ in the paraphrases of similes, M = 0.69 vs 0.73). Furthermore, as would be expected from the dual reference hypothesis, the mean number of superordinate (metaphorical) properties for metaphors was significantly higher than for similes, suggesting that metaphors are a more effective means of communicating than are similes. Similes are, in effect, more 'literal', conveying more mundane information than do metaphors. Clearly, metaphors and similes are not equivalent, either in form, or in their effect. As Richard Russo wrote in his scathing novel of academic life 'Straight Man':

Sophomoric student: I like the clouds. They're, like, a metaphor.
Sarcastic professor: They *are* a metaphor... if they were *like* a metaphor they'd be, like, a simile.

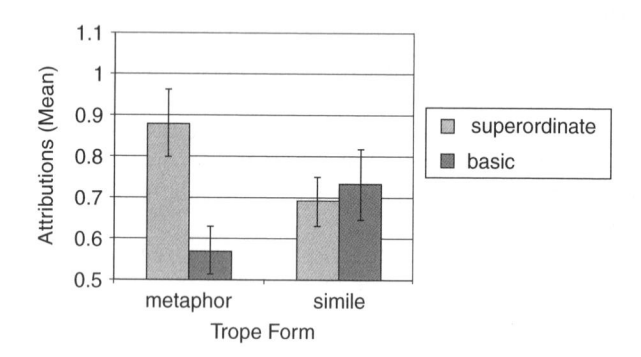

Significantly more basic-level (literal) attributions for similes than for metaphors. Significantly more superordinate level (metaphorical) attributions for metaphors than for similes.

Figure 4.2 Mean number of attributions as a function of trope form

Clearly, similes are *like* metaphors, but they're not metaphors themselves. We turn now to a different instantiation of nominal metaphors, implicit metaphors in the surface form of noun-noun and adjective-noun phrases.

4 Conceptual combination

How do people understand adjective-noun and noun-noun combinations such as *blind lawyer* and *shark lawyer*? The primacy of the literal would lead us to expect that literal interpretations would be preferred to non-literal ones: (a) whenever the literal makes sense in context, that is, is not 'defective' in any discernible way; and (b) that the process by which such constructions are understood would be governed, at least initially, by syntactic and semantic analysis. In other words, 'post-comprehension' (read 'post-lexical' or 'post-syntactic') processes such as pragmatic inferences would be operative only after a literal interpretation turned out to be defective. There is strong evidence against both of these predictions. First, just as with explicit nominal metaphors, noun-noun compounds such as *shark lawyer* and *sieve memory* are interpreted metaphorically, even though non-defective literal interpretations are available. Second, emergent properties of adjective-noun compounds that cannot be generated compositionally are not only common, but can even be generated faster than compositionally derived properties. I will briefly describe each of these phenomena in turn.

4.1 When are metaphorical interpretations generated?

In English, noun-noun combinations, such as *rock star* and *moon rock*, pose problems for interpretation. Unlike languages such as Russian, which have explicit case marking, noun-noun combinations in English provide no explicit information on the roles of the two nouns. To compound the problem (pun intended), noun-noun combinations can be interpreted both literally and metaphorically. For example, if the compound *shark lawyer* is interpreted as a lawyer who is predatory and aggressive, then the noun 'shark' is used to refer to a metaphorical rather than literal shark. If the compound is interpreted as a lawyer who represents an environmental group dedicated to protecting sharks from over-fishing, then 'shark' is used to refer to the literal shark. What determines whether a compound will be interpreted literally or metaphorically?

Noun-noun combinations are interpreted primarily via two alternative processing strategies: property attribution and relational linking (Goldvarg and Glucksberg, 1998; Wisniewski, 1997). For example, the combination *medicine music* is typically interpreted via property attribution to refer to music that can be used for healing purposes. In such cases, one or more properties of the modifier noun (medicine) are attributed to the head noun (music). Wisniewski (1997) points out that in many cases of property attribution the modifier is used metaphorically rather than literally.[2] In contrast, *mourner musician* is typically interpreted via relational linking as meaning a musician who plays for mourners. In such cases, the compound is interpreted in terms of a relation between the head noun (musician) and modifier noun (mourner). Such relational interpretations are overwhelmingly literal. What will people do when presented with noun-noun compounds that can be interpreted either metaphorically (via property attribution) or literally (via relational linking)? If literal meanings have priority, then people should opt for the literal for two reasons. Literal meanings are automatically generated first, and so they should become available first. Then, because the literal meanings are not defective in any sense, there should no reason to infer additional non-literal meanings. If, however, metaphorical meanings are also automatically generated, then such meanings should be generated whenever they are available, irrespective of whether or not non-defective literal meanings are also available.

Yevgenya Goldvarg and I tested these two alternative predictions to answer the question: are the operations required for metaphor comprehension stimulus driven just as our syntactic and semantic mechanisms are stimulus driven? We gave college students noun-noun compounds that could be interpreted metaphorically via property attribution, or literally via relational linking. Such compounds have a modifier noun that can provide properties that are relevant to a head noun. In addition, these noun-noun compounds, if presented in the form *X is a Y*, would be apt and comprehensible metaphors (see note 2). The criteria for such modifier and head nouns were derived from Glucksberg, McGlone and Manfredi's (1997) interactive property attribution model: the modifier noun had salient properties that were relevant to the head noun concept. *Shark lawyer* is an example of such a compound. A relevant dimension of *lawyer* is degree of aggressiveness. Some lawyers are quiet, scholarly and gentle; others are vicious and predatory. Salient properties of the *shark* concept include the properties *aggressive, vicious* and *predatory*. If people are sensitive to such relations between modifier and head noun concepts, then such compounds should tend to be interpreted metaphorically via property attribution, despite the availability of literal, relational interpretations. In contrast, compounds that cannot be paraphrased as comprehensible metaphors (i.e., do not have the appropriate relationship between modifier and head noun concepts) should be interpreted literally via relational linking. *Witch parade* and *murder brochure* are two examples of such compounds.

---

[2] One test of whether a noun-noun compound has been interpreted metaphorically was proposed by Levy (1978). If the compound becomes a metaphor when paraphrased as a categorical assertion, as in *music is medicine*, and yields the same interpretation as the compound (music can be healing), then the noun-noun interpretation can be considered metaphorical.

We gave college students both types of compounds, and asked them to provide interpretations of them. The interpretations were then scored by two independent judges. Each interpretation was assigned to one of three categories: 'property attribution', 'relation linking' and 'other'. The judges used the following criteria: if an interpretation consisted of an adjective in place of modifier and unchanged head-noun, as in *strong arms* for *steel arms*, the strategy was classified as property attribution and metaphorical. If an interpretation consisted of an unchanged head-noun and an unchanged modifier, and a verb between them, as in *arms made-of steel*, the interpretation was classified as relation linking and literal. All other interpretations were assigned to the 'other' category.

As expected, most (75 per cent) of the noun-noun combinations that could be paraphrased as metaphors were interpreted metaphorically. In contrast, 82 per cent of the combinations that could not be paraphrased in this way were interpreted relationally and literally (Goldvarg and Glucksberg, 1998, experiment 1), indicating that people had no problems in producing relational interpretations in general. These results speak directly to the hypothesis that metaphor comprehension is stimulus driven. When a metaphorical interpretation is available, even in implicit noun-noun form, people generate that interpretation. This finding is analogous to Glucksberg, Gildea and Bookins's (1982) demonstration that explicit metaphors are understood non-optionally. Metaphors cannot be ignored, whether they are explicit or implicit.

4.2 Are compounds understood compositionally or pragmatically?

The initial attempts to explain how people deal with adjective-noun and noun-noun compounds reflected the 'primacy-of-the-literal' assumption, in this case in the form of compositional semantics. To generate the meaning of, say, 'red apples', compositional models posit a two-stage process in which the features of each member of the conceptual combination are first accessed independently and then, in a second stage, combined to yield the features of the combination (Springer and Murphy, 1992). This compositional process is appealing and could sometimes work. For a phrase such as *red apples*, the feature 'red' and the features of 'apples' can be combined to yield *apples that have the colour red*. However, there are many conceptual combinations for which the sequential compositional model fails. For example, the feature 'white' of *peeled apples* cannot come from either the modifier *peeled* or the head noun *apples* because neither peeled things nor apples are generally white. Similarly, for the combination *pet bird*, most people agree that this could mean a bird that talks, even though neither pets nor birds typically talk (Hampton, 1987). These features emerge from the interpretation of the combined concept as a whole, not from an understanding of its constituent parts. Such features are called *phrase features*, because they are true of the phrase but are not true of either the head noun or the modifier in isolation. That is, peeled apples are white, though neither apples nor peeled things in general are white. Phrase features can be contrasted with *noun features*, which are true of both the combined concept and the head noun in isolation. For instance, 'round' is a noun feature in that both peeled apples and apples in general are round.

To account for emergent phrase features such as 'white' for 'peeled apples', Smith, Osherson, Rips and Keane (1988) proposed a three-stage model rather than the two-stage model for combinatorial features (e.g., red for red apples). Like the two-stage compositional model, the first and second stages involve an initial spreading activation process that activates the features of each member of the combination, followed by a feature combination stage. However, after the activation and combination of features, world knowledge is used to construct more elaborate representations. Thus, for the combination *leather seats*, the spreading activation process would yield the features of 'leather' and of 'seats', which would then be combined. Finally, a slower elaborative process could generate emergent features such as 'found in luxury automobiles' (Weber, 1989, cited in Springer and Murphy, 1992). The elaboration process may be in the form of inference from world knowledge or, for conventional expressions such as 'social x-ray',[3] retrieval from a phrasal lexicon (Jackendoff, 1995). This kind of multi-stage model for processing conceptual combinations is of a piece with standard views of sentence processing: linguistic decoding followed by pragmatic inferences to, as Sperber and Wilson (2003) put it, 'bridge the gap between sentence meaning and speaker's meaning'. Furthermore, it makes eminent intuitive sense. How else derive the property 'white' from the phrase 'peeled apples'? Surely, the features of a combination's constituents must be accessed first, followed by an inference that produces the emergent features of the combination.[4] As intuitive as this might be, experimental evidence suggests otherwise. Consider, again, how we might come to the understanding that peeled apples are white. The sequential models posit that features of the constituents 'peeled' and 'apples' are accessed first, and so the feature 'round' should be accessible before the phrase feature 'white'. This is because 'round' is accessed in the initial feature activation stage from the constituent *apples*, while 'white' can only be generated in a subsequent elaboration stage from our knowledge of apples and what they look like after being peeled. To test this hypothesis, Springer and Murphy (1992) asked people to verify noun features such as *peeled apples are round* and phrase features such as *peeled apples are white*. Their results were as clear as they were surprising: phrase

---

[3] The term social x-ray is used to refer to incredibly thin, often anorexic women who virtually starve to conform to the American elite's standards of beauty.
[4] Note that the very use of the term 'emergent' presupposes a sequential process beginning with constituent features, leading to features that are true only of the combination.

features are verified more quickly and more accurately than noun features. Similar findings were reported by Potter and Faulconer (1979) and by Hampton and Springer (1989).

What accounts for this differential accessibility of phrase and noun features, that is, the phrase feature superiority effect? Relevance may be the critical factor. We suggest that, in the absence of a context that indicates otherwise, people automatically consider phrase features to be more relevant than noun features. Why should people do this? When a phrase such as 'peeled apples are...' is encountered, people assume that *peeled apples* (instead of just *apples*) were specified in order to highlight some way in which peeled apples might differ from unpeeled apples. In other words, phrase features become relevant because they differentiate the combined concept from other members of the head noun category. Thus, phrase features such as 'white' and 'sticky' would be relevant and appropriate completions of the phrase 'peeled apples are...'. In contrast, the feature 'round' does not distinguish peeled apples from unpeeled ones, and therefore would be an inappropriate and unexpected completion. This default, automatic assumption of relevance might produce the phrase feature superiority effect.

We already know that relevant contexts facilitate access to the features of simple concepts (e.g., *apples*) from a variety of paradigms, including sentence verification (McKoon and Ratcliff, 1988, 1982; Tabossi and Johnson-Laird, 1980), lexical decision (Tabossi, 1988) and naming tasks (Hess, Foss and Carroll, 1995). We used a similar paradigm, but tested combined concepts rather than simple concepts. We expected that relevant contexts would also facilitate feature verification in combined concepts, and that this will hold true irrespective of whether it is a noun feature or a phrase feature. If noun features are relevant, then they should be more accessible than phrase features. If phrase features are relevant, then they should be more accessible than noun features. And if relevance is responsible for phrase feature superiority in neutral contexts, then this superiority should not only be eliminated but should be reversed by contexts that make noun features relevant and phrase features irrelevant.

Zachary Estes and I (Glucksberg and Estes, 2000) tested this hypothesis in an experiment in which people verified statements about either noun or phrase features, for example, 'peeled apples are round' versus 'peeled apples are white'. However, the statements to be verified were not presented in isolation, but rather in contexts that made either noun (e.g., 'round') or phrase (e.g., 'white') features relevant. One of the contexts that made a noun feature relevant was:

> Alan and Susan were bored one Sunday afternoon, and they decided to play lawn bowling in their back yard. But they didn't have any lawn balls, so they searched around the house. The first things they found

was a pair of peeled apples that were going to be used with dinner. They were a little sticky, but they worked just fine.

The corresponding phrase-feature-relevant context was:

Alan was a famous French chef who used fresh fruit to garnish his meals. Each night, he spent half an hour selecting the perfect fruit for the centerpiece. Last night, Alan decided to make a colorful centerpiece. He used orange slices, kiwi and peeled apples. The centerpiece was gorgeous, until the guests began to eat it.

We expected phrase features to be verified faster than noun features when they were relevant, and the reverse to be true when noun features were relevant. For instance, when colour is relevant, the phrase feature 'white' of *peeled apples* should be more accessible. But when shape is relevant, then the noun feature 'round' should be more accessible. Our results were crystal clear. When noun features were relevant, noun features were verified faster than phrase features, 1980 msec versus 2117. They were also verified more accurately, 89 per cent versus 82 per cent. The reverse was true when phrase features were relevant. Phrase features were verified faster and more accurately than noun features, 1921 msec versus 2222 msec, and 89 per cent versus 81 per cent, respectively. Furthermore, the effects of relevance were symmetrical, suggesting that given relevance, noun and phrase features are equally accessible. Could the elimination of the phrase superiority effect be attributable to the extra effort of processing target items in particular contexts rather than in isolation? The levels of accuracy and response times in this experiment suggest not; they were comparable to those obtained in other experiments that found phrase feature superiority using either no contexts (Springer and Murphy, 1992) or neutral contexts (Gagné and Murphy, 1996). This suggests that the contexts that we used did not require extra effort or processing beyond that ordinarily required for items presented in isolation or in neutral contexts. In short, the effects of making one or another feature-type relevant is not attributable to any additional, integrative processing beyond what is generally required for sentence verification without any context.

Relevant information is thus more accessible than irrelevant information, irrespective of whether it is a phrase or noun feature. When noun features are relevant, then they are accessed more quickly than phrase features. When phrase features are relevant, then they are accessed more quickly than noun features. Apparently, when people understand conceptual combinations in which any number of features are potentially available, feature accessibility is selective, favouring those features that are relevant in the particular context. If the context does not make any particular feature relevant, then an automatic default strategy is deployed. In adjective-noun and noun-noun combinations, the automatic default is to treat the information provided by the modifier as relevant. Hence, when such combinations are encountered in isolation, phrase features are accessed preferentially over noun features. But as our data indicate, this default can be completely overridden by appropriate contexts.

5 Pragmatics and automaticity reconsidered

I have argued that many, if not all, pragmatic processes are automatically engaged when people process any verbal input. Put another way, they are stimulus driven, just as phonological, syntactic and semantic processes are stimulus driven. In this sense, one can postulate a pragmatic module in the most general sense intended by Sperber and Wilson (2003), as a domain- or task-specific autonomous computational mechanism. Sperber and Wilson view the 'pragmatic module' as a sub-module of a more general mind-reading module. The general mind-reading module automatically computes other people's intentions, including what others are thinking, attending to and trying to communicate. Their sub-module computes 'relevance-based procedure(s) to ostensive stimuli, and in particular to linguistic utterances'. While my concept of modularity is certainly not incompatible with theirs, it is both more restrictive and more general. It is more restrictive in that it claims only automatic elicitation of pragmatic procedures, without a commitment to specific procedures. It is more general in that it is not confined to 'mind-reading' computations on the one hand, nor to relevance-based procedures on the other. That said, our respective approaches are certainly compatible, and concerned with the same sets of issues. Among those issues are: (a) the automaticity of pragmatic processes; and (b) how one might specify the boundaries between linguistic decoding on the one hand, and inferential processes on the other. I first turn to the automaticity issue.

5.1 Automatic engagement of pragmatic processes

Standard pragmatic theory treated metaphor processing as optional, that is, not obligatory. We tested this claim experimentally, and found it to be false, both with respect to explicit nominal and predicative metaphors, and to implicit metaphors in conceptual combinations. How general is our claim for automaticity? Are there pragmatic processes that are not obligatory, that are, in Gricean terms, defeasible? One candidate for a non-obligatory process is scalar implicature. Normally, people interpret *some* as implicating *not all*. For example, people tend to disagree with statements such as *some elephants have trunks*. Interestingly, children seem perfectly happy with such statements. Noveck (2001) asked children (ages 7–11) and adults if they agreed with logically true but pragmatically infelicitous statements quantified by *some*. The children tended to agree about 87 per cent of the time, while adults agreed only 41 per cent of the time (Experiment 3), leading Noveck to surmise that children are more logical than adults. More germane to the automaticity issue, Noveck found that adults were capable of responding 'logically' if they were trained to do so. Both adults and children were shown a box containing a toy parrot. The box was then closed and then they were shown a puppet who said, 'there might be a parrot in the box'. They were then asked whether or not the puppet was right. Without any specific training in this task, children tended to say 'yes', while adults overwhelmingly said 'no', presumably because the adults interpreted 'might' as implicating 'might not'. After intensive training, adults improved, responding 'logically' 75 per cent of the time, but still did not match 7-year-olds, who responded logically 94 per cent of the time.

Noveck interpreted these results as indicating that scalar implicatures are non-obligatory, but another interpretation is possible. Recall our experiments with metaphors: people could not inhibit metaphor processing even when asked to attend solely to literal truth conditions. However, they could respond correctly on the basis of literal truth value: it just took them longer (Glucksberg et al., 1982). The adults in Noveck's experiments also could respond correctly, that is, 'logically', at least 75 per cent of the time. However, in order to demonstrate that scalar implicatures were not drawn, one would need more sensitive measures such as response latencies. If people need more time to respond logically when such responses conflict with pragmatic ones, then the most likely conclusion is that scalar implicatures are not blocked, but instead simply aren't expressed overtly. This issue is clearly important and, fortunately, open to experimental investigation (see Chapters 12 and 14).

A second issue concerns the boundary between decoding and inference. In the case of scalar implicature, one could argue that, for adults, the meaning of *some* includes the implication of *not all*. A weaker form of this view is that the scalar implicature has become automatized, and so is triggered whenever we encounter the quantifier *some*. Either the outcome of decoding or the outcome of the obligatory implicature can be inhibited, allowing adults to be 'logical' in Noveck's terms, but the decoding or implicature itself is obligatory. This issue – how to draw the line between decoding and inference – goes well beyond scalar implicature. It is central to our understanding of the nature of conceptual representations.

5.2 Conceptual representation: when is inference required?

Most treatments of linguistic decoding assume that linguistic input is represented propositionally, that is, in abstract symbolic form (e.g., Kintsch, 1998). Such representations may be quite rich. Carston (1998), for example, argues that number terms can be represented in at least two different ways. For example, when someone asks 'were there a hundred people at your lecture', and I reply 'oh yes', I could be understood as saying that there were exactly 100 people, or that there were at least that many. Number terms can thus be used to communicate exact amount (precisely *n*) or an as an interval (at least *n*). Which of these two representations is generated by an addressee will be contextually determined, but in either case would be generated by an enrichment process. The ensuing representation would be 'part of the proposition expressed, that is, truth conditional' (Carston, 1998, p. 21), and so would not require any additional, inferential processes. The necessary 'enrichment process' is relevance-driven and, presumably, obligatory.

How far might such enrichment processes go, and are they limited to propositional representations? A growing body of literature argues that language comprehension involves much richer conceptual representations than those provided by amodal propositional structures. Barsalou (1999) argues that people use perceptual representations rather than (or perhaps in addition to?) amodal propositions to represent meanings. These perceptual symbols have an analogue relationship with their referents, and can be thought of as schematic, imaginal forms of the mental models proposed by Johnson-Laird (1983) in the domain of deductive reasoning.

What kinds of information are included in people's conceptual representations of linguistic input? Put in terms of Wilson and Sperber's (1993) discussion of linguistic form and relevance, what types of information are on the decoding side of the borderline between decoding and inference? Recent research suggests that far more is available than has generally been assumed. Among the perceptual attributes that people seem to encode automatically are object shapes and orientations. For example, the perceived shape of an eagle differs when it is flying from when it is in a nest. Do people encode the specific shapes of eagles as a function of linguistic context, that is, described as flying or as sitting in a nest? Zwaan, Stanfeld and Yaxley (2002) examined this issue by giving people sentences such as *The ranger saw the eagle in the sky* or *The ranger saw the eagle in its nest*. Note that the different shapes are not explicitly described, only implied. A picture of an eagle was then shown, and people had to name the picture, for example by saying 'eagle'. Naming responses were faster following a shape match (e.g., wings spread out following eagle-in-the sky sentence) than a shape mismatch, suggesting that people did indeed routinely encode shapes implied by linguistic input. Whether the encoding was imaginal-perceptual or in propositional form cannot be decided by these data, but this does not matter for the argument. What does matter is that shape information was automatically encoded even though it was not needed for the task at hand. Shape information in this instance seems to be on the decoding side of the borderline between decoding and inference. Analogous results have been reported for object orientation, for example, a pencil put in a cup versus a pencil put on a table (Stanfield and Zwaan, 2001).

If object shapes and orientations are routinely encoded from linguistic inputs, then it seems safe to assume that object colours might also be encoded. If so, then the information that peeled apples are white might not require a pragmatic inference after all. This colour information, like shape and orientation, might also be on the decoding side of the decoding–inference borderline. Recall that the colour of peeled apples is more accessible than the shape. We argued earlier that this was attributable to the default relevance of colour over shape in neutral contexts. However, when shape is relevant in a given context, then it is more accessible than colour (Glucksberg and Estes, 2000). If indeed the shape and colour of objects can be included in the coded linguistic input, and if the choice of colour or shape can be influenced by relevance considerations, then a central claim of Relevance Theory is supported. Not only pragmatic inferences, but the selection of information to be decoded will be a function of relevance in any given context (cf. Carston, 1998).

In light of these considerations, the task of drawing a borderline between decoding and inference becomes more subtle. As with cognitive theory in general, there is always an interaction between representation and process. In any given domain, the richer the representation, the simpler the process. In the domain of language comprehension, the richer the information that is linguistically encoded, the less the need for enriching inferences. As Wilson and Sperber put it: 'Linguistic decoding provides input to the inferential phase of comprehension' (1993, p. 1). If relevant information is already in the linguistic decoding, then that information need not be inferred, it is already available. Clearly, one task for experimental pragmatics will be to determine, for all important classes of linguistic expressions, what information is linguistically encoded, and what needs to be inferred.

6 Concluding remarks on pragmatic modules

We began this chapter with the observation that pragmatic processes might be stimulus driven, in the same way that syntactic and semantic processes are stimulus driven. If so, then we might be licensed to posit a pragmatic module, a specialized set of procedures that are automatically engaged by linguistic input. Such a module need not have all of the characteristics of the paradigmatic input modules proposed by Fodor (1983). For example, pragmatic modules should not be 'impenetrable', and indeed they seem not to be. As we demonstrated in our study of conceptual combinations and relevance, the pragmatic process of assigning salience to one or another property of noun-phrase referents was context-dependent (Glucksberg and Estes, 2000). Pragmatic modules may or may not have dedicated neurological substrates. My intuition is that they would not. To my knowledge, no specific brain areas have been implicated in pragmatic functions, either through studies of brain damage, or by brain imaging. Most likely, prefrontal cortex is involved via either language-specific or general executive functioning. The one important characteristic of a pragmatic module (or modules) is that pragmatic operations are obligatory: they are driven by linguistic input.

Furthermore, they are intrinsically involved in both linguistic encoding and post-encoding inferences. Pragmatics is thus no less primary than syntax or semantics to language processing.

References

Barsalou, L. (1983). Ad hoc categories. *Memory and Cognition* 11: 211–27.
Barsalou, L. W. (1999). Perceptual symbol systems. *Behavioral and Brain Sciences* 22: 577–660.
Blasko, D. G., and Connine, C. M. (1993). Effects of familiarity and aptness on metaphor processing. *Journal of Experimental Psychology: Learning, Memory and Cognition* 19: 295–308.
Brown, R. (1958). *Words and Things*. New York: The Free Press.
Camac. M., and Glucksberg, S. (1984). Metaphors do not use associations between concepts, they are used to create them. *Journal of Psycholinguistic Research* 13: 443–55.
Carston, R. (1998). Informativeness, relevance and scalar implicature. In R. Carston and Seiji Uchida (eds), *Relevance Theory: Applications and Implications*: 179–236. Amsterdam and Philadelphia: John Benjamins.
Fodor, J. A. (1983). *The Modularity of Mind*. Cambridge, MA: Bradford Books.
Gagné, C. L., and Murphy, G. L. (1996). Influence of discourse context on feature availability in conceptual combination. *Discourse Processes* 22: 79–101.
Gentner, D., and Wolff, P. (1997). Alignment in the processing of metaphor. *Journal of Memory and Language* 37: 331–55.
Gernsbacher, M. A., Keysar, B., Robertson, R. R., and Werner, N. K. (2001). The role of suppression in understanding metaphors. *Journal of Memory and Language* 44: 1–18.
Gildea, P., and Glucksberg, S. (1983). On understanding metaphor: The role of context. *Journal of Verbal Learning and Verbal Behavior* 22: 577–90.
Glucksberg, S. (2001). *Understanding Figurative Language: From Metaphor to Idiom*. New York: Oxford University Press.
Glucksberg, S., and Estes, E. (2000). Feature accessibility in conceptual combination: Effects of context-induced relevance. *Psychonomic Bulletin and Review* 7: 510–15.
Glucksberg, S., Gildea, P., and Bookin, H. A. (1982). On understanding nonliteral speech: Can people ignore metaphors? *Journal of Verbal Learning and Verbal Behavior* 21: 85–98.
Glucksberg, S., and Keysar, B. (1990). Understanding metaphorical comparisons: Beyond similarity. *Psychological Review* 97: 3–18.
Glucksberg, S., McGlone, M. S., and Manfredi, D. (1997). Property attribution in metaphor comprehension. *Journal of Memory and Language* 36: 50–67.
Glucksberg, S., Newsome, M. R., and Goldvarg, Y. (2001). Inhibition of the literal: Filtering metaphor-irrelevant information during metaphor comprehension. *Metaphor and Symbol* 16: 277–93.
Goldvarg, Y., and Glucksberg, S. (1998). Conceptual combinations: The role of similarity. *Metaphor and Symbol* 13: 243–55.
Goodman, N. (1972). Seven strictures on similarity. In N. Goodman, *Problems and Projects*. New York: Bobbs-Merril.
Grice, H. P. (1975). Logic and conversation. In P. Cole and J. Morgan (eds), *Syntax and Semantics: Vol. 3. Speech Acts*: 41–58. New York: Academic Press.
Hampton, J. A. (1987). Inheritance of attributes in natural concept conjunctions. *Memory and Cognition* 15: 55–71.

Hampton, J. A., and Springer, K. (1989). Long speeches are boring: Verifying properties of conjunctive concepts. Paper presented at the thirtieth meeting of the Psychonomic Society, Atlanta, Georgia.

Hasson, U., Estes, Z., and Glucksberg, S. (2001). Metaphors communicate more effectively than do similes. Paper presented at the forty-second annual meeting of the Psychonomic Society, Orlando, Florida.

Hess, D. J., Foss, D. J., and Carroll, P. (1995). Effects of global and local context on lexical processing during language comprehension. *Journal of Experimental Psychology: General* 124: 62–82.

Jackendoff, R. (1995). The boundaries of the lexicon. In M. Everaert, E van den Linden, A. Schenk and R. Schreuder (eds), *Idioms: Structural and Psychological Perspectives*: 133–66. Hillsdale, NJ: LEA.

Johnson, A. T. (1996). Comprehension of metaphors and similes: A reaction time study. *Metaphor and Symbolic Activity* 11(2): 145–59.

Johnson-Laird, P. N. (1983). *Mental Models: Towards a Cognitive Science of Language, Inference, and Consciousness*. Cambridge: Cambridge University Press.

Keysar, B. (1989). On the functional equivalence of literal and metaphorical interpretations in discourse. *Journal of Memory and Language* 28: 375–85.

Kintsch, W. (1998). *Comprehension: A Paradigm for Cognition*. New York: Cambridge University Press.

Levy, J. N. (1978). *The Syntax and Semantics of Complex Nominals*. New York: Academic Press.

Malgady, R. G., and Johnson, M. G. (1976). Modifiers in metaphor: Effects of constituent phrase similarity on the interpretation of figurative sentences. *Journal of Psycholinguistic Research* 5: 43–52.

McKoon, G., and Ratcliff, R. (1988). Contextually relevant aspects of meaning. *Journal of Experimental Psychology: Learning, Memory, and Cognition* 14: 331–43.

Medin, D. L., and Shoben, E. J. (1988). Context and structure in conceptual combination. *Cognitive Psychology* 20: 158–90.

Miller, G. A., (1979). Images and models, similes and metaphors. In A. Ortony (ed.), *Metaphor and Thought*: 202–50. Cambridge: Cambridge University Press.

Miller, G. A., and Johnson-Laird, P. (1976). *Language and Perception*. Cambridge, MA: Harvard University Press.

Murphy, G. L. (1988). Comprehending complex concepts. *Cognitive Science* 12: 529–62.

Murphy, G. L. (1990). Noun phrase interpretation and conceptual combination. *Journal of Memory and Language* 29: 259–88.

Noveck, I. A. (2001). When children are more logical than adults: Experimental investigations of scalar implicature. *Cognition* 78: 165–88.

Ortony, A. (1979). Beyond literal similarity. *Psychological Review* 86: 161–80.

Ortony, A., Schallert, D., Reynolds, R., and Antos, S. (1978). Interpreting metaphors and idioms: Some effects of context on comprehension. *Journal of Verbal Learning and Verbal Behavior* 17: 465–77.

Potter, M. C., and Faulconer, B. A. (1979). Understanding noun phrases. *Journal of Verbal Learning and Verbal Behavior* 18: 509–21.

Russo, R. (1997) *Straight Man*. New York: Vintage.

Searle, J. (1979). Metaphor. In A. Ortony (ed.), *Metaphor and Thought*: 92–123. New York: Cambridge University Press.

Smith, E. E., Osherson, D. N., Rips, L. J., and Keane, M. (1988). Combining prototypes: A selective modification model. *Cognitive Science* 12: 485–527.

Sperber, D., and Wilson, D. (1986). *Relevance: Communication and Cognition.* Oxford: Blackwell.

Sperber, D., and Wilson, D. (2003). Pragmatics, modularity and mind reading. *Mind in Language* 17: 3–23.

Springer, K., and Murphy, G. L. (1992). Feature availability in conceptual combination. *Psychological Science* 3: 111–17.

Stanfield, R. A., and Zwaan, R. A. (2001). The effect of implied orientation derived from verbal context on picture recognition. *Psychological Science* 12: 153–6.

Stroop, J. R. (1935). Studies of interference in serial verbal reactions. *Journal of Experimental Psychology* 18: 643–62.

Tabossi, P. (1982). Sentential context and the interpretation of unambiguous words. *Quarterly Journal of Experimental Psychology* 34A: 79–90.

Tabossi, P. (1988). Effects of context on the immediate interpretation of unambiguous words. *Journal of Experimental Psychology: Learning, Memory, and Cognition* 14: 153–62.

Tabossi, P., and Johnson-Laird, P. N. (1980). Linguistic context and the priming of semantic information. *Quarterly Journal of Experimental Psychology* 32: 595–603.

Torreano, L. (1997). *Understanding Metaphorical Use of Verbs.* Unpublished doctoral dissertation, Princeton University, Princeton, New Jersey.

Tversky, A. (1977). Features of similarity. *Psychological Review* 85: 327–52.

Wilson, D., and Sperber, D. (1993). Linguistic form and relevance. *Lingua* 90: 1–25.

Wisniewski, E. J. (1997). When concepts combine. *Psychonomic Bulletin and Review* 4(2): 167–84.

Zwaan, R. A., Stanfield, R. A., and Yaxley, R. H. (2002). Do language comprehenders routinely represent the shapes of objects? *Psychological Science* 13: 168–171.

5
Reasoning, Judgement and Pragmatics
Guy Politzer

1 Introduction

In psychological experiments on reasoning, participants are typically presented with premises which refer to general knowledge or which are integrated in an original scenario; then, either they are asked to derive what follows from the premises or they are provided with one or several conclusions and asked to decide whether or not these conclusions follow from the premises. There is always a logical argument underlying the premises and the conclusion, and the aim of such experiments is to study participants' performance with respect to a theoretical model, either normative or, as is more usual nowadays, descriptive. The experiments on judgement do not differ much, except that they look more like a problem to solve, where the final question is a request for a comparison, a qualitative or a quantitative evaluation, and so on. The experiment may be administered orally during an interview with the experimenter, but more often it is administered in a written form, using paper and pencil or a computer. Given that there are two interlocutors engaged in a communication, a conversational analysis is appropriate, whether the presence of the experimenter is physically real or mediated by the support of the written messages.

After he has been provided with the instructions and the information that supports the question (the scenario, the argument, the problem statement, etc.) the participant is presented with the target question. Like any utterance, this question must be interpreted. Its meaning generally is not straightforwardly identifiable because the information may be more or less long, complicated (and occasionally conceptually hard). It may also be vague or ambiguous. As for any question, its interpretation is determined by the content of the putative answer: the answer should satisfy the expectation of relevance attributed by the participant to the experimenter. Now, in experimental settings (as well as in instructional settings and more generally in testing situations) the participant is aware that the question put to him is a higher order question, that is, does not implicate 'the experimenter does not know how to find the answer' but rather 'the experimenter knows how to find the answer and she wants to know whether I know how to find it'.

The interpretation of the question is determined in part and revealed by the specific kind of knowledge that the participant chooses to exhibit through his response: this choice is made on the assumption that what is relevant to the experimenter is to know whether the participant has that kind of knowledge. This choice and the underlying assumption reveal in turn the participant's *representation of the task*. This is why knowledge of the population tested is essential. The range of questions of interest which participants are likely to attribute to the experimenter must be anticipated by the experimenter (another, higher order, attributional process) in the light of the participants' educational and cultural backgrounds. This requires a macroanalysis of the information provided, including the non-verbal experimental material (e.g., does the material used suggest that reaction times will be measured?). Social psychologist had related concerns quite some time ago, albeit more limited and focused on the transparency of the experiment; for example, Orne (1962) defined the notion of *demand characteristics* as 'the totality of cues which convey an experimental hypothesis to the subject'. Only recently did a few investigators of thinking and reasoning (Hilton, 1995; Schwarz, 1996, and co-workers) apply the so-called 'conversational' approach to the relationship between experimenter and participant, in order to study how participant's expectations and attributions affect their responses.

There is, in addition, another kind of analysis, based on pragmatic theory, which needs to be applied to the sentences used to state the argument or the problem. The output of this analysis is the determination of the interpretation of the premises, conclusion or question which the participant is likely to work out; in other words, it delivers the actual proposition(s) which will be processed during the inferential treatment, taking into account (as will be exemplified below) the frame of the task representation. The reason for performing this microanalysis is that it is an essential step to guarantee the validity of the experimental task. Indeed, the experimenter is interested in the processing of specific propositions which she expects the participant to recover from the sentences used in the argument or problem statement. Unluckily (at least in the early days of the experimental investigation of thinking), these sentences used to be either awkward and artificial formulations inspired by logic textbooks or sentences expressed in very impoverished contexts; and it was assumed that some kind of literal meaning was communicated and then the associated propositions processed. It is clear that a formal logical argument can be deemed to have been followed or not followed only to the extent that the propositions which constitute it are those which the participant has actually processed. For example, in the study of deduction, the endorsement of a conclusion which does not follow validly from the premises, or the non-endorsement of a conclusion which follows validly can be declared reasoning errors only if it can be ascertained that the participant did construe the propositions (premises and conclusion) in a way that coincided with the formal logical description of the argument.

In brief, knowledge of how people represent reasoning and judgmental tasks and of how they interpret the premises or the questions is an indispensable prerequisite for the investigation of the inferential process proper. The recommendation that experimental tasks should be submitted to a macro- and a microanalysis is made with hindsight. For a long period which ended in the late 1970s, psychologists showed little concern about such problems. The reason is that most of them were not yet familiar with the tools offered by pragmatic theory (and at an earlier time pragmatic theory itself was not developed enough to offer such tools). As a result, many erroneous evaluations of the performance observed in experiments and many unfounded claims about human rationality were made. This will be illustrated by reviewing a number of tasks, some of which have been extremely influential, and by describing some of the experimental work carried out in support of the pragmatic approach just outlined. Studies that concern reasoning (deduction and induction) and judgement (probabilistic and classificatory) will be considered in turn.

2 Studies of deduction

2.1 Quantifiers

It will be useful to begin with a prototypical case, namely the deductions called *immediate inferences*. They are elementary one-premise arguments in which the premise and the conclusion are quantified sentences which belong to Aristotle's square of opposition. In experiments, participants are presented with one premise such as, for example, *[on the blackboard] some squares are white*, and asked to evaluate (by 'true', 'false' or 'one cannot know') one or several conclusions provided to them, such as *all squares are white; no square is white*, and so on. Whereas performance for contraries (*all...are...*, to *no...are...*, and vice versa) and for contradictories (*all...are...*, to *some...are not...*, and vice versa; *no...are...*, to *some...are...*, and vice versa) is nearly perfect, performance for subalterns (*all...are...*, to *some...are...*, and vice versa; *no...are...*, to *some...are not...*, and vice versa) is apparently very poor (around one quarter of the responses coincide with the formal logical response, that is, 'true' from universal to particular sentences, and 'one cannot know' from particular to universal sentences, while a strong majority opt for the response 'false' in both directions. The same obtains for subcontraries (*some...are...*, to *some... are not...*, and vice versa) to which most people respond with 'true' instead of the formal logical response 'one cannot know' which logic textbooks would prescribe (Begg and Harris, 1982; Newstead and Griggs, 1983; Politzer, 1990).

It would be a mistake to attribute poor logicality to participants in such experiments. Assuming that participants process the sentences as if they were uttered in a daily conversation (rather than using the conventions of logicians which require a literal interpretation), the microanalysis applied to quantifiers suggests that people add the scalar implicature *not all* to *some*. If this is so, all the data are coherent. A universal sentence (e.g., *all...are...*) and its particular counterpart (*some...are...*) being contradictory under the interpretation of the latter as *some...but not all are...*, the inferences that involve these two sentences will lead the reasoner to the conclusion 'false'. And similarly, both particular sentences being equivalent to *some...are...but some...are not...*, the reasoner concludes 'true' when one is a premise and the other the conclusion.

As this example shows, pragmatic theory provides the conceptual tools to identify the propositions actually processed by participants in psychological experiments. It could be argued that, in return, the tasks used by psychologists can provide useful tools to test some claims made by pragmatic theory. As far as quantifiers are concerned, one of these claims is that the hearer's awareness of the speaker's epistemic state can affect his interpretation of *some*. If the speaker is known to be fully informed, the choice of the weaker item on the scale does convey an implicature based on the fact that the stronger item which is more informative or more relevant was not chosen; but if he is known to be not fully informed, then the choice of the weaker item may as well be attributed to lack of knowledge, and the implicature is less likely to be generated. Consider now the following situation. A radar operator is describing the screen. Some participants are told that the operator is working without time pressure and with certainty, that is, she is omniscient, and some others that she is working with time pressure and uncertainty (non-omniscient). Consider the statement, *some spots are large*. When she is omniscient, the use of *some* may license the implicature *not all* for the reasons seen above. But when she is not, it cannot be ruled out that all the spots are large. In an experiment (Politzer, unpublished) that used this scenario, the frequency of restrictive interpretations of *some* could be inferred on the basis of the conclusions that participants endorsed (such as *all spots are large*). When the speaker was assumed to be omniscient, the rate of restrictive interpretations was around 75 per cent; but when she was assumed to be non-omniscient, it dropped on average to 50 per cent. This difference was reliable and it was observed in a within-subjects as well as in a between-subjects design, which bears out the general pragmatic prediction. One might wonder why the restrictive interpretations did not collapse altogether. This seems to illustrate one limitation of the paper-and-pencil methodology, namely the difficulty for participants to exploit mental states attributed to fictitious characters. Given the artificiality of the manipulation, one might even regard its effect as impressive.

2.2 Conditional reasoning

For many years, studies of propositional reasoning have focused on 'conditional reasoning', that is, two deductively valid arguments:

- Modus Ponendo Ponens (MP): *if A then C; A; therefore C*; and
- Modus Tollendo Tollens (MT): *if A then C; not-C; therefore not-A*,

and two invalid arguments, which are the fallacies of:

- Affirming the Consequent: *if A then C; C; therefore A*; and
- Denying the Antecedent: *if A then C; not-A; therefore not-C.*

Nearly everyone endorses the conclusion of MP. For example (instantiating A with *it rains*, and C with *Mary stays at home*), given *if it rains Mary stays at home* and *it rains*, most people instructed to consider the premises as true endorse the conclusion *Mary stays at home*. However, not everyone endorses the conclusion of MT: knowing for sure that *if it rains Mary stays at home*, and that *Mary does not stay at home*, only about two-thirds conclude *it does not rain*. Performance on the two invalid arguments seems even less satisfactory: given that *if it rains Mary stays at home*, and that *it does not rain*, around one half of the people endorse the conclusion *Mary does not stay at home*, although this does not follow deductively. And similarly, from the premisses *if it rains Mary stays at home*, and *it does not rain*, around one half of the people incorrectly endorse the conclusion *Mary does not stay at home*. These are robust observations (Evans, Newstead and Byrne, 1993).

Invalid arguments. Do all people who endorse the conclusion of invalid arguments commit a fallacy? Let us first consider the microanalysis of the task.

Ducrot (1971) proposed a principle (similar to Grice's first Maxim of Quantity), which he called the *law of exhaustivity*, 'give your interlocutor the strongest information that is at your disposal and that is supposed to be of interest to him', from which it follows that there is a tendency to comprehend a limited assertion as the assertion of a limitation; in particular, *if it rains Mary stays at home* suggests that it is *only* in case it rains that Mary stays home, which explains the interpretation of *if* as a sufficient and necessary condition (or *biconditional* for short).

Geis and Zwicky (1971) used the now often-quoted example, *if you mow the lawn, I'll give you five dollars* to show that in some contexts a conditional sentence suggests an *invited inference*, in the present case the obverse of the original sentence, *if you don't mow the lawn, I will not give you five dollars*. This inference was hypothesized to follow from a *principle of conditional perfection*, but Lilje (1972) questioned whether there is such a principle. He objected that the inference crucially depends on the circumstances, as shown by the example in which the target sentence would be a reply to *'How can I earn five dollars?'*. In such a context, there are alternative antecedents (clean up the garage or whatever) that prevent mowing the lawn from being a necessary condition. Nevertheless, Geis and Zwicky's paper was very influential, so that the conditional reasoning task was the first reasoning task to be examined from a pragmatic point of view (Taplin and Staudenmayer, 1973; Staudenmayer, 1975; Rips and Marcus, 1977). There are more recent theoretical treatments of conditional perfection (Horn, 2000; Van der Auwera, 1997); without entering the technical debate, it will be assumed that the interpretation of *if* as a biconditional stems from an implicature which the hearer may generate on the basis of his knowledge base, given the aim of the conversational exchange (Politzer, 2003).

This leads us to the macroanalysis. Braine (1978) was among the first psychologists to stress the differences between 'practical reasoning' which uses premises as they are comprehended in daily verbal exchange, and formal reasoning which requires a special attitude in order to set aside implicatures. That there are individual differences in interpretation of the conditional which can be related to educational background (among other factors) was demonstrated by the results of a truth-table task (Politzer, 1981). In such a task, given a conditional sentence *if A then C*, participants are asked to choose which of the four possible contingencies (A and C; A and not-C; not-A and C; not-A and not-C) they judge to be compatible with the sentence. The choices made by Arts students were characteristic of a biconditional interpretation (A and C; not-A and not-C) more often than the choices made by Science students; these in turn more often had the formal interpretation (all cases except A and not-C). Clearly, the Science students (even though they were untutored in formal logic) were more apt to represent the task as a formal game using literal meaning.

Now an important point is that under a biconditional interpretation of the conditional premise the two fallacious arguments become valid: from *if A (and only if A) then C; not-A*, the conclusion *not-C* follows; and similarly from *if A (and only if A) then C; C*, the conclusion A follows. Consequently, if a participant endorses the conclusions of the two invalid arguments while construing the conditional sentence as a biconditional, one cannot talk any more of committing a fallacy because under such an interpretation the arguments become valid. It follows that the only way to know whether people commit a fallacy, and if so, how often, is to present a conditional premise in which the implicature is cancelled. In order to do so, Rumain, Connell and Braine (1983) presented a control group of participants with the invalid arguments made of a major premise such as *if there is a dog in the box, then there is an orange in the box* and the appropriate minor premise, *there is no dog in the box* (for the argument of Negation of the Antecedent) or *there is an orange in the box* (for the argument of Affirmation of the consequent); the fallacies (namely concluding *there is not an orange in the box* and *there is a dog in the box*, respectively) were commited 70 per cent of the time. The experimental group was presented with the same two premises together with an additional conditional premise such as *if there is a tiger in the box, then there is an orange in the box* indicating that there may be an orange without a dog. This aimed at cancelling the implicature *if there is not a dog, then there is not an orange* that is held responsible for the biconditional interpretation and therefore for the fallacies. Indeed, participants in this group committed the fallacies only 30 per cent of the time, presumably because the cancellation of the implicature gave way to the conditional interpretation. (The question of the residual 30 per cent of fallacies is beyond the scope of this chapter.) This kind of manipulation has been widely replicated and generalized to various contexts (Byrne, 1989; Manktelow and Fairley, 2000; Markovits, 1985).

Valid arguments and credibility of the premises. While it is established that performance in the invalid conditional arguments crucially depends on the interpretation of the major conditional premise, in the past 12 years a number of experimental manipulations have revealed interesting effects on the endorsement of the conclusion of the two valid arguments.

Cummins (1995; Cummins, Lubart, Alksnis and Rist, 1991) studied these arguments with causal conditionals. She and her colleagues demonstrated that the acceptance rate of the conclusion depends on the domain referred to in the major premiss. For example, of the two following arguments:

1. If the match was struck, then it lit; the match was struck; therefore it lit.
2. If Joe cut his finger, then it bled; Joe cut his finger; therefore it bled.

people are less prone to accept the conclusion of the first. The variable which was manipulated is the number of 'disabling conditions' that are available. Disabling conditions are such that their satisfaction is sufficient to prevent an effect from occurring (and their non-satisfaction is therefore necessary for the effect to occur, e.g., dampness of the match, and superficiality of the cut, respectively): the acceptance rate was a decreasing function of their number.

Thompson (1994, 1995) obtained differences in the endorsement rate of the conclusion with causals as well as non-causal rules such as obligations, permissions and definitions by using conditionals that varied in 'perceived sufficiency' (estimated by judges). A sufficient relationship was defined as one in which the consequent always happens when the antecedent does; for example, the following sentences were attributed high and low sufficiency, respectively: *If the licensing board grants them a licence then a restaurant is allowed to sell liquor. If an athlete passes the drug test at the Olympics then the IOC can give them a medal.* She observed that the endorsement rate of the conclusion was an increasing function of the level of perceived sufficiency.

Newstead, Ellis, Evans and Dennis (1997) and Evans and Twyman-Musgrove (1998) used as a variable the type of speech act conveyed by the major conditional premise; they observed differences in the rate of endorsement of the conclusion: promises and threats on the one hand, and tips and warnings on the other hand constituted two contrasted groups, the former giving rise to more frequent endorsements of the conclusion than the latter.

(These classes of conditionals were investigated in the seventies by Fillenbaum, 1975, 1978.) They noted that the key factor seems to be the extent to which the speaker has control over the occurrence of the consequent, which is higher for promises and threats than for tips and warnings.

George (1995) manipulated the credibility of the conditional premise of MP arguments. Two groups of participants received contrasted instructions. One group was asked to assume the truth of debatable conditionals such as *If a painter is talented, then his/her works are expensive*, while another group was reminded of the uncertain status of such statements. As a result, 60 per cent in the first group endorsed the conclusion of at least three of the four MP arguments, but only 25 per cent did in the second group.

While each of these authors has an explanation for his or her own results separately, it will be proposed that there is a single explanation along the following lines.

1. Conditionals are uttered in a background knowledge within which they explicitly link two units (the antecedent and the consequent), keeping implicit the rest of it, which will be called a *conditional field*.

2. The conditional field has the structure of a disjunctive form, as proposed by Mackie (1974) for causals. The mental representation of a conditional *if A then C* (excluding analytically true conditionals) in its conditional field can be formulated as follows:

$$[(A \& A_1 \& A_2 \& \ldots) \vee (B \& B_1 \& B_2 \& \ldots) \vee \ldots] \rightarrow C$$

A is the antecedent of the conditional under consideration; B is an alternative condition that could justify the assertion of *if B then C* in an appropriate context. (The fact that alternative antecedents like B and its conjuncts may not exist, or may be assumed to not exist, is at the origin of the *if not-A, then not-C* implicature considered above, but this is not our current concern.) We focus on the abridged form:

$$(A \& A_1 \& A_2 \& \ldots) \rightarrow C$$

While $(A \& A_1 \& A_2 \& \ldots)$ is a sufficient condition as a whole, each conjunct $A_1, A_2, \ldots$ is separately necessary with respect to A. These conjuncts will be called *complementary necessary conditions* (henceforth CNC). Each of the CNC's has its own availability, and this availability is part of what specifies the conditional field.

3. It is hypothesized that in asserting the conditional *if A then C*, the speaker assumes that the necessity status of the conditions $A_1, A_2, \ldots$ is part of the cognitive environment, and most importantly that the speaker

has no reason to believe that these conditions are not satisfied. The formula can be rewritten as:

$\{A_1 \ \& \ A_2 \ \& \ \ldots\} \ \& \ A \rightarrow C$

where the braces indicate that the CNCs are tacitly assumed to hold. This is justified on the basis of relevance: in uttering the conditional sentence, the speaker guarantees that the utterance is worth paying attention to. But this in turn requires that the speaker has no evidence that the CNCs are unsatisfied, failing which the sentence would be of little use for inferential purposes. (In making this assumption, one must accept that the implicature concerns not a single constant, such as A_1, but a variable A_i.)

In brief, conditionals are typically uttered with an implicit *ceteris paribus* assumption to the effect that the normal conditions of the world (the satisfaction of the CNCs that belong to the cognitive environment) hold to the best of the speaker's knowledge. Suppose now that for some reason the satisfaction of the CNC can be questioned. This typically occurs when it has high availability. The conditional sentence no longer conveys a sufficient condition and consequently the conclusion of the argument does not follow any more. This explains the results of the foregoing manipulations. For the sake of simplicity the formula can be rewritten as:

$\{A_1\} \ \& \ A \rightarrow C$

Formally, from

if ($\{A_1\}$ & A) then C; A

C follows, whereas from

if (A_1 & A) then C; A

C does not follow.

Compare two arguments defined by different conditionals such that one has less available CNCs (or disabling conditions in terms of causality) than the other, like *If Joe cut his finger, then it bled* against *If the match was struck, then it lit*: in the first case, the low availability of the CNCs makes it more likely that their satisfaction goes unchallenged than in the second case. This analysis generalises to the non-causal sentences like the 'licensing board' or the 'athlete' scenarios above. In fact, it makes a step towards the formalisation of the concept of credibility of a conditional sentence: once the antecedent and the consequent have been identified as related to each other, the conditional is all the more credible as there are fewer CNCs whose satisfaction is questionable.

There are close links between this claim and the classic view that belief in a conditional is measured by the conditional belief of the consequent on the antecedent, and it can be formally demonstrated that the former is a specification of the latter (Politzer and Bourmaud, 2002).

In the experiments mentioned above, there is an interesting case where the epistemic implicature is reinforced. This is the case with the Evans and colleagues manipulation mentioned earlier: the speaker of a promise or a threat warrants the satisfaction of CNCs, which he is not in a position to do when uttering a tip or a warning. The difference is one between a warrant 'to the best of one's knowledge' and a warrant of full knowledge that renders the conditional more credible.

Finally, George's manipulation (mentioned earlier) of the level of credibility of the conditional is another way of questioning the satisfaction of CNCs: by being asked to asssume the truth of such conditionals, participants were invited to dismiss CNCs acting as possible objections like *the painter must be famous*, whereas stressing the uncertainty of the statement is a way of inviting them to take such objections into account.

Valid arguments and non-monotonic effects. There are other means of cancelling the implicature and this is what gives rise to non-monotonic effects to which we now turn. Non-monotonic deduction is defined by the following property: consider a proposition Q that is deducible from P; Q is not necessarily deducible from the conjunction of P with another proposition R, contrary to the case of classic deduction.

Byrne (1989) asked one control group of participants to solve standard arguments such as, for MP:

If Mary meets her friend, then she will go to a play;
Mary meets her friend;
therefore:
  (a) *Mary will go to a play.*
  (b) *Mary will not go to a play.*
  (c) *Mary may or may not go to a play.*

As is commonly observed, nearly every participant chose option (a). An experimental group was asked to solve the same arguments modified by the addition of a third premise, *if Mary has enough money, then she will go to a play*. The result is that fewer than 40 per cent in this group chose option (a) and the others chose option (c). A similar effect was observed with MT. Notice the special structure of the argument: the third (additional) premise was a conditional that had a necessary condition in its antecedent; since it had the same consequent as the major premise, it contained a necessary condition for the consequent of the major premise (in fact, a CNC) and served as a means of introducing it in the context. The result has been replicated many times with rates of non-endorsement varying from one-third to two-thirds, depending on sentence type and population.

Within the proposed framework, the additional premise raises doubt on the assumption of satisfaction of a CNC in the main conditional. This is made possible by using the CNC in the antecedent of another conditional: in uttering 'if Mary has enough money...' the speaker implicates that she does not know whether or not Mary has enough money, so cancelling the implicature that accompanies the main conditional. This now has decreased credibility and the conclusion follows with a level of credibility inherited from the premises. This is why in an all-or-none format of response, a majority of people choose option (c).

This explanation has testable consequences. First, by replacing the additional conditional sentence with a categorical sentence that expresses doubt, such as *it is not certain that she has enough money*, it should be possible: (i) to simulate the effect (a decrease in the rate of endorsement of the conclusion); and (ii) to bring this rate of endorsement in fact to zero since the doubt stems from an explicit statement and no longer from an implicature that may not always be generated. This is precisely what was observed (Politzer, in press). Second, when participants are given a chance to evaluate the conclusion, the proportion who find it doubtful should be about the same as the proportion who chose option (c) above; again this is what was observed.

Another consequence is that it should be possible to manipulate the credibility of the major conditional premise by introducing various degrees of satisfaction of the CNCs and observe correlated degrees of belief in the conclusion. This was tested by Politzer and Bourmaud (2002) who used different MT arguments such as:

If somebody touches an object on display then the alarm is set off;
the alarm was not set off;
therefore: *nobody touched an object on display* (to be evaluated on a five-point scale ranging from certainly true to certainly false).

This was a control; in the three experimental conditions, degrees of credibility in the conditional were defined by way of an additional premise that provided information on a CNC:

High credibility: *there was no problem with the equipment;*
Low: *there were some problems with the equipment;*
Very low: *the equipment was totally out of order.*

The coefficients of corrrelation between level of credibility and belief in the truth of the conclusion ranged between 0.48 and 0.71 and were highly significant. This result supports the proposed theoretical approach all the more as the kind of rule used was not limited to causals but included also means-end, remedial and decision rules.

Non-monotonicity is highly difficult to manage by Artificial Intelligence systems because of the necessity of looking for possible exceptions through the entire data base. What I have suggested is some kind of reversal of the 'burden of the proof' for human cognition: at least for conditionals (but this could generalize) looking for exceptions is itself an exception because the conditional information comes with an implicit guarantee of normality.

3 Hypothesis testing

Some people are professionally trained to test their hypotheses; they may be scientists or practitioners such as detectives, medical doctors or technicians specialized in trouble-shooting. But how do lay people behave when they have to put a hypothesis to the test? One of the classic laboratory tasks used to answer this question was designed by Wason (1960). The situation resembles a game played between the experimenter and the participant. The experimenter chooses a rule to generate sequences of three numbers. The aim of the game for the participant is to discover this rule. In order to do so, the participant can use two kinds of information. The main source of information is the result of tests which he carries out as follows: he submits triples to the experimenter who replies every time by 'yes' (the three numbers obey the rule) or 'no' (they do not). The second source of information is an initial example of a sequence conforming to the rule provided by the experimenter at the beginning of the game: this sequence is *2, 4, 6*. When the participant thinks he has discovered the rule, he states it; in case he is wrong, the game may continue for another cycle until the rule stated is correct or the participant gives up. The rule which the experimenter follows is *three increasing numbers* (integers). It is usually observed that the majority of participants state at least one incorrect rule and that failure is not uncommon. More strikingly, the incorrect rules proposed by participants often express one of the salient features of the initial exemplar (2, 4, 6), such as *even numbers, increasing by the same interval, or increasing by two* and it seems difficult for them to eliminate such hypotheses. This is especially interesting from a pragmatic point of view because the triple *2, 4, 6* has very salient features; given that it has been specially selected and presented as an instance by the experimenter, participants are thereby invited to assume that its features are relevant; but unluckily for the participant, these features overdetermine the rule (the numbers need not be even, they need not increase by two, and so on in order to follow the rule actually used), so that one can consider the whole situation to be deceptive. As every teacher knows, it is misleading to offer an example of a concept that is too specific. This analysis made on theoretical grounds (Politzer, 1986) has received support from the results of a recent experiment performed by Van der Henst, Rossi and Schroyens (2002). In their experimental procedure, the 2, 4, 6 instance was not presented to participants as resulting from a deliberate choice made by the experimenter, but rather as the output of a computer program which randomly generated instances of the rule: the authors observed that the erroneous first solutions diminished by one-half, and that the mean number of rules proposed as solutions diminished by one-third, presumably because the salient features are not presumed to be relevant if they are the result of a random, non-intentional process.

The 2, 4, 6 task is not the only inductive task that deserves pragmatic scrutiny. One of the most extensively investigated tasks in the psychology of reasoning, also due to Wason and also designed to study hypothesis testing behaviour, is the four-card problem (or selection task) in which participants are required to select the information that they think is necessary in order to test whether a conditional rule is true or false. Studies by Sperber, Cara and Girotto (1995) and by Girotto, Kemmelmeir, Sperber and Van der Henst (2001) show that the task, as understood by the experimenter, is rather opaque to participants. Ironically, the comprehension mechanisms pre-empt any domain-specific reasoning mechanism, so that the task cannot be considered as one of reasoning in the strict sense.

4 Studies of probabilistic judgement

There is a huge psychological literature on probabilistic judgement that dates back to the 1960s. The conclusion which has been retained, especially among philosophers and economists, is that performance is poor and often reveals irrational judgments. This widely shared opinion is essentially due to the work of Tversky and Kahneman (1982 for an overview). Whether they are right or wrong is not an issue to debate here; instead, it can be argued that their demonstration is often unconvincing because in too many cases they grossly neglected the pragmatic analysis of their experimental paradigms. Two of these tasks, possibly the most famous ones, the Linda problem and the Lawyer-Engineer problem, will be discussed.

4.1 The conjunction fallacy (the Linda problem)

In a typical version of the experimental paradigm, participants are presented with the following description:

> Linda is 31 years old, single, outspoken and very bright; she majored in philosophy. As a student, she was deeply concerned with issues of discrimination and social justice, and also participated in anti-nuclear demonstrations.
>
> (Tversky and Kahneman, 1982)

They are then asked to decide which of the following statements is the most probable:

- Linda is a bank teller (B)
- Linda is active in the feminist movement (F)
- Linda is a bank teller and active in the feminist movement (B + F)

Whatever the response format (multiple choice, rank ordering, etc) over 80 per cent judge B + F to be more probable than B, in apparent violation of a fundamental axiom of probability theory which requires that the probability of a conjunction be no more probable than that of any one of its conjuncts. The authors take this result as evidence for the use of the representativeness heuristic, that is, an assessment of the degree of correspondence between a model and an outcome: being a 'feminist bank teller' (B + F) is more representative of the description because it has one common feature with the description, which 'bank teller' (B) is lacking. This explanation is appealing if only because of its simplicity, but it cannot be accepted before a pragmatic analysis of the task has been made. Now, from this point of view, there are two main problems with the task.

The first problem is that the crucial options have an obvious anomaly: in comparing two items B versus (B and F), there are two permissible construals for B in the first option, namely an inclusive construal *(B whether or not F)*, and an exclusive one that carries an implicature *(B but not F)*.

The claim that the implicature is licensed by the juxtaposition of the two options was supported by the results of the following manipulation (Politzer and Noveck, 1991; see Dulany and Hilton, 1991 for a similar approach). Keeping constant a scenario that depicted a very brilliant and determined student, two formulations of the options were presented to two experimental groups as follows.

The first group had clearly nested options (and for this reason it was hypothesized that conjunction errors would be less frequent than in a Linda-type control):

1. Daniel entered Medical School.
2. Daniel dropped out of Medical School for lack of interest.
3. Daniel graduated from Medical School.

The second group had the same options, but with the explicit mention of the inclusion structure of the questions introduced by *and*, which was predicted to trigger an implicature attached to option one:

1. Daniel entered Medical School.
2. Daniel entered Medical School and dropped out for lack of interest.
3. Daniel entered Medical School and graduated.

Indeed, while 77 per cent committed the error on the Linda-type control the rate of errors collapsed to 31 per cent for the first control, but as predicted it increased significantly to 53 per cent for the second control.

The second problem with the task is even more basic; it revolves around the task representation. From a computational point of view, Linda's profile is useless: all the necessary and sufficient logical information is given in the options. But participants normally assume the description to be relevant and one obvious way to satisfy this is to consider the task as a test of one's sociological or psychological skills and the description as a source of information that provides a theme together with the necessary evidence for or against the answer to a question (the possibility that Linda is a feminist): the *and-not* interpretation of option (B) is then constrained.

This point is important in relation to the between-subjects task. In this variant of the task, only one statement is presented: B to one group, and B + F to the other, and participants are asked to estimate the probability of the statement. As B + F is rated as more probable than B, many investigators have been convinced in favour of the representativeness theory. But what this demonstrates is only that participants are inclined to try to render the description relevant to the question asked: they identify the kind of activity which provides greater relevance to the description of the character and like when one has to imagine what could be the best end of a story, it does not have to be the most probable event – rather it generally is not.

4.2 The base-rate fallacy (the Engineer-Lawyer problem)

In this paradigm, participants are told that a panel of psychologists has written personality descriptions of 30 engineers and 70 lawyers (the associated proportions provide what is called the *base rates*). A description that is assumed to have been chosen at random and that coincides with the stereotype of an engineer is presented; one group of participants is asked to estimate the probability that the person described is an engineer; another group is asked to do the same based on the reversed base rates: 70 engineers and 30 lawyers. Provided some technical assumptions are satisfied, standard probability theory requires that the estimate given by the first group should be lower than that of the second group. The first study reported by Kahneman and Tversky (1973) showed no difference, hence the widely held belief that 'people are insensitive to the base rates'; however, more recent studies have shown that people do take base rates into account, although 'not optimally or even consistently' (Koehler, 1996). Tversky and Kahneman's explanation for their results is again based on the representative heuristic: people would base their judgment exclusively on the extent to which the description fits the stereotype. This explanation is again problematic because it does not take into account the participants' representation of the task. In a recent series of experiments (Politzer and Macchi, in press) it was hypothesized that people view the task as a request to exploit a psychological description that is assumed to be relevant. If that is the case, the neglect of base rates should be relative and could be suppressed in an experimental condition where no psychological description is provided, but instead the psychological characterization is provided in a single statement to the effect that the person's description is typical of an engineer: in this way, the outcome is available (in order to let the representativeness heuristic operate, if at all) but the details are missing in order to suppress the interpretation of the task as one of extraction of a psychological profile from such data. In being told that the description is typical, these participants receive a near answer to the question, which makes it lack relevance; consequently, they reinterpret the question as a request for an unconditioned probability, which enables them to render both the statement of typicality and the base-rate information relevant and to fulfil the task, so that most of them should give the base rate as their response. This is what was observed (85 per cent used the base rate exclusively while the rate of its use in a control group was 17 per cent). It seems therefore that the paradigm could be better described as showing that people have difficulty in combining information from two sources, the base rates and the individuating information, and that they focus on the one that maximizes relevance (see also Baratgin and Noveck, 2000). Previous research has shown that when the psychological description is uninformative (i.e., completely non-diagnostic between the engineer and the lawyer stereotypes), they rely entirely on the base rates.

5 Class inclusion and categorization

5.1 Class inclusion in children

One of the most thoroughly investigated paradigms in developmental psychology during the period that runs from the 1960s to the 1980s, and which nowadays is still subject to debate is class inclusion, initially created by Piaget (Piaget and Inhelder, 1959; Piaget and Szeminska, 1941). In a typical experiment, the child is presented with a picture of five daisies and three tulips, and then asked, 'Are there more daisies or more flowers?'. The rate of what is considered the correct response, 'more flowers', reaches the 50 per cent value only around 8 or 9 years of age. This highly robust result is puzzling given the well-documented precocity in the acquisition of lexical hierarchies. We will consider in turn the interpretation of the interrogative sentence and the representation of the task.

First, the microanalysis indicates that the relation of hyperonymy–hyponymy between *flower* and *tulip/daisy* licenses the use of *flower* to refer to either all the flowers or a subclass of them. Indeed, it can be demonstrated that in the experimental setting, *flower* is indeterminate between an inclusive sense (all the flowers) and an exclusive sense (tulip). This was done as follows. Two groups of 6- and 7-year-old children were presented with the picture. The control group was just asked: (i) to first point to the flowers; and then (ii) to the daisies. In contrast, the experimental group was asked the same questions *in the reversed order*. Whereas 90 per cent of the children in the control group, who were asked to point to the flowers, pointed to all the flowers, half of the children in the experimental group pointed to *all* the flowers and the other half pointed to the tulips. This demonstrates that *flower* apparently had become completely indeterminate in the context of *daisy*. Half of the children decided that *flower* must refer to the flowers that are not daisies – presumably because the word *daisy* had just been used; the other half were not able or not willing to make this decision.

Consequently, the standard class-inclusion question is ambiguous because the lexeme *flower* can receive either its inclusive/hyperonym or its exclusive/hyponym interpretation. It follows that many children may compare the daisies with the tulips (which is well documented), a comparison that is not intended by the experimenter though semantically permitted, and pragmatically justifiable under one representation of the task as we will see shortly.

If this explanation is correct, it should be possible to enhance performance by disambiguating the question. This was done in another experiment that used a double disambiguation procedure. First, 5- to 8-year-old children were requested to 'point to the flowers' and then to 'point to the daisies' (as in the previous experiment). Secondly, they were asked a modified class-inclusion question in which all three terms appeared: 'Are there more tulips, or more daisies, or more flowers?' The 5-year-olds reached the 50 per cent rate of success (control: 6 per cent) and the 7- and 8-year-olds were very close to the 100 per cent rate (control: 30 per cent). Two other experiments showed that each disambiguating procedure is effective separately but less than in combination. In brief, the disambiguation of the question has revealed that children acquire inclusion three to four years earlier than previously claimed.

But still a major question remains to be answered: Why do children change their response to the standard question when they are about 8 or 9 years old? The answer is that the younger choose the exclusive interpretation of *flowers* (tulips) and the older the inclusive interpretation (all the flowers). But again, why? This question leads us to the macroanalysis and the representation of the task. So long as the child attributes to the experimenter an interest in knowing whether he can count (one of the great achievements during that period) the relevant comparison is between the tulips and the daisies (this response is likely to produce the more cognitive effects: you will know that I know how to count). But when the child has progressed enough in the development of metacognitive skills such as logical necessity (Cormier and Dagenais, 1983; Miller, Custer and Nassau, 2000) and awareness of semantic ambiguities (Gombert, 1990), he can attribute to the experimenter an interest in these abilities, and the relevant comparison shifts to comparing all the flowers and the daisies, which yields the 'correct' response. In brief, this overview of an old paradigm in the study of logical development shows once again that the verbal material and the speaker/experimenter–hearer/ participant relationship must be pragmatically scrutinized.

5.2 Categorization: mathematical hierarchies

Although the approach taken here is focused on laboratory tasks, the analysis that has been proposed can help identify some sources of difficulty in learning mathematical concepts; more specifically the application of the foregoing analysis of the inclusion question to lexical hierarchies reveals a tension between the use that is made of them by the lay person/student on the one hand and the scientist/teacher on the other.

We noticed earlier that the standard class-inclusion question is ambiguous because the lexical unit *flower* can receive either its inclusive/hyperonym interpretation or its exclusive/hyponym interpretation. This case is reminiscent of markedness: opting for the inclusive rather than the exclusive meaning amounts to opting for an unmarked rather than a marked interpretation. This is at the basis of riddles such as *'What animal barks but is not a dog?'*, the solution of which is blocked if *dog* is interpreted as unmarked, but transparent if *dog* is interpreted as contrasting with *bitch*. (This ambiguity is sometimes referred to as *privative*.) Now, as far as mathematical hierarchies are concerned, the speaker's freedom to use an ambiguous lexical unit is constrained by the register of the communication. In daily life, it seems that the items on such hierarchies are essentially used exclusively; for instance, *square* contrasts with *rectangle* (which in turn contrasts with *parallellogram*, etc.), which means that for a 'naïve' person no square is a rectangle.[1] On the contrary, in the mathematical vocabulary, items on the same hierarchy are used inclusively: a square is a special rectangle (which in turn is a special parallellogram, etc.); hence technically all squares are rectangles. Similarly, for the layman integers are not decimal numbers although mathematically they are. In brief, whenever two items are compared, the subset to set relations generated by the folk hierarchy and the mathematical hierarchy are logical contraries. It follows that a crucial difficulty in the learning of these classifications lies in the student's capability to shift appropriately from his familiar classification to the technical

---

[1] One might argue that this phenomenon is but a particular case of scalar phenomenon, by which the use of *rectangle* on the scale implicates *not square*, a higher item on the scale (Horn, 1972). However, while it is easy to imagine or observe in daily life utterances that exhibit literal meaning on various scales (quantifiers, modals, frequency terms, etc.) it seems debatable that this happens with mathematical classifications. Whether *rectangle* can refer to a square in a non-mathematical context is an open question that could be answered empirically. In the absence of evidence to the contrary, it is assumed that there is no lexical unit in ordinary English to refer to the set of figures that conjoins the squares and the rectangles. That it is so is understandable: in daily life, it is the exclusive contrast that is useful; the inclusive contrast has only a metacognitive theoretical interest, which justifies its scholarly use.

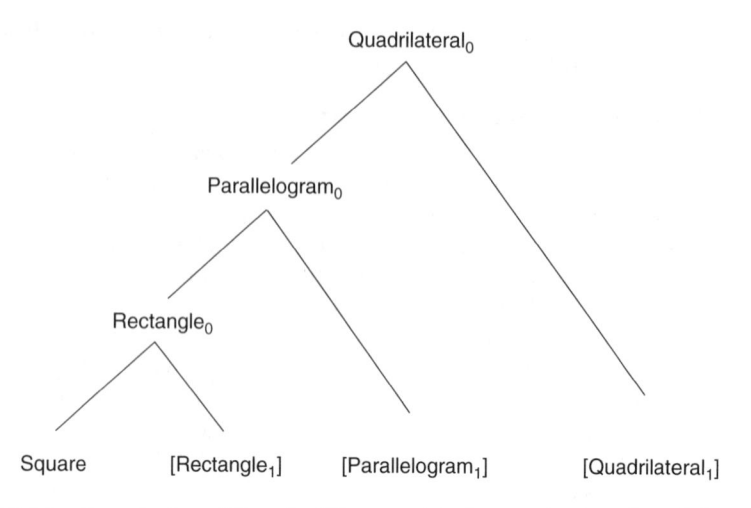

Figure 5.1 Organization of the scientific geometrical categorization of quadrilaterals

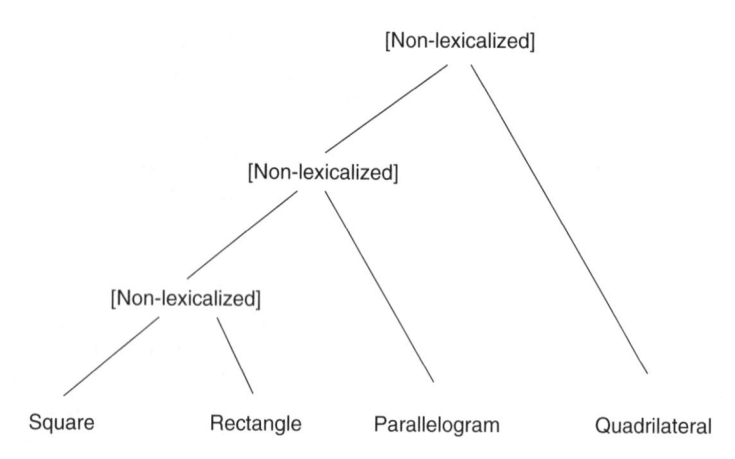

Figure 5.2 Organization of the lay geometrical categorization of quadrilaterals

one (Politzer, 1991). The cognitive difficulty is illustrated in Figures 5.1 and 5.2 which show both hierarchies for elementary geometry.

6 Conclusions

From a methodological point of view, the experimental study of thinking is among the most difficult in cognitive psychology to carry out. This is the area where the representation of the task interferes the most with the thought process under study, to such an extent that the task may be devoid of validity if no precautions are taken. It has been argued that precautionary measures should include two kinds of analysis. One, which has been called *macroanalysis*, aims to determine the task representation, that is, the participant's/student's attributions to the experimenter/teacher about the latter's expectations regarding the former's knowledge or performance. This is based on the content of each task, taking into account the specificity of the relationship between experimenter/teacher and participant/student which creates a special element of pretence in their communication. The other, that has been called *microanalysis*, takes into account the result of the first and aims to determine the disambiguations, referential assignments and implicatures which the participant/student works out on the way to his final interpretation of the premises, questions, problem statement and the like. When such analyses yield interpretations that are at variance with the experimenter's intended meaning, it is possible to write up an alternative formulation or to design an alternative task whose validity is no more questionable and to compare performance on this new task with the initial one. In the past, many unwarranted conclusions in terms of human irrationality have been drawn from participants' seemingly poor performance. The experimental method that compares initial and modified materials on the basis of pragmatic theory plays a crucial role to redress the balance.

References

Baratgin, J., and Noveck, I. A. (2000). Not only base rates are neglected on the Engineer-Lawyer problem: An investigation of reasoners' underutilization of complementarity. *Memory and Cognition* 29(1): 79–91.

Begg, I., and Harris, G. (1982). On the interpretation of syllogisms. *Journal of Verbal Learning and Verbal Behavior* 21: 595–620.

Braine, M. D. S. (1978). On the relation between the natural logic of reasoning and standard logic. *Psychological Review* 85: 1–21.

Byrne, R. M. J. (1989). Supressing valid inferences with conditionals. *Cognition* 31: 61–83.

Cormier, P., and Dagenais, Y. (1983). Class-inclusion developmental levels and logical necessity. *International Journal of Behavioral Development* 6: 1–14.

Cummins, D. D. (1995). Naive theories and causal deduction. *Memory and Cognition* 23: 646–58.

Cummins, D. D., Lubart, T., Alksnis, O., and Rist, R. (1991). Conditional reasoning and causation. *Memory and Cognition* 19: 274–82.

Ducrot, O. (1971). L'expression en français de la notion de condition suffisante. *Langue Française* 12: 60–7.

Dulany, D. E., and Hilton, D. J. (1991). Conversational implicature, conscious representation, and the conjunction fallacy. *Social Cognition* 9: 85–110.

Evans, J. St B. T., Newstead, S. E., and Byrne, R. M. J. (1993). *Human Reasoning: The Psychology of Deduction*. Hove: Lawrence Erlbaum.

Evans, J. St B. T., and Twyman-Musgrove, J. (1998). Conditional reasoning with inducements and advice. *Cognition* 69: B11–B16.

Fillenbaum, S. (1975). IF: Some uses. *Psychological Research* 37: 245–60.

Fillenbaum, S. (1978). How to do some things with *if*. In J. W. Cotton and R. L. Klatzky (eds), *Semantic Factors in Cognition*: 169–214. Hillsdale, NJ: Lawrence Erlbaum.

Geis, M. L., and Zwicky, A. M. (1971). On invited inferences. *Linguistic Inquiry* 2: 561–6.

George, C. (1995). The endorsement of the premises: Assumption-based or belief-based reasoning. *British Journal of Psychology* 86: 93–111.

Girotto, V., Kemmelmeir, M., Sperber, D., and Van der Henst, J.-B. (2001). Inept reasoners or pragmatic virtuosos? Relevance and the deontic selection task. *Cognition* 81: B69–B76.

Gombert, E. (1990). *Le développement métalinguistique*. Paris: Presses Universitaires de France. [English translation: *Metalinguistic Development*. University of Chicago Press, 1992].

Hilton, D. J. (1995). The social content of reasoning: Conversational inference and rational judgement. *Psychological Bulletin* 118: 248–71.

Horn, L. R. (1972). *On the Semantic Properties of Logical Operators in English*. Bloomington: Indiana University Linguistics Club.

Horn, L. R. (2000). From *if* to *iff*: Conditional perfection as pragmatic strengthening. *Journal of Pragmatics* 32: 289–386.

Kahneman, D., and Tversky, A. (1973). On the psychology of prediction. *Psychological Review* 80: 237–51.

Koehler, J. J. (1996). The base rate fallacy reconsidered: Descriptive, normative, and methodological challenges. *Behavioral and Brain Sciences* 19: 1–17

Lilje, G. W. (1972). Uninvited inferences. *Linguistic Inquiry* 3: 540–2.

Mackie, J. L. (1974). *The Cement of the Universe*. Oxford: Oxford University Press.

Manktelow, K. I., and Fairley, N. (2000). Superordinate principles in reasoning with causal and deontic conditionals. *Thinking and Reasoning* 6: 41–65.

Markovits, H. (1985). Incorrect conditional reasoning among adults: Competence or performance? *British Journal of Psychology* 76: 241–7.

Miller, S. A., Custer, W. L., and Nassau, G. (2000). Children's understanding of the necessity of logically necessary truths. *Cognitive Development* 15: 383–403.

Newstead, S. E., Ellis, M. C., Evans, J. St B. T., and Dennis, I. (1997). Conditional reasoning with realistic material. *Thinking and Reasoning* 3: 49–76.

Newstead, S. E., and Griggs, R. A. (1983). Drawing inferences from quantified statements: A study of the square of opposition. *Journal of Verbal Learning and Verbal Behavior* 22: 535–46.

Orne, M. T. (1962). On the social psychology of the psychological experiment: With particular reference to demand characteristics and their implications. *American Psychologist* 17: 776–83.

Piaget, J., and Inhelder, B. (1959). *La genèse des structures logiques élémentaires*. Neuchâtel: Delachaux et Niestlé. [English translation: *The Early Growth of Logic in the Child: Classification and Seriation*. London: Routledge and Kegan Paul, 1964].

Piaget, J., and Szeminska, A. (1941). *La genèse du nombre chez l'enfant*. Neuchâtel: Delachaux et Niestlé. [English translation: *The Child's Conception of Number*. London: Routledge and Kegan Paul, 1952].

Politzer, G. (1981). Differences in interpretation of implication. *American Journal of Psychology* 94: 461–77.

Politzer, G. (1986). Laws of language use and formal logic. *Journal of Psycholinguistic Research* 15: 47–92.

Politzer, G. (1990). Immediate deduction between quantified sentences. In K. J. Gilhooly, M. T. G. Keane, R. H. Logie and G. Erdos (eds), *Lines of Thinking: Reflections on the Psychology of Thought*, vol. 1: 85–97. London: John Wiley.

Politzer, G. (1991). L'informativité des énoncés: contraintes sur le jugement et le raisonnement. *Intellectica* 11: 111–47.
Politzer, G. (1993). *La psychologie du raisonnement: Lois de la pragmatique et logique formelle*. [The psychology of reasoning: Laws of pragmatics and formal logic. Unpublished Ph.D. thesis. University of Paris VIII].
Politzer, G. (2003). Premise interpretation in conditional reasoning. In D. Hardman and L. Macchi (eds), *Thinking: Psychological Perspectives on Reasoning, Judgment, and Decision Making*: 79–93. London: Wiley.
Politzer, G. (in press). Uncertainty and the suppression of inferences. *Thinking and Reasoning*.
Politzer, G., and Bourmaud, G. (2002). Deductive reasoning from uncertain premises. *British Journal of Psychology* 93: 345–81.
Politzer, G., and Macchi, L. (in press). The representation of the task: The case of the Lawyer-Engineer problem in probability judgment. In V. Girotto and P. N. Johnson-Laird (eds), *The Shape of Reason: Essays in Honor of P. Legrenzi*. Hove: Psychology Press.
Politzer, G., and Noveck, I. (1991). Are conjunction rule violations the result of conversational rule violations? *Journal of Psycholinguistic Research* 20: 83–103.
Rips, R. J., and Marcus, S. L. (1977). Suppositions and the analysis of conditional sentences. In M. A. Just and P. A. Carpenter (eds), *Cognitive Processes in Comprehension*: 185–220. Hillsdale, NJ: Lawrence Erlbaum.
Rumain, B., Connell, J., and Braine, M. D. S. (1983). Conversational comprehension processes are responsible for reasoning fallacies in children as well as adults: *If* is not the biconditional. *Developmental Psychology* 19: 471–81.
Schwarz, N. (1996). *Cognition and Communication*. Mahwah, NJ: Lawrence Erlbaum.
Sperber, D., Cara, F., and Girotto, V. (1995). Relevance theory explains the selection task. *Cognition* 52: 3–39.
Staudenmayer, H. (1975). Understanding conditional reasoning with meaningful propositions. In R. J. Falmagne (ed.), *Reasoning: Representation and Process in Children and Adults*: 55–79. Hillsdale, NJ: Lawrence Erlbaum.
Taplin, J. E., and Staudenmayer, H. (1973). Interpretation of abstract conditional sentences in deductive reasoning. *Journal of Verbal Learning and Verbal Behavior* 12: 530–42.
Thompson, V. A. (1994). Interpretational factors in conditional reasoning. *Memory and Cognition* 22: 742–58.
Thompson, V. A. (1995). Conditional reasoning: The necessary and sufficient conditions. *Canadian Journal of Experimental Psychology* 49: 1–60.
Tversky, A., and Kahneman, D. (1982). Judgments of and by representativeness. In D. Kahneman, P. Slovic and A. Tversky (eds), *Judgment under Uncertainty: Heuristics and Biases*: 84–100. Cambridge: Cambridge University Press.
Van der Auwera, J. (1997). Conditional perfection. In A. Athanasiadou and R. Dirven (eds), *On Conditionals Again*: 169–90. Amsterdam: John Benjamins.
Van der Henst, J.-B., Rossi, S., and Schroyens, W. (2002). When participants are not misled they are not so bad after all: A pragmatic analysis of a rule discovery task. *Proceedings of the 24th Annual Conference of the Cognitive Science Society*. Mahwah, NJ: Lawrence Earlbaum.
Wason, P. C. (1960). On the Failure to eliminate hypotheses in a conceptual task. *Quarterly Journal of Experimental Psychology* 12: 129–40.

6
Exploring Quantifiers: Pragmatics Meets the Psychology of Comprehension

A. J. Sanford and Linda M. Moxey

1 Introduction

If I hear that few students in Glasgow understand Japanese, what does this mean? For the past 15 years (Moxey, 1986; Moxey and Sanford, 1987), we have been trying to establish the psychological (processing) properties of natural language quantifiers, and how quantified statements are understood. This includes how they are used, how they are represented in the minds of producers and listeners, and how to capture their meaning in a psychologically plausible description. The venture was motivated by the frequency of quantity statements in everyday life (e.g., *People often find statistics difficult; Few of our students know more than two languages*), and by the obvious difficulties in working out just what quantifiers denote. Although we are primarily psychologists with an interest in language comprehension, working in a Cognitive Science environment guaranteed contact with both formal linguistics and logic. Since the most comprehensive accounts of the meanings of natural language quantifiers were formal (e.g., Barwise and Cooper, 1981; Keenan and Stavi, 1986; Westersthål, 1989; Zwarts, 1996) and not psychological, inevitably it was desirable that our psychological data made contact with formal theories. Equally, we believed that it was important for formal theories to be brought into contact with psychological data. Colleagues of a formal persuasion were interested in whether formal properties of generalized quantifiers, such as downward entailment, could explain the results of psychological experiments. Equally, since the fit of natural language quantifiers to generalized quantifier theory depends upon the intuitive validity of certain inferences (see Barwise and Cooper, 1981) we became concerned with judgements of logical necessity as a processing problem. In short, we became interested in how well psychological data meshed with formal ideas (and vice versa), as well as in our main goal of developing a psychological process model of quantifier understanding.

Although there have been substantial treatments within psychology of so-called logical, or standard quantifiers (*all, some, none, some-not*; e.g., Johnson-Laird, 1983), the treatment of the larger set of non-standard quantifiers (such as *few, many, most*, etc.) was restricted to the question of what they denoted, in terms of number and proportions. However, Moxey and Sanford, (1987; 1993a, 1993b; Sanford, Moxey and Paterson, 1996) established other communicative functions of non-standard quantifiers, and claimed that these went some way to explaining the very large number of possible quantifying expressions given a very large overlap in numerical denotation.

Two of these functions define the boundaries of the present chapter: focus, the set that a particular quantifier brings to mind, and denial of some supposition. Both are revealed through an investigation of negative quantifiers, and how they contrast with positive quantifiers.

2 Focus and negation in quantified statements

There has been a great deal of interest in the concept of focus as it relates to discourse anaphora. In psycholinguistics (e.g., Garnham, 2001; Sanford and Garrod, 1989), as in computational linguistics (e.g., Walker, Joshi and Prince, 1998), elements of a sentence that are in focus have typically been taken as those that are easily accessed by pronominal anaphors. Observations by Moxey and Sanford (1987) showed that the patterns of easy anaphoric reference were different for negative and positive quantifiers. For instance, in (1), it is normally considered to be the set of boys who went to the park who are being referred to by *They*:

(1) A few of the boys went to the park. They had a good time.

Within standard formal treatments of anaphora, such as Discourse Representation Theory (DRT; Kamp and Reyle, 1993), the only sets that are available as a result of the first sentence of (1) are the set of boys in general, and the set of boys who went to the park. However, Moxey and Sanford (1987; Moxey, 1986) found that patterns with negative quantifiers were not the same. For instance, consider (2):

(2) Not many of the boys went to the park. They stayed at home instead.

Not only does *They* clearly not refer to the boys who went to the park, it is taken by many subjects to refer to the ones who did NOT go to the park (Moxey and Sanford, 1987; Sanford, Moxey and Paterson, 1996). Furthermore, in tasks where subjects were presented with sentences like the first in (2), and invited to continue with a further sentence beginning with *They*, they produced many sentences conforming to this pattern: that is, one in which the producers themselves took the pronoun to refer to the set who did not

Table 6.1 Incidence in percentage of reference types in a continuation task

Quantifier	Compset	Refset	Other
Not quite all (N)	63	7	30
Less than half (N)	60	23	17
Not Many (N)	83	7	10
Few (N)	66	27	7
Not quite 10% (N)	50	30	20
Not quite 50% (N)	76	7	17
Nearly all (P)	0	97	3
More than half (P)	0	83	17
Many (P)	0	100	0
A few (P)	0	97	3
Nearly 10% (P)	7	83	10
Nearly 50% (P)	0	90	10

Notes to Table: Compset is explained in the text. Refset is the 'normal' pattern of reference, where the pronoun is taken to refer to those cases of which the predicate of the quantified sentence holds. (N) denotes negative quantifiers, (P) denotes positive quantifiers.
Source: Data from A. J. Sanford, L. M. Moxey and K. B. Paterson (1996), Attentional focusing with quantifiers in production and comprehension, *Memory and Cognition* 24: 144–55, with permission of the authors and thanks to the Psychonomic Society.

go to the park (from now on, we shall call this **complement set reference**). The incidence of these continuations from a variety of studies is shown in Table 6.1. For the positive items, complement set reference is virtually absent, while for the negatives, it is present and in some cases strongly predominates.

Traditionally, such patterns have been considered impossible. Thus the second sentence in (3) has been widely cited as evidence that the complement set cannot be accessed:

(3) Eight of the ten marbles are in the bag. They are under the sofa.

However, we simply point out that *eight of the ten (x)* is not negative.
Studies of reading times for sentences show that if a sentence mismatches the preferred pattern of focus for a quantified statement, then that sentence will take longer to read than one that matches (Sanford, Moxey and Paterson, 1996; Paterson, Sanford, Moxey and Dawydiak, 1998). Thus, following (4), (6) takes longer to read than does (5), while for (7), (6) is read in less time than (5):

(4) A few of the MPs attended the meeting.
(5) Their presence helped the meeting run smoothly.
(6) Their absence helped the meeting run smoothly.

(7) Few of the MPs attended the meeting.
(5) Their presence helped the meeting run smoothly.
(6) Their absence helped the meeting run smoothly.

While the different patterns of focus seem indisputable, the precise nature of the effect has been a topic of some debate. For instance, the question arose as to whether complement set references aren't references to the superset (the *minimally abstracted* set, in DRT terms). For instance, in (2), it might be taken as the case that the *boys in general* stayed at home instead. Sanford, Moxey and Paterson (1996; see also Moxey and Sanford, 2000) found that certain continuations could not be accounted for in this way. For instance, the following types of continuation occurred in our data:

(8) Not quite all of the MPs were at the meeting. They were back in their constituencies trying to muster support.

The *They* could not possibly refer to MPs in general, since the proportion of MPs here is very much less than half (so a generalization is not possible; see Moxey and Sanford, 2000, for further arguments). In addition, other work shows that the attachment of individual to sets follows the same pattern:

(9) Not many people went to the meeting, including Fred.

Did Fred attend the meeting? The dominant response is 'no' (Sanford, Williams and Fay, 2001; Dawydiak, Sanford and Moxey, submitted 2004).

Given the rather large body of evidence for different focus patterns for negative and positive quantifiers, and the indisputable practical significance of this difference (see Sections 4 and 5) our attention was fixed on why the difference holds. Furthermore, because complement set focus was taken to be a somewhat odd phenomenon by the linguistic community (e.g., Corblin, 1997; Geurts, 1997; Percus, Gibson and Tunstall, 1997), the question of how it fitted into linguistic theory had to be confronted. This in turn led to the issue of which properties of quantifiers 'license' complement set reference. The fact that the focus patterns were not absolute for the negative expressions seemed to substantiate the general feeling that there was something not quite respectable about the focusing differences, or at least something that was not central to an understanding of how quantifier meaning might be understood. From a psychological perspective, continua are just as commonplace as categories, however.

3 Establishing the truth of inferences: the case of quantifiers

What properties of quantifiers are important determinants of focus patterns? Although we have referred to quantifiers as negative or positive up to now, clearly negativity is a multi-faceted property (Horn, 1986). Thus the search for 'licensing' conditions led us to investigate the status of quantifiers within theoretical frameworks. In turn, we became concerned at how well natural language quantifiers fitted the categories of the formal frameworks. Within linguistics, the use of intuitive judgements of grammaticality as primary data has a long tradition. Sentences are judged as grammatical, or ungrammatical, in which case they have been marked with an asterisk *. This represents a typically binary contrast between the grammatical and the ungrammatical, rather like the binary distinction between true and false utterances. On occasion, the question sign ? has been used in situations where a judgement seems to be difficult to make. Even less often, ratings of grammaticality have been used, or some other sort of continuous scoring system, such as the use of magnitude estimation by Bard, Robertson and Sorace (1996).

Once one moves away from the binary position to continua, one at once admits that judgements of grammaticality constitute more complex processing than simply checking a sentence against a readily applicable grammar. It may also be difficult to disentangle strict grammaticality from ugliness of style, felicity and perhaps meaningfulness. One approach is to take the view that judgements of grammaticality depend upon people being properly trained to make the judgements (so that typical speakers of a language may be poor at judging). This is in many ways similar to the view that logical judgements of truth or falsity may be better made by experts in logic, because they know how to rule out strong pragmatic constraints on typical usage, or know how to keep extraneous variables constant when there is temptation for the untrained observer not to.

Turning to quantifiers, the problem of judgement emerges as in the guise of how well judgements of sentences containing natural language quantifiers fit sentences that are admissible within the theory (e.g., Barwise and Cooper, 1981; Moxey and Sanford, 1993a).

3.1 Judgements of downward entailment

Generalized quantifiers have been categorized with respect to the category monotonicity by a number of semanticists (e.g., Barwise and Cooper, 1981; Keenan and Stavi, 1986). Monotonicity refers to the preservation of the truth of a property into larger (upwards) or smaller (downwards) sets. Thus, for monotone increasing quantifiers, of which *more than 10 (x)* is an analytic example, the following holds:

(10) If more than ten men went to the party early, then more than ten men went to the party.

That is, what holds for a subset still holds for the superset of the left argument. For monotone decreasing expressions, like *less than 10 (x)*, this subset → superset frame does not hold (11), but the superset → proper subset frame does (12):

(11) *If less than ten men went to the party early, then less than ten men went to the party.
(the crucial thing is that more could have turned up later)

(12) If less than ten men went to the party, then less than ten men went to the party early.
(crucially, no more than the number who went to the party at all could have arrived early)

Downward monotonicity in the left argument (Mdec) requires that two of the four De Morgan conditions hold that define classical logical negation, and so downward monotonicity has been viewed as a 'weak' form of negation: an Mdec quantifier is weakly negative (see especially Zwarts, 1996, 1998; but also Atlas, 1997, for complications). We noted that, in general, it seemed to be monotone decreasing expressions that licensed compset focus, and held this a rule-of-thumb for a while, finally testing the idea from 2001 onwards.

The possibility that the property Mdec could formally license complement reference was explored by Kibble (1997). Kibble's approach was to show how, for some Mdec quantifiers, a formula containing the compset could be derived from internal negations of their Minc counterparts. To jump the gun a little, it turns out that Mdec is not a sufficient property to explain complement set reference (Moxey and Sanford, 1998b; Moxey, Sanford and Dawydiak, 2001; Sanford, Williams and Fay, 2001). But in parenthesis, we explored the way in which NLQs fitted the logical test-frames for judgements of monotonicity. Some NLQs appear to be analytically Mdec. For example, *less than 10(x), or at most 10(x)*, since the null set is obviously entailed within the semantics of the expressions. However, it is the case that many quantifiers are not analytically Mdec, even though they are often treated as though they were Mdec. Consider the quantifier *few(x)*, for example, within a downward-entailment frame:

(14) If few students went to the meeting, then few students went to the meeting early.

Many naïve and informed respondents think that this is a true conclusion (see, e.g., Barwise and Cooper, 1981; Zwarts, 1996; 1998), even though it only holds if the null set is assumed to be included in the meaning of *few*. Yet such an inference is based purely on intuition, and leads to difficulties. If one imagines that *few* translates into 'a small number or less, including the null set' then it holds, but if it does not, for instance by assuming that there is at least one case (i.e., *some but not many*), then the quantifier is not monotone decreasing. There may be several interpretations of *few*, some

Table 6.2 Judgements within frames for testing for monotonicity

	Minc Index	Mdec Index	NonM
Almost none	0.28	0.36	0.12
Hardly any	0.20	0.64	0.04
Not quite all	0.24	0.48	0.28
Less than 10%	0.20	0.28	0.28
At most 10%	0.28	0.28	0.28
few	0.08	0.68	0.20
Not many	0.12	0.64	0.20
Almost all	0.64	0.12	0.24
all	0.64	0.12	0.12
many	0.32	0.20	0.36

Notes to Table: The Minc index shows the proportion of people (N = 24) who found the Minc frame acceptable for each quantifier. The Mdec score shows the proportion who found the Mdec frame acceptable. The NonMon index shows the proportion who found NEITHER acceptable. In some cases, the same person found both Minc and Mdec to be acceptable.
Source: Unpublished data from Dawydiak, Sanford and Moxey (submitted 2004).

being monotone decreasing, and some not. This should lead to variability in the judgement of downwards entailment using frames like (14).

There has been little work on the acceptability of inference frames like (14) for 'negative' quantifiers. However, work carried out in our laboratory has revealed a number of interesting issues. Table 6.2 shows the percentage of 'acceptable' responses for judgements of negative and positive quantifiers within upwards and downwards entailing frames. Above all, judges were not uniform in their decisions about the validity of inferences. Second, they did not always accept the validity of inferences even where they are analytically necessary, as with *at most N (x)* being Mdec, and they did not always reject Minc inferences when they are analytically excluded. It is also interesting that the purely *intuitive* Mdec cases, like *few* and *not many*, are more frequently judged as Mdec than are the analytic cases *less than X per cent (x)*, and *at most X per cent (x)*.

Of course, much of this reflects peoples' failure to stick by the laws of logic in the analytic cases, and our plans include determining the basis of peoples' responses. But for the intuitive cases, it is difficult to know how to evaluate and use peoples' responses, because there is nothing analytic to go on.

3.2 Judgements of null set inclusion in quantifier denotation

A related and important distinction amongst quantifiers is whether or not their semantics includes the null set. Indeed, for a quantifier to be Mdec, it must allow the null set. For instance, the quantifier *hardly any (x)* is typically taken to mean something like a very small number by any expectations, but semantically it may arguably include the null set. It may be tested for inclusion of the null set by using one of Horn's (1984) standard suspenders of implicature, as in:

(15) Hardly anybody went to the meeting, if anybody did.

To the extent that (15) seems reasonable, it indicates that the assumption of one or more people going is defeasible, and hence pragmatic. That some quantifiers include the null set is analytic. For instance, the denotation of *less than N* plainly includes 0, because 0 is less than N, provided N is positive. But N can take any positive value. For instance, it could be 5 per cent, or it could be 95 per cent. One would expect this to influence the strength of the implicature that *some* is included in the meaning of the expression. As part of testing properties of quantifiers for establishing their meaning from a user's viewpoint, Majid and Sanford (unpublished data) investigated the acceptability of using the frame:

(16) [Quant] of the people went to the meeting, if any.

The results of the judgements of a group of participants for a set of quantifiers are shown in Table 6.3 where a high rating indicates a high degree of acceptability.

A rather interesting pattern occurred. First, there were some findings that might well be expected. So, a high acceptability was found for *few, hardly any, not many* and *less than 30 per cent*. Also, *less than 80 per cent* was relatively unacceptable; the contrast with *less than 30 per cent* is exactly what we expected in terms of how easy it should be for semantics to take precedent over pragmatics in making an acceptability judgement. Positive expressions, whether monotone increasing or non-monotone, give low acceptability ratings, with two exceptions: *some* gives a middle rating, and *a few* gives

Table 6.3 Averaged acceptability ratings for the frame [*Quant*] *of the fans went to the match, if any did.* Maximum (completely acceptable) has a rating of 60; completely unacceptable has a rating of 0.

Quantifier	Rating	Quantifier	Rating
Exactly 80%	2	Not quite all	15
All	2.5	Less than 80%	15
More than 30%	2.5	Less than 30%	42
More than 80%	5	Not many	42
Many	3	Few	58
Some	31	Hardly any	48
A few	48		
Nearly all	4		

as high a rating as *not many*. These anomalies are hard to explain. In the present context, however, we are simply illustrating how reliance upon intuition to classify quantifiers suffers the problem of a lack of uniformity of judgement, and so testing the fit of quantifiers to a formal theory is intrinsically bound up with human performance. Empirically, the question is what makes people judge things the way they do, and we would suggest that these judgements reflect how the quantifiers are normally interpreted (i.e., what their functional significance is, rather than their formal semantics within an abstract framework).

4 Denial, or not?: judgements of tags

The same observations of continuities in classifications rather than categorical data occur when one considers the negative-related property of denial of supposition, but on this occasion, we were able to use this fact in exploring what underpins complement set focus.

4.1 Denial and negation

Several observations suggested that the property Mdec was not sufficient to explain the appearance of complement set focus. For instance, it had been suggested (without test) that complement set references would not occur with the analytically Mdec NLQ *At most N of the (x)* (Kibble, 1997), and a very low incidence was obtained for this expression in an experiment by Moxey and colleagues (2001, Experiment 2). However, in the same experiment, a very high incidence of complement set reference was obtained for the NLQ *No more than N of the (x)*. What is the difference between these two Mdec expressions? The difference relates to a distinction between denial and affirmation, itself related to the idea of sentence versus predicate negation (Horn, 1989).

Denial is a property associated with negation (Horn, 1989), and is usually considered to be pragmatic. If I assert *I didn't have soup for lunch*, you may reasonably suppose that I had reason to expect to have soup for lunch (or that I thought you might expect that). The negation presupposes a state of affairs (the affirmative proposition), and then effectively denies that state of affairs. The claim that negations are interpreted as denials is supported by the argument that it is in some way not informative to assert a negation (So, to know I didn't have soup does not say what I did have, while to say *I had a sandwich for lunch* narrows the possibilities without introducing any presuppositions). To make the negation informative, it must deny a presupposition (see Horn, 1989, for a discussion). This case has been extensively championed in psycholinguistics by Wason, (1959, 1961). He refers to the use of negation as bringing about a Context of Plausible Denial, in which negative sentences emphasize that a fact is contrary to an expectation.

Sanford, Williams and Fay, (1996; Moxey et al., 2001) suggested that quantifiers that produce negations in simple declarative sentences may be the ones that produce complement set focus. The argument was that a denial raises the question of why the denial had to be made. For instance, with *Not many people went to the park*, the question is *why did less than was expected go to the park?*. The answer is in terms of reasons why this was the case, and the most obvious explanation is in terms of what happened to those who could not go to the park (the complement set). To test this idea, we had to devise ways of testing which NLQs formed denials.

What methods are available to test for denial? One way, described in Section 3 (above) is direct: test for what people infer was being expected. Another way is to use a variety of linguistic tests. One family of tests, described by Klima (1964), is the use of tags. As part of the phenomena surrounding negation, Klima observed that different sentences can take different tags. For instance, in (17), the negative sentence takes *does he?* as a tag, but not *doesn't he?*. The opposite is true for its affirmative counterpart, (18):

(17) John doesn't like cheese, does he?/* doesn't he? (* denotes unacceptability)
(18) John likes cheese, doesn't he?/* does he?

The pattern with (17) has been taken as reflecting sentence negation, whether the statement is an affirmation of a state of affairs, or a denial of a (pre)supposed state of affairs (Clark, 1976; Horn, 1989). On this basis, *not many (x)* comes out as forming denials:

(19) Not many people like cheese, do they?/* don't they?

Consider next *either* versus *too* tags:

(20) Not many men are happy and not many women are happy either/* too.
(21) Many men are happy and many women are happy too/* either.

Essentially, the use of *either* connects two denial statements.
A final one we shall discuss here is the *neither* versus *so* case:

(22) Many men are happy, and so/* neither are the women in the group. (so, *many* does not form a denial)
(23) Not many men are happy, and neither/* so are the women in the group. (so, *not many* forms a denial)

These tags have been argued as indicating denial rather than the semantic element behind negation, namely downward monotonicity. This point was noted by Moxey, Sanford and Dawydiak (2001). For instance, the quantifier

At most N (x) is analytically monotone decreasing, yet it seems to form assertions, not denials, for instance:

(24) At most ten of the students showed up, didn't they?/* did they?

In contrast, *No more than N (x)*, also analytically monotone decreasing, forms denials:

(25) No more than ten of the students showed up, did they?/* didn't they?

In continuation data, Moxey and her colleagues (2001) found that while *At most N (x)* produced few complement set continuations, *No more than N (x)* produced many. This suggested that denial was indeed important.

We have presented these details as though they are categorical (denial/assertion), yet some judgements of which tag questions might fit sentences made with particular quantifiers yield continuous answers over a group of participants asked to make the judgements. For instance, for the analytically monotone decreasing expression *less than N (x)*, about half of the responses to the three sorts of tags described above indicated that it formed an assertion, while the other half indicated that it formed a denial (Moxey et al., 2001, p. 440).

4.2 The predictive power of the denial continuum

In more recent studies, we have obtained data showing that monotone decreasing quantifiers could be classed on a continuum of forming denials, using the proportion of participant's choices to the 'denial' tags as a denial index. A number of subjects were asked to make judgements of the acceptability of the six different tag options shown above. A composite score was produced of the percentage of judgements fitting the denial (sentence negation) pattern, and this we termed the Denial Index. The same subjects also completed a single judgement of the 'including' type:

(26) Not many people went to the meeting, and that includes John.
    Did John go to the meeting? YES NO

The proportion of NO responses was taken as the complement set index. The close relationship is shown in Table 6.4. In fact, in a regression equation, the denial index accounts for over 90 per cent of the variance in complement attachment (Dawydiak, Sanford and Moxey, submitted).

According to the **Inference Theory of Complement Focus**, which gradually evolved over the course of these studies, denial, associated with negation, is at the heart of the complement reference phenomenon (Moxey, Sanford and Dawydiak, 2001; Sanford, Moxey and Paterson, 1996). Negative quantifiers that form denials lead the processor to ask the (implicit) question *why was*

Table 6.4 The relation of complement set rate to denial index rate

Quantifier	Compset index	Denial index
almost none	0.96	0.973
almost all	0.08	0.04
hardly any	0.92	0.92
not quite all	0.96	0.947
less than 10%	0.2	0.307
less than 50%	0.48	0.373
less than 90%	0.64	0.44
at most 10%	0.04	0.107
at most 50%	0.16	0.067
at most 90%	0.08	0.067
few	0.52	0.613
not many	1	0.893
many	0.04	0.027

Source: Unpublished data selected from Dawydiak, Sanford and Moxey (submitted).

the amount less than expected? In a continuation task, the sentence produced will then be an attempt to answer this implicit cognitive question. Indeed, throughout our research, we and our colleagues have found that negatives do give rise to explanations of why the number or amount was less. For instance, the following continuation is typical:

(26) Not many of the fans went to the match. They preferred to watch it at home on TV instead.

Here, preferring to watch at home is an explanation for not going to the match. Once an explanation is formulated, its natural form of expression is by reference to the complement set, in the case above, in terms of the fans who did not go to the match. This theory contrasts with other attempts to provide 'licensing conditions' for complement set focus based on the semantic property of downward entailment (Kibble, 1997).

If, as we suggest, focus depends on denial, then an expression that is an assertion should not produce complement set references, even if it has a downward entailing quantifier. Given the earlier discussion, a case in point is the form *At most N (x)*, that produces assertions, in contrast to *No more than N (x)*, that produces denials. In a pair of continuation studies, Moxey and colleagues (2001) showed that the *At most* version produced far fewer complement set continuations than did than did the *No more than* version.

The relationship between the various tests used to obtain the denial index, and the complement set focus index is, of course correlational. Furthermore, as far as the authors can tell, there is no demonstrated formal way of expressing the link between the tests used in the denial index, and the notion of denial. Clearly there is still much to explore here, which is scarcely surprising, since these phenomena are at the heart of how negation itself is to be defined and understood. However, the work described in this section suggests to us that it is promising to pursue the relations amongst focus and denial.

From the perspective of the current chapter, what is especially interesting is that the denial results lie on a continuum, suggesting that interpretations themselves vary, and that the extent of variation is a function of quantifier. One must also ask whether the variation could be brought under contextual control. Too often, interpretations of sentences are taken out of context (especially in dealing with formal treatments, but also in the present psychological work). Our current efforts are directed at this issue.

5 Who has the expectations?

A further point of contact with pragmatic issues concerns who has expectations based on language input. The notion of denial assumes that someone (listener, speaker or both) has an expectation that is being denied. But who? Within the framework of communication, this is an important issue. If *Not many people came to the talk* presupposes there was reason to expect that more might have come, just who is doing the presupposing and believing? From a psychological perspective, this is a perfectly good question (see also Horn, 1989, p. 180, *et seq.*, for a wide-ranging general discussion). For instance, it might be supposed that the quantifier *not many (x)* leads an interpreter simply to suppose that there is reason for someone to expect that, that the expectation is in some abstract sense applicable to the situation being depicted. This may be the case, but other work by Moxey and Sanford (1993b) suggests a possibly more complex situation, in which different expectations may be attributed to different people depending upon the quantifiers used in making the utterances.

Moxey and Sanford (1993b) measured the interpretations given by people at three 'levels' in relation to quantified statements. The questions can be understood best in terms of Figure 6.1.

The interpretation of quantifiers (Level 1 in Figure 6.1) is known to be influenced by base-rate expectation, and three settings were used to generate three different base-rate expectations: the number of local doctors who are female (low expectation), the number of people likely to be influenced by a political speech (medium expectation), and the number of people who enjoyed a party (high expectation). These expectations were confirmed by a pre-test. In the experiment, one group of participants were asked what percentage a given quantifier was intended by the speaker to denote, that is, Level 1 (in all cases, participants made only one judgement). A second group

> SPEAKER SAYS: 'Few of the fans went to the football match.'
>
> HEARER INTERPRETS:
>
> *Level 1* Speaker must mean that x% went to the match.
>
> *Level 2* Speaker must have expected y% would go to the match.
>
> *Level 3* Speaker must think that the LISTENER expected z% would go to the match.

Figure 6.1 Levels of interpretation as proportions investigated by Moxey and Sanford (1993a)

was asked to say what proportion they thought the speaker expected to be the case before the speaker discovered the facts (Level 2). A third group had to say what proportion the speaker thought the listener might have expected before the listener heard the quantified statement (Level 3). Again, participants had to make only one judgement, to avoid contaminating contextual effects.

If quantifiers convey nothing about expectations at Levels 2 and 3, then there should have been no effect of quantifier on the responses to these questions. The only effects that guide and influence judgements would be the independently assessed base-rates, on the assumption that base-rate represents the norm. However, it turned out that not just base-rate, but also quantifiers had an effect. The following expressions produced relatively high proportions to the Level 2 question:

- Very few, few, not many, and only a few.

For Level 3, however, the picture was different, with only the following producing relatively high proportions:

- Not many, only a few.

The proportions estimated at Levels 2 and 3 for these negative expressions are shown in Figure 6.2.

These observations are interesting, in that they suggest that while *not many* and *only a few* are taken as revealing things about the speaker and the speaker's expectations of the listeners beliefs, *very few* and *few* only reveal reliable information about the speaker's expectations. This subtle difference, along with our earlier observations regarding focus, led us to examine how a speaker's choice of quantifier might be made.

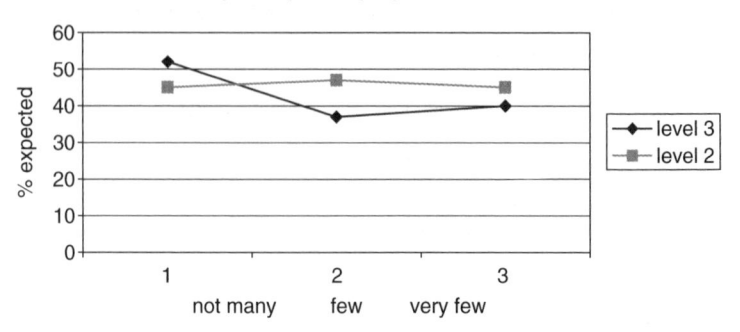

Relationship of expected proportions to levels

Figure 6.2 How expected proportion varied with level of interpretation in the Moxey and Sanford (1993a) study

6 Communication: choosing the right quantifier

If semantics is concerned with the formal properties of meaning, then pragmatics is plainly concerned with the interface with usage and communication. In the domain of quantifiers, one of our key questions has been how the vast variety of quantifiers is used for effective communication. There have been rather few experimental studies of the optimal choices between alternative ways of saying things, although there has been a rather large psychological literature on how to express risk and uncertainty (see, e.g., Moxey and Sanford, 2000, for a review). As part of our studies, Moxey and Sanford (1997) reported a direct investigation into choice of quantifier. This was theory-driven, being based on the idea that for different communicative situations, focus (perspective) and expectation at different levels would contribute to choice.

The method used was to present to participants vignettes that depicted situations, which ended with a choice of quantifiers, with the instructions to rank order these in terms of how well they fit the situation. We tested the idea that choice would be a function of focus and presupposition, as discussed separately above. Three vignettes were presented to readers, designed to capture the critical contrasts under test. In the first, we contrasted complement focus with reference set focus. In the scenario, we presented a depiction of a hospital patient facing a necessary but dangerous operation. It is the surgeon's duty to inform the patient of the low chances of coming through unharmed in some way. He has the choice of saying:

A few
Only a few
Few   patients survive for long afterwards.
Not many
Hardly any

Hardly any, not many and *few* were all unacceptable, in contrast to *a few* and *only a few*, which were equally highly acceptable. This is to be expected. The desirable communication is to signal to the patient that there are survivors. There are other situations where the focus should be on the complement set. In a second scenario, we depicted a car sale scenario. The customer asks about the reliability of a particular model, which the salesman in anxious to sell. The choices for the salesman response were:

A few
Only a few
Few   need more than a basic service in the first three years.
Not many
Hardly any

The acceptability ratings gave the following ordering, from acceptable to unacceptable:

(Few = hardly any) > not many > only a few > a few

This confirms the utility of complement set focusing: by focusing on the complement set, the customer should be set to think about the large proportion of cars that do not require more than a basic service. However, there was a reliable difference between *few* and *not many*, with *few* being preferred over *not many*. Why should this be? We proposed (Moxey and Sanford, 1997) that to use *not many* would introduce the supposition that the customer might have expected more to be the case, and that is something that is undesirable in the present arena of discourse, from the point of view of the salesman. He does not want to allow the customer to think that he acknowledges the possibility of them expecting a large number of problems! Most readers get the intuition that *few* is better than *not many*, paralleling the findings of the experiment.

If this argument is right, in situations where the prior expectations of the listener should NOT be ignored, the preference should be reversed. This was tested through the constructions of another scenario. In this one, we depicted a situation where a keen football fan could not get to a match by an amateur side he is keen on supporting. His friend went, but nobody else did. His friend has to break the news of a low turnout, and he hedges this, but has to take his friends expectations into account. The choices were:

A few
Hardly any
Not many   people went.... Well, in fact, nobody did.
Only a few
A few

The results showed the following ordering (from acceptable to unacceptable):

Hardly any > not many > few >> only a few = a few

As expected, not only are the complement focusers the only ones that are acceptable; *not many* is preferred over *few*.

Taken together, these results seem to support the position that patterns of supposition, as well as patterns of focus, influence the choice of appropriate quantifier. More fundamentally, we believe that these studies show how preference for a form of communication can be understood in terms of underlying mechanisms, where in both cases experimental data substitutes for pure intuition.

7 Communication: focus on fat

In our ongoing programme of work, we have put an applied spin on our ideas of explaining intuitions about language. Sanford, Fay, Stewart and Moxey (2002) investigated the well-known intuition that (27) is a better thing to say about your product than (28) if you want to sell it:

(27) This food product is 95 per cent fat-free.
(28) This food product contains 5 per cent fat.

The fat-free formulation is the one that is most prevalent on products on the supermarket shelves. In psychology, there have been studies showing that describing beef as 75 per cent lean rather than 25 per cent fat led participants to rate the beef as leaner and less greasy. This effect persisted even after the participants had tasted the beef! (Levin and Gaeth, 1988). Such effects have been seen as part of the framing problem (e.g., Kahneman and Tversky, 1979; Levin, Schneider and Gaeth, 1998). However, this way of thinking about things does not explain the mechanism by which the preference operates.

In line with the earlier arguments, the fat-free formulation puts focus on the amount of the food product that is not fat. For instance, in one experiment, Sanford and colleagues (2002) showed that people are more likely to endorse (31) following (29) than (30):

(29) Bloggs lite yoghurt is 95 per cent fat-free.
(30) Bloggs lite yoghurt contains 5 per cent fat.
(31) This is a healthy product.

But how does this really work? Does concentrating on fat-freeness actually prevent the reader from seeing things from the point of view of how much fat the product **does** contain? Consider what happens at different fat levels. If the fat content is increased, then the product should be viewed as less healthy, thus (32) is endorsed as healthy with a much lower frequency than (30):

(32) Bloggs lite yoghurt contains 25 per cent fat.

However, this difference is very much reduced when (33) is contrasted with (29):

(33) Bloggs lite yoghurt is 75 per cent fat-free.

Thus, the impact of knowledge about health and fat levels differs in the two situations. The fat-free formulation allows higher levels of fat to be introduced without influencing the healthiness judgement so much. In a further experiment, Sanford and colleagues (2002) showed that these effects are seen at the very time the sentences are comprehended. Materials like (34) were used. The first sentence depicts the product setting, the second the options in the way the fat information may be presented, and the third a target sentence (sentence of interest). The question was whether the ease of reading the target sentence would depend upon the fit of the healthiness statement to the level of fat and the way it was depicted (*per cent fat* or *per cent fat-free*):

(34) A new home-made style yoghurt is to be sold in supermarkets.
    The yoghurt {contains 5 per cent fat/contains 25 per cent fat}/{is 95 per cent fat-free/is 75 per cent fat-free}.
    It is widely believed to be a {healthy}/{unhealthy} product.

Consider first the percentage of fat depiction. For the 5 per cent level, the target was easier to integrate for the healthy condition than for the unhealthy condition, as indexed by reading speed. However, at the 25 per cent level, this differential disappeared, in line with the idea that the product is no longer clearly healthy. The story was very different for the fat-free depiction, however. Here the *healthy* target was integrated more easily regardless of fat level: there was no difference in ease of integration of the *healthy* target. Sanford and colleagues (2002) concluded that rapid access to general knowledge about the relationship of amount of fat to healthiness was inhibited under the fat-free formulation, and that this is the reason for the fat-free formulation sounding 'healthier', or at least, it is a major part of the story.

These kinds of perspective phenomena are general, and may be applied to a variety of quantity statements. For instance, note the difference in perspective on risk associated with using the quantifying expressions a *small risk* and *little risk* (Moxey and Sanford, 2000):

(35) There is a small risk of side-effects occurring with this drug. (which is a bad thing)
(36) There is little risk of side-effects occurring with this drug. (which is a good thing)

The examples with fat belong to the same family of effects as those discussed from a more theoretical perspective under complement set focus. In each case, a quantity statement cannot be made without implicitly adopting some sort of perspective. In the fat cases discussed in this section, the introduction of perspective is through different patterns of assertion, while in the quantifier case, it is through the mechanism of denial if our theory is correct.

8 Conclusions

In this chapter we have presented a summary of our experiences, experiments and general attempts to understand a phenomenon relating to how quantifiers are understood by people. Quantifiers have played a central role in formal treatments of semantics and natural language, so it is scarcely surprising that we were drawn into thinking about both formal semantics and pragmatics. The fact that we were concerned with negative quantifiers simply exacerbated these things. The problem is to establish sensible links between the insights of different disciplines. Perhaps the biggest single problem is the relation of data to formalisms. Whether the data come from linguistics in the form of acceptability judgements, or from psychology in the form of reading-times and so on, any discrepancy between a formalism and data may cause a problem. From the formalists' perspective, as psychologists we suspect that the formalist would view the problem as being one of the psychologist obtaining data that was in some way imperfect. We believe that one problem is the way in which the relationship of the formal to the empirical is formulated. If the psychologist is seen as providing tests of the ideas resulting from formalisms, then the imperfect data formulation of discrepancies seems to follow quite naturally. If the psychologist is seen as providing data (empirical facts) that need to be described within a formal framework, the notion of imperfect data vanishes, to be replaced by the question of whether the psychologist is happy that the measures being taken reflect the kind of process she or he is trying to tap.

Similar problems arise with the relationship of sentence-linguistic data to formalisms. If a string is acceptable grammatically and yet should not occur according to a formalism, does this reflect some kind of human fallibility, or does it reflect a weakness in the formalism? The empirical investigation of cases where decisions of acceptability show variation, as with the denial indices in the present chapter, suggests that formal treatments that allow continuities of acceptability might fair well in explaining the interrelationship amongst linguistic phenomena. We see the problem as parallel to that with psychological data. The case of judgements of monotonicity also raises a problem of the fit of formalisms to data: many quantifiers (such as *few*) that are not analytically monotone decreasing are often judged to be so. Such cases seem to us to require formal treatments that explain quasi-monotonicity, and there are plenty of those in natural language.

Pragmatics sits neatly at the centre of these arguments. Judgements of what is logically necessary, when made by humans, can easily be influenced by pragmatic considerations. Good examples abound in the psychological literature on syllogistic reasoning, where pragmatic implicatures may lead to technically distorted views of logical necessity. However, from a human-user's (performance) perspective, expressions in language primarily serve a rhetorical function. Selecting the right one allows the speaker to control what it is the listener is thinking about. Perhaps a proper approach to the meanings of expressions should be more about what the expressions put into focus, that is, what aspects of their semantics or pragmatics is most important when they are used in communication. In linguistics, a similar view has been put forward by Horn (personal communication) that under some circumstances semantically entailed material is often disregarded for the purposes of linguistic diagnosis (in particular polarity licensing when the entailment is outside of the scope of what is asserted). It takes genuine experimental pragmatics (applying psychology to pragmatics) to work out what is most important when particular quantifiers are used in communication. For instance, many people judged *at most N (x)* as not being monotone decreasing, even though, analytically, it embraces none. This would make sense if the focal part of the assertion hinges on the number N ('why would a speaker pick N?'): for the null set, this issue would have to be blocked. Instead, it looks as if the semantic entailment is blocked.

Our programme of research into the function and meaning of quantifiers has brought us into contact with all of these issues, and as we have attempted to solve various problems, we have had to develop techniques that we believe belong to the emerging discipline of Experimental Pragmatics.

References

Atlas, J. D. (1997). Negative adverbials, prototypical negation, and the De Morgan Taxonomy. *Journal of Semantics* 14: 349–67.
Bard, E., Robertson, D., and Sorace, A. (1996). Magnitude estimation and linguistic acceptability. *Language* 72: 32–68.
Barwise, J., and Cooper, R. (1981). Generalized quantifiers and natural language. *Linguistics and Philosophy* 4: 159–219.
Clark, H. H. (1976). *Semantics and Comprehension*. The Hague: Mouton.
Corblin, F. (1997). Quantification et anaphore discursive: la référence aux complimentaires. *Languages* 123: 51–74.
Dawydiak, E. J., Sanford, A. J., and Moxey, L. M. (submitted 2004). A cognitive theory of quantifier perspective effects.

Garnham, A. (2001). *Mental Models and the Interpretation of Anaphora*. Hove, UK: Psychology Press.
Guerts, B. (1997). Review of L. M. Moxey and A. J. Sanford, (1993) Communicating Quantities. *Journal of Semantics* 18: 87–94.
Horn, L. R. (1984). Toward a new taxonomy for pragmatic inference: Q-based and R-based implicature. In D. Schiffrin (ed), *Meaning, Form, and Use in Context: Linguistic Applications*. Georgetown University round table 1984. Washington, DC.
Horn, L. R. (1989). *A Natural History of Negation*. Chicago: University of Chicago Press.
Johnson-Laird, P. N. (1983) *Mental Models*. Cambridge University Press, Cambridge.
Kahnemen, D., and Tversky, A. (1979). Prospect Theory: An analysis of decision under risk. *Econometrika* 47: 263–91.
Kamp, H., and Reyle, U. (1993). *From Discourse to Logic: Introduction to Model Theoretic Semantics of Natural Language, Formal Logic, and Discourse Representation Theory*. Dordrecht: The Netherlands.
Keenan, E. L., and Stavi, J. (1986). A semantic characterization of natural language determiners, *Linguistics and Philosophy* 9: 253–326.
Kibble, R. (1997). Complement anaphora and monotonicity. In G. J. M. Kruijff, G. V. Morrill and R. T. Oehrle (eds), *Proceedings of a Conference on Formal Grammar*: 125–36. Aix-en-Provence, August 1997.
Klima, E. S. (1964). Negation in English. In J. A. Fodor and J. J. Katz (eds), *The Structure of Language*. Englewood Cliffs, NJ: Prentice-Hall.
Levin, I. P., and Gaeth, G. J. (1988). How consumers are affected by the framing of attribute information before and after consuming the product. *Journal of Consumer Research* 15: 374–8.
Levin, I. P., Schneider, S. L., and Gaeth, G. J. (1998). All frames are not created equal: A typological and critical analysis of framing effects. *Organizational Behavior and Human Decision Processes* 76: 149–88.
Levin, I. P., Schittjer, S. K., and Thee, S. L. (1988). Information framing effects in social and personal decisions. *Journal of Experimental Social Psychology* 24: 520–9.
Moxey, L. M. (1986) *A Psychological Investigation of the Use and Interpretation of English Quantifiers*. Unpublished Ph.D. Thesis, University of Glasgow, Scotland.
Moxey, L. M., and Sanford, A. J. (1987). Quantifiers and Focus. *Journal of Semantics* 5: 189–206.
Moxey, L. M., and Sanford, A. J. (1993a). Prior expectation and the interpretation of natural language quantifiers. *European Journal of Cognitive Psychology* 5: 73–91.
Moxey, L. M., and Sanford, A. J. (1993b). *Communicating Quantities: A Psychological Perspective*. Hove: Lawrence Erlbaum Associates.
Moxey, L. M., and Sanford, A. J. (1997). Choosing the right quantifier. In T. Givon (ed.), *Conversation*. Philadelphia: John Benjamins.
Moxey, L. M., and Sanford, A. J. (1998a). Choosing the right quantifier: Usage in the context of communication. In T. Givon (ed.), *Conversation*: 207–31. Amsterdam: John Benjamins
Moxey, L. M., and Sanford, A. J. (1998b). Complement set reference and quantifiers. In M. Gernsbacher and S. J. Derry (eds), *Proceedings of the Twentieth Annual Conference of the Cognitive Science Society, Madison, WS*: 734–9. Mahwah, NJ: Lawrence Erlbaum Associates.
Moxey, L. M., and Sanford, A. J. (2000). Focus effects associated with negative quantifiers. In M. Crocker, M. Pickering and C. Clifton (eds), *Architectures and Mechanisms of Language Processing*. Cambridge: Cambridge University Press.

Moxey, L. M., Sanford, A. J., and Dawydiak, E. J. (2001). Denial as controllers of negative quantifier focus. *Journal of Memory and Language* 44: 427–42.
Paterson, K. B., Sanford, A. J., Moxey, L. M., and Dawydiak, E. J. (1998). Quantifier polarity and referential focus during reading. *Journal of Memory and Language* 39: 290–306.
Percus, O., Gibson, T., and Tunstall, S. (1997). *Antecedenthood and the Evaluation of Quantifiers.* Poster presented at the tenth CUNY conference, Santa Monica, California, March 20–22.
Sanford, A. J., Fay, N., Stewart, A. J. and Moxey, L. M. (2002). Perspective in statements of quantity, with implications for consumer psychology. *Psychological Science* 13: 130–4.
Sanford, A. J., and Garrod, S. C. (1989). What, when and how? Questions of immediacy in anaphoric reference resolution. *Language and Cognitive Processes* 4: 235–62.
Sanford, A. J., Moxey, L. M., and Paterson, K. B. (1996). Attentional focusing with quantifiers in production and comprehension. *Memory and Cognition* 24: 144–55.
Sanford, A. J., Williams, C., and Fay, N. (2001). When being included is being excluded: A note on complement set focus and the inclusion relation. *Memory and Cognition* 29(8): 1096–101
Walker, M. A., Joshi, A. K., and Prince, E. F. (1998). *Centring Theory in Discourse.* Oxford: Clarendon Press.
Wason, P. C. (1959). The processing of positive and negative information. *Quarterly Journal of Experimental Psychology* 11: 92–107.
Wason, P. C. (1961). Response to affirmative and negative binary statements. *British Journal of Psychology* 52: 133–42.
Wason, P. C. (1965). The contexts of plausible denial. *Journal of Verbal Learning and Verbal Behavior* 4: 7–11.
Westerståhl, D. (1989). Quantifiers in formal and natural languages. In D. Gabbay and F. Guenthner (eds), *Handbook of Philosophical Logic* 4: 1–131.
Zwarts, F. (1996). Facets of negation. In J. van der Does and J. van Eijk (eds), *Quantifiers, Logic and Language.* Stanford, CA: CSLI.
Zwarts, F. (1998). Three types of polarity. In E. Hinrichs and F. Hamm (eds), *Plural Quantification*: 177–238. Dordrecht: Kluwer.

Part II
Current Issues in Experimental Pragmatics

7
Testing the Cognitive and Communicative Principles of Relevance

Jean-Baptiste Van der Henst and Dan Sperber

1 Introduction

A general theory is testable not directly but through consequences it implies when it is taken together with auxiliary hypotheses. The test can be weaker or stronger depending, in particular, on the extent to which the consequences tested are specifically entailed by the theory (as opposed to being mostly entailed by the auxiliary hypotheses and being equally compatible with other general theories). The earliest experimental work based on Relevance Theory (Jorgensen, Miller and Sperber, 1984; Happé 1993) tested and confirmed Sperber and Wilson's (1981) echoic account of irony (and much experimental work done since on irony has broadly confirmed it and refined it further). While this account of irony is part and parcel of Relevance Theory, it is nevertheless compatible with different pragmatic approaches. The experimental confirmation of this account, therefore, provides only weak support for Relevance Theory as a whole. More recent experimental work has made explicit, tested and confirmed other and more specific and central consequences of Relevance Theory (e.g. Sperber, Cara and Girotto, 1995; Politzer, 1996; Gibbs and Moise, 1997; Hardman, 1998; Nicolle and Clark, 1999; Matsui, 2000, 2001; Girotto, Kemmelmeir, Sperber and Van der Henst, 2001; Noveck, 2001; Noveck, Bianco and Castry, 2001; Van der Henst, Sperber and Politzer, 2002; Van der Henst, Carles and Sperber, 2002; Noveck and Posada, 2003; Ryder and Leinonen, 2003). Here we review experiments that test consequences of the most central tenets of the theory, namely the Cognitive and the Communicative Principles of Relevance.

2 The basic tenets of Relevance Theory

Relevance, as characterized in Relevance Theory, is a property of inputs to cognitive processes. These inputs include external stimuli (for instance utterances) and internal representations (for instance memories or conclusions from inferences that may then be used as premises for further inferences). When is an input relevant? An input is relevant to an individual when processing it in a context of previously available assumptions yields positive cognitive effects, that is, improvements to the individual's knowledge that could not be achieved from processing either the context on its own, or the new input on its own. These improvements may consist in the derivation of contextual implications, in the confirmation of uncertain assumptions, in the correction of errors, and also, arguably, in the reorganization of knowledge so as to make it more appropriate for future use.

Inputs are not just relevant or irrelevant; when relevant, they are more or less so. A relatively high degree of relevance is what makes some inputs worth processing. Many of the potential inputs competing for an individual's processing resources at a given time may offer a modicum of relevance, but few are likely to be relevant enough to deserve attention. What makes these worth processing is, to begin with, that they yield comparatively higher cognitive effects. However, two inputs yielding the same amount of cognitive effect may differ in the amount of processing effort[1] required to produce this effect. Obviously, the less the effort, the better. If relevance is what makes an input worth processing, then the relevance of an input is not just a matter of the cognitive effect it yields but also of the mental effort it requires. Hence, the characterization of relevance in terms of *effect* and *effort*:

(1) *Relevance of an input to an individual*

   (a) Other things being equal, the greater the positive cognitive effects achieved by processing an input, the greater the relevance of the input to the individual at that time.
   (b) Other things being equal, the greater the processing effort expended, the lower the relevance of the input to the individual at that time.

Here is a simplified illustration of how the relevance of alternative inputs might be compared in terms of effort and effect. Suppose you want to take the next train to Bordeaux and compare statements (2)–(4) (assumed to be uttered by a reliable informer):

(2) The next train to Bordeaux is at 3:24pm.
(3) The next train to Bordeaux is after 3pm.
(4) The next train to Bordeaux is 36 minutes before 4pm.

---

[1] 'Effort' as used here refers here to any expenditure of energy in the pursuit of a goal. It is not restricted to conscious effort.

All three statements would be relevant to you, but (2) would be more relevant than either (3) or (4). Statement (2) would be more relevant than (3) for reasons of cognitive effect: (2) entails (3), and therefore yields all the conclusions derivable from (2), and more besides, and these extra conclusions themselves have practical consequences for the planning of your trip. Statement (2) would be more relevant than (4) for reasons of processing effort: although (2) and (4) are logically equivalent, and therefore yield exactly the same cognitive effects, these effects are easier to derive from (2) than from (4), which requires an additional effort of calculus with no additional benefit whatsoever (in the ordinary situation envisaged). More generally, when similar amounts of effort are required by two alternative inputs, the effect factor is decisive in determining degrees of relevance, and when similar amounts of effect are achievable, the effort factor is decisive. In experimental work, as we will illustrate, this makes it relatively easy to manipulate the relevance of stimuli across conditions by keeping the effort factor constant and modifying the effect factor or, conversely, by keeping the effect factor constant and modifying the effort factor.

Relevance theory claims that, because of the way their cognitive system has evolved, humans have an automatic tendency to maximize relevance. As a result of constant selection pressure towards efficiency, perceptual mechanisms tend automatically to pick out potentially relevant stimuli, memory mechanisms tend automatically to store and, when appropriate, retrieve potentially relevant pieces of knowledge, and inferential mechanisms tend spontaneously to process these inputs in the most productive way. This universal tendency is described in the First, or Cognitive, Principle of Relevance:

(5) *Cognitive Principle of Relevance*

Human cognition tends to be geared to the maximization of relevance.

This spontaneous tendency to maximize relevance makes it possible to predict to some extent to which available stimuli people will pay attention and how they will process them.

There is a wealth of evidence in the experimental study of attention and memory that could be re-analysed in order to see to what extent it supports the Cognitive Principle of Relevance. This is not our field of expertise, but the challenge there, we surmise, would be not so much to find support as to find support that is specific enough to relevance theory, in other words to find predictions that follow from the Cognitive Principle of Relevance but not – or not as directly – from standard psychological approaches to attention and memory. In other areas, the study of inference and that of communication in particular, the cognitive principle does have consequences that are far from trivial. Some of these consequences in the domain of category-based induction have been explored by Medin, Coley, Storms and Hayes

(2003). In Section 3 of this chapter we will present experimental tests of consequences based on work by Van der Henst and his collaborators on relational reasoning.

Relevance Theory has been mostly an exploration of the implications of the Second, Communicative Principle of Relevance for human verbal communication. The human tendency to maximize relevance makes it possible not only to predict some of other people's cognitive processes, but also to try to influence them – how indeed could you aim at influencing people if you had no way to predict how your behaviour would affect their thought? Human intentional communication, and in particular verbal communication, involves the attribution, by the communicator and the addressee, of mental states to one another. This attribution is greatly helped by the relative predictability of relevance-guided cognitive processes. In particular, a speaker must intend and expect that the hearer will pay attention to the utterance produced. If attention tends automatically to go to inputs that seem relevant enough to be worth processing, then it follows that, to succeed, the speaker must intend and expect her utterance to be seen as relevant enough by the hearer she is addressing. By the very act of speaking to him, the communicator therefore encourages the hearer to presume that the utterance is so relevant. This is the basis for the Communicative Principle of Relevance:

(6) *Communicative Principle of Relevance*

Every utterance conveys a presumption of its own optimal relevance.

An utterance, so the theory claims, conveys not just a vague expectation, but a precise presumption of relevance, which the notion of 'optimal relevance' captures:

(7) *Optimal relevance*

An utterance is optimally relevant to the hearer just in case:
(a) It is relevant enough to be worth the hearer's processing effort.
(b) It is the most relevant one compatible with the speaker's abilities and preferences.

According to clause (7a) of this definition, the hearer is entitled to expect the utterance to be at least relevant enough to be worth processing, which means (given the Cognitive Principle of Relevance) that the utterance should be more relevant than any alternative input available at the time.

Is the hearer entitled to higher expectation than this (already high) minimum level spelled out in clause (7a)? The speaker wants to be understood. It is therefore in her interest to make her utterance as easy as possible to understand, and to provide evidence not just for the cognitive effects she aims to achieve in the hearer but also for further cognitive effects which, by holding his attention, will help her achieve her goal. Speakers, however, are not omniscient, and they cannot be expected to go against their own interests and preferences in producing an utterance. There may be relevant information that they are unable or unwilling to provide, and wordings that would convey their meaning more economically, but that they are unable to think of at the time, or are unwilling to use (for reason of propriety for instance). All this is spelled out in clause (7b) of the definition of optimal relevance, which states that the ostensive stimulus is the most relevant one (i.e., yielding the greatest effects, in return for the smallest processing effort) that the communicator is *able and willing* to produce.

The Communicative Principle of Relevance justifies a specific inferential procedure for interpreting an utterance, that is, for discovering what the speaker meant by uttering it:

(8) *Relevance-guided comprehension procedure*
   (a) Follow a path of least effort in constructing and testing interpretive hypotheses (regarding disambiguation, reference resolutions, implicatures, etc.).
   (b) Stop when your expectations of relevance are satisfied.

Given clause (7b) of the definition of optimal relevance, it is reasonable for the hearer to follow a path of least effort because the speaker is expected (within the limits of her abilities and preferences) to make her utterance as easy as possible to understand. Since relevance varies inversely with effort, the very fact that an interpretation is easily accessible gives it an initial degree of plausibility. It is also reasonable for the hearer to stop at the first interpretation that satisfies his expectations of relevance, because there should never be more than one. A speaker who wants her utterance to be as easy as possible to understand should formulate it (within the limits of her abilities and preferences) so that the first interpretation to satisfy the hearer's expectation of relevance is the one she intended to convey. An utterance with two apparently satisfactory competing interpretations would cause the hearer the unnecessary extra effort of choosing between them, and, because of this extra effort, the resulting interpretation (if there were one) could never satisfy clause (7b) of the definition of optimal relevance. Thus, when a hearer following the path of least effort arrives at an interpretation that satisfies his expectations of relevance, he should, in the absence of contrary evidence, adopt it. Since comprehension is a non-demonstrative inference process, this interpretation of the speaker's meaning may be erroneous. Still, it is the most plausible interpretation in the circumstances.

The hypothesis that hearers spontaneously follow the relevance-guided comprehension procedure spelled out in (8) can be experimentally tested by manipulating the effort factor and, in particular, by changing the order of accessibility of various interpretations. It can also be tested by manipulating the effect factor and thereby making a specific interpretation more or less likely to satisfy the hearer's expectations of relevance. This, as we will illustrate in Section 4, is what Girotto, Sperber and their collaborators have done in a series of experiments with the Wason Selection Task.

Most work in Relevance Theory so far has been focused on utterance interpretation rather than on utterance production. The theory, however, has testable implications regarding the production process. Speakers often fail to be relevant to their audience, and sometimes do not even make the effort to be relevant. Still, utterances couldn't effectively convey the presumption of their own relevance unless speakers were, most of the time, aiming at optimal relevance and achieving it often enough. In Section 5, we describe a series of experiments that were aimed at testing to what extent speakers were actually aiming at optimal relevance.

3 Testing the Cognitive Principle of Relevance with relational reasoning tasks

In most studies on reasoning, psychologists analyse participants' successful or unsuccessful performance in reasoning tasks. They look at the percentages of correct conclusions or at the time taken to draw such a conclusion. They investigate factors that impede or enhance correct performance, such as the premises' content, the premises' complexity, task instructions or IQ. They use this evidence to test various theories of the inferential machinery that underlies our reasoning ability. Some argue that people reason by constructing mental models of the premises (Johnson-Laird and Byrne, 1991). Others support the idea that people reason by applying general inference rules (Rips, 1994; Braine and O'Brien, 1998). Yet others have proposed that reasoning relies on domain-specific procedures (Cheng and Holyoak, 1985; Cosmides, 1989).

Relevance Theory claims that comprehension is based on a domain-specific inferential procedure, but it is not, in and by itself, a theory of human reasoning. It is, in fact, compatible with the view that an important role is played in reasoning by mental models, or by inference rules, or by both, or by yet other kinds of procedures in a domain-general or in a domain-specific way.[2] Nevertheless, Relevance Theory may make a direct contribution to the study of reasoning by suggesting testable claims not on the procedures (except in the case of comprehension) but on the goals of reasoning processes.

---

[2] Sperber, however, has been defending the view that the human mind is 'massively modular' (Sperber, 1994), and Sperber and Wilson (2002) have argued that linguistic comprehension is modular.

Standard approaches to the study of reasoning have had little to say on what causes people to engage in reasoning – when they are not, that is, requested to do so by an experimenter – what expectations they have in doing so, and what kind of conclusions satisfy these expectations, bringing the process to a close.[3] What guides reasoners to infer a specific conclusion? At first sight, one might argue that people aim at inferring a conclusion that *logically follows* from the premises. However, from any given set of premises, an infinity of conclusions logically follows. Most of these valid conclusions are of no interest at all. For instance, nobody would burden one's mind by inferring from the single premise P the logical conclusion Not (not (not (not P))). Harman has formulated this idea as a *principle of clutter avoidance*: 'It is not reasonable or rational to fill your mind with trivial consequences of your beliefs, when you have better things to do with your time, as you often do' (Harman, 1995, p. 186).

It is not sufficient for a conclusion to be logically valid in order to be worth inferring. Some valid conclusions are too trivial ever to be derived, and others may be derived in some circumstances and not in others. From the same set of premises, we might derive one particular conclusion in one situation, another conclusion in a second situation, or no conclusions at all in a third. In a recent study, we proposed that the conclusions that people are inclined to draw are those, if any, that seem relevant enough in the context (Van der Henst et al., 2002). This, of course, is a direct consequence of the Cognitive Principle of Relevance.

In this study, we compared so-called 'determinate' and 'indeterminate' relational problems such as these:

A determinate problem   *An indeterminate problem*
    A is taller than B   A is taller than B
    B is taller than C   A is taller than C

Such relational problems have been empirically investigated in many studies (see Evans, Newstead and Byrne, 1993, for a review). Determinate problems are so called because the one relation between the three terms A, B and C which is not explicitly described in the premises, that between A and C, is nevertheless inferable from them: in our example, A is taller than C. Indeterminate problems are so called because the one relation which is not described in the premises is not inferable from them: in our example, B might be taller than C or C might be taller than B. Hence, nothing follows from the premises about the relation between B and C. The goal of most studies on relational problems has been to describe the way in which

---

[3] See Johnson-Laird and Byrne (1991: 20–2) for a notable exception.

the premises are being mentally represented and processed by reasoners. Typically, participants have had to answer a specific question like 'What is the relation between A (or B) and C?' and the evidence consists in the rate of correct answers. The correct answer for the determinate problem above would be: A is taller than C. The correct answer for the indeterminate problem would be: it is impossible to tell. Indeterminate problems tend to yield a lower rate of correct answers than determinate problems.[4]

In our study, our aim was not to assess and explain the relative difficulty of determinate and indeterminate problems. Instead of asking a question about a specific relation between two terms mentioned in the premises, we just asked *what, if anything, follows from the premises?* We were interested in what causes some participants, particularly with indeterminate problems, to answer *nothing follows*.

Not only is it always possible to infer conclusions from a given set of premises, but what is more, some of these conclusions are quite obvious: for instance, from two premises P and Q, their conjunction P-and-Q trivially follows. So when people answer that nothing follows from a given set of premises, either they just fail to see the obvious, or, we suggest, they mean that nothing *relevant* follows. If so, *nothing follows* answers are evidence of people's intuitions of relevance. In particular, if a problem creates the expectation that the most relevant conclusion to be derived should be of a certain type and, at the same time, does not warrant any conclusion of this particular type, people may be tempted to answer that nothing follows. This, we tried to show, is what happens with indeterminate relational problems.

What conclusion could participants expect to infer from two relational premises in the context of a reasoning task? In determinate and indeterminate relational problems such as the examples above, there are three terms, A, B and C, one type of asymmetric and transitive relation, for example *taller than*, and therefore three possible relations of this type, in the pairs A–B, B–C, A–C. Two of these relations are described in the premises. Given the Communicative Principle of Relevance, these relations are presumed to be relevant in the context of the task, and, more specifically, the two relations given in the premises are expected be relevant in allowing the inference of the third relation. Of course, it could be rightly pointed out that, in these experimental situations, the premises on which participants are asked to reason are arbitrary and without relationship to their real-life concerns. Therefore, neither the premises nor the conclusions that can be derived from them have any genuine relevance. Still, we would argue, just as participants

---

[4] Supporters of mental model theory explain this fact by pointing out that the mental representation of indeterminate problems calls for two mental models (to represent the two possible relations between B and C) as opposed to one model for the determinate problems (Byrne and Johnson-Laird, 1989).

reason under the pretence that the premises are true (that, say, the premise 'Jim is taller than Paul' is about two actual people), they reason under the pretence that the premises, and the conclusions they are expected to derive from them, might be relevant in some ordinary context of knowledge about the individuals or the entities described in the premises. It is not hard, for instance, to pretend that it might be relevant to know that Jim is taller than Paul and that Paul is taller than Dick, and to assume then that it would be relevant to draw the inference that Jim is taller than Dick.

Participants' expectations of relevance are easily satisfied in the case of determinate problems but not in the case of indeterminate ones, where the relation that is not specified in the premises cannot be inferred from them. Hence, with indeterminate problems, participants may be tempted to answer that nothing follows. This is indeed what we observed. In our study, 43 per cent of the participants gave a 'nothing follows' response to indeterminate problems, while only 8 per cent did so with determinate problems. This difference in the rate of 'nothing follows' answers between determinate and indeterminate problems is, of course, not surprising. However, it had never been demonstrated before, and, more importantly, only Relevance Theory provides a simple and direct explanation of this difference. When participants say that nothing follows, what they mean, we surmise, is not that it is impossible to infer anything at all from the two premises, but that it is impossible to derive a conclusion relevant enough to be worth deriving, namely a conclusion about the third undescribed relation among the three items mentioned in the premises.

Nevertheless, facing a situation where what would be the most relevant conclusion cannot be inferred, about one-half of the participants do offer some positive conclusion. Are they giving up on relevance and aiming just for any logically valid conclusion, or are they still guided by considerations of relevance? As we will show, one can find out by examining the specific conclusions they actually derive.

Consider the determinate conclusion *A is taller than B and C*, or equivalently, *A is the tallest*, derived from our indeterminate problem. This conclusion is merely a linguistic integration of the premises. It may seem trivial, especially in the context of a reasoning experiment where, generally, participants are eager to demonstrate their reasoning skills to the experimenter. However, a conclusion such as *A is the tallest* may have some relevance of its own. There are ordinary situations where it would be relevant to know which item in a set is above the others with respect to some given property (e.g. who is the tallest?). Actually, in many situations, knowing which item in a set is above all the others with respect to some comparative property is more relevant than knowing the relative position of two other items in the set that are lower on the comparison scale. For instance, suppose you have the choice among three different cars all of which would satisfy your needs and you just want to buy the cheapest. You will probably be more interested in knowing which is the cheapest of the three than in knowing which is the cheaper of the other two. Hence, inferring *A is more...than B and C* has some relevance since, assuming a quite ordinary context, it can be a step towards inferring further contextual implications (e.g. about which car to buy).

One might query: how can deductively deriving a conclusion and adding it to, or substituting it for, an initial set of premises yield a more relevant point of departure for further reasoning, given that nothing can be derived from this conclusion that wasn't already derivable from the initial premises? In other words, how can such a conclusion be relevant at all, in a context where the premises from which it is derived are given? The fact that relevance is defined not just in terms of effect but also in terms of processing effort provides a simple answer. A set of premises with some deductively derived conclusion added could not carry more cognitive effects than the initial set and thus be more relevant *on the effect side*, but it can be more relevant *on the effort side* by allowing the same effects to be derived with less effort. The deduction of some specific conclusion from a set of premises may be a preliminary and effort-costly necessary step towards deriving cognitive effects from this set of premises. In that case, the conclusion is as relevant as the premises on the effect side and more relevant than the premises on the effort side.

We frequently encounter information which we think is likely to prove useful in the future. We then retain this information, and often process it in such a way as to optimize its potential usefulness. Suppose, for instance, that you arrive in a holiday resort where you plan to spend a month with your family. You learn that there are three doctors in the resort, Smith, Jones and Williams. You also learn the following two pieces of information: {*Smith is a better doctor than Jones, Jones is a better doctor than Williams*}. At the time, you don't need a doctor, but you might in the future, and would then want to visit the best doctor in town. So the information is potentially relevant to you. You might just store the two pieces of information above, but from a cognitive point of view it would be more efficient to draw the conclusion: *Smith is the best doctor* straight away. By drawing this conclusion now, you prepare for future circumstances in which you would need a doctor. By adding this conclusion to the two initial premises, you are left with a set of premises for future inference with a greater expected relevance, since its exploitation will require fewer inferential steps. Moreover, if you expect not to need information about the other two doctors, it may be sufficient to remember just the conclusion *Smith is the best doctor*, replacing the initial two-premise set with the single derived conclusion, thus reducing the memory load.

If what makes a conclusion seem relevant is that it spares effort for the possible derivation of cognitive effects, then it follows that the more effort it spares for such possible derivations, the greater will be its perceived relevance. In our initial study (Van der Henst et al., 2002), we manipulated the relevance of a relational conclusion of the form *A is more...than B and C* by formulating the premises so as to make the derivation of such a conclusion more or less effortful. In one type of problem, the derivation of this conclusion was very easy and thus the effort saved for the possible derivation of cognitive effects was quite low, whereas with another type of problem, deriving the conclusion was harder, and thus the effort saved was greater. The problems we used were the following:[5]

>   Problem 1   Problem 2
>   A is taller than B   B is taller than A
>   A is taller than C   C is taller than A

In both problems, the relation between B and C is indeterminate. Still, from either problem one can derive a variety of conclusions. For instance, from Problem 1, one can infer conclusion (9a) and (9b), and from Problem 2, one can infer conclusion (10a) and (10b):

(9) (a) 'A is taller than B and C'
    (b) 'B and C are shorter than A'
(10) (a) 'A is shorter than B and C'
    (b) 'B and C are taller than A'

With the usual element of pretence involved in the experimental study of reasoning, such conclusions can be seen as having some relevance in that they may facilitate the derivation of further cognitive effects, given some plausible context.

Deriving the single-subject conclusion (9a) from the premises of Problem 1 hardly involves any inferential effort. Since the grammatical subject (*A*) and the comparative term (*taller than*) are the same in the conclusion and in the premises, it amounts just to merging the two premises in a single sentence. Deriving the single-subject conclusion (10a) from the premises of Problem 2, on the other hand, involves some genuine inferential effort: the grammatical object in the premises (*A*) has to be put in subject position, and the comparative term (*taller than*) has to be converted into its opposite (*shorter than*). It is rather the double-subject conclusion (10b) that amounts to a mere merging of the premises. If participants just went for the less effort-demanding conclusion, they should choose (9a) and (10b). However, if they are guided by considerations of relevance, they should choose (9a) and (10a).

Conclusions (9a) and (9b) are logically equivalent and therefore, in any context, would yield the same effects, and so are and would conclusions (10a) and (10b). However, in most contexts, deriving these effects by using

---

[5] All the experiments reported in this section were carried out in French.

the single-subject conclusions (9a) and (10a) as premises is likely to cause less effort than by using the double-subject conclusions (9b) or (10b) as premises. Why? Because most pieces of knowledge transmitted, constructed and stored in human cognition have as their topic a single entity or a single category rather than a pair of entities or categories (for fairly obvious reasons having to do with cognitive efficiency). One is more likely, for instance, to encounter a contextual conditional premise of the form (11a) than of the form (11b):

(11) (a) *'If A is taller than B and C, then...'*
(b) *'If B and C are shorter than A, then...'*

From either of (9a) and (9b) and either of (11a) and (11b) as premises, the same conclusions can be derived, but the derivation will be more direct if the minor premise, that is, (9a) or (9b), of this conditional syllogism matches the antecedent of the major premise, that is, (11a) or (11b). In other words, in most realistic contexts, single-subject conclusions such as (9a) and (10a) are likely to prove more relevant than double-subject conclusions such as (9b) and (10b). We predicted therefore that, in both Problem 1 and Problem 2, participants, guided by considerations of relevance, would derive more single-subject that double-subject conclusions.

There is a further reason, specific to the premises of Problem 2, why (10a) should be perceived as more potentially relevant than (10b). It is that the extra effort involved in deriving (10a) as compared to (10b) is effort expanded in the right direction. It can be seen as preparatory for the derivation of cognitive effect. This argument does not apply to (9a) and (9b) in Problem 1. As we mentioned, the derivation of (9a), unlike that of (9b), involves almost no effort. In other terms, the derivation of both (9a) and (10a) are steps in the right direction, but the derivation of (10a) is a much bigger step, and therefore a more useful one. This suggests that Problem 2 should be seen as yielding a relevant enough conclusion more frequently than Problem 1.

For the reasons just developed, we expected that participants who produced a conclusion with Problem 1 and 2 would predominantly produce a single-subject conclusion and that there would be more such conclusions, and fewer 'nothing follows', with Problem 2 than with Problem 1. Note that there is nothing intuitively obvious about these predictions, which follow quite directly from the Cognitive Principle of Relevance applied to this particular reasoning problem, and from no other approach we are aware of. Our findings, presented in Table 7.1, confirmed these predictions.

Another way to increase the relevance of a conclusion *A is more...than B and C* inferred from indeterminate relational premises is to act on the effect side. As we pointed out, a conclusion cannot yield more cognitive effects than the premises from which it is deductively derived. However, the information contained in the premises of a problem can yield greater or lesser cognitive effects, depending on the wider context. The greater are

Table 7.1 Percentage of conclusion types for Problems 1 and 2

	Problem 1 A is taller than B A is taller than C	Problem 2 B is taller than A C is taller than A	Total
Single-subject conclusions	26	45	35
Double-subject conclusions	14	15	14
Nothing follows	54	31	43
Other	6	9	8

these effects, the more useful it is to derive a conclusion which is a step towards the production of these effects, and therefore the more relevant is this conclusion. Acting on the effect side here means providing or suggesting a context in which a conclusion derived from the premises of a problem might yield greater or lesser cognitive effects.

In Problem 2, the conclusion *A is taller than B and C* has a modicum of potential relevance. The cognitive effects that this conclusion might yield remain vague since no context is given. The relevance of such a conclusion can be increased by manipulating the effect factor in the way we have just suggested. This can be done, in particular, by providing a context in which this conclusion will have clear contextual implications. Imagine, for instance, that the premises of Problem 2 are processed with the knowledge that the tallest person of A, B and C, is the tallest person in the world. In this context, deducing that *A is taller than B and C* is a necessary step towards inferring that *A is the tallest person in the world*.

We predict that people should be more inclined to produce the conclusion *A is more...than B and C* when an appropriate context is given than when no context is given, or than when a less or non-appropriate context is given. We tested this prediction in three experiments of an unpublished study done with Guy Politzer.

In the first experiment, participants received either a problem without explicit context (Problem 3) or a problem with an explicit context (Problem 4) and had to produce a conclusion:

Problem 3

Premises: A is ahead of B
A is ahead of C

Problem 4

Context: *A, B and C were the top three finishers in the race last Sunday.*
Premises: A is ahead of B
A is ahead of C

For both problems, it follows from the premises that *A is ahead of B and C*. However, in the race context, inferring the logical conclusion *A is ahead of B and C* is a step towards inferring the contextual implication *A won the race*. The possibility of deriving this contextual implication endows the logical conclusion *A is ahead of B and C* with greater relevance than in the absence of any explicit context. Since the inference that A is ahead of B and C has a greater relevance in the race context, it should be more frequently performed, and participants should formulate more determinate conclusions and fewer 'nothing follows' answers. Our results indeed show that Problem 4 resulted in a higher rate of determinate conclusions than Problem 3 (54 % vs 70 %, χ^2 (1) = 5.59, p < 0.02). Moreover, in the race context, there were three times as many determinate conclusions referring to the race context like *'A is the first'* or *'A is the winner'* than conclusions simply integrating the two premises like *'A is ahead of B and C'* or *'B and C are behind A'*.

In a second experiment, we manipulated the effect factor by using two different explicit contexts both of which increased the relevance of the conclusion *A is more...than B and C*. However the context of Problem 5 (almost identical to that of Problem 4 above) produced a greater increase in relevance than that of Problem 6:

Problem 5

Context: A, B and C were the first three finishers in the race last Sunday.
Premises: A arrived before B
A arrived before C

Problem 6

Context: A, B and C were the last three finishers in the race last Sunday.
Premises: A arrived before B
A arrived before C

In Problem 5, the context explicitly focuses on people who were the first three in a race; if this is relevant at all, knowing who was *the* first should be even more relevant. The premises of the problem can thus achieve relevance by making it possible to infer who precisely arrived first and who did not. Deriving that A arrived before B and C enables one to infer three contextual implications: *A won the race, B did not win the race*, and *C did not win the race*. In Problem 6, the context focuses on people who arrived last in an athletics race. In contrast with Problem 5, deriving that A arrived before B and C makes it possible to infer only one contextual implication: *A did not arrive last*. Because the relation between B and C is indeterminate, it is impossible to infer who arrived last. The conclusion that A arrived before B and C has some relevance in Problem 6, but less so than in Problem 5 and should therefore be produced less often. Our results (see Table 7.2) show that people indeed derived more determinate conclusions in Problem 5 than in Problem 6 (94.4 % vs 74.7 %, χ^2 (1) = 13.45, p < 0.001).

Table 7.2 Percentage of conclusion types for Problems 6 and 7

	Problem 5	Problem 6
	N = 90	N = 91
Determinate conclusions	94.4	74.7
Nothing follows	3.3	18.7
Errors and weird answers	2.2	6.6

Any explicit context evokes a wider *implicit* context of general knowledge. For instance, the explicit context of Problem 5, '*A, B and C were the first three finishers in the race last Sunday*', evokes background knowledge about racing, about the value attributed to winning, prizes or medals given to winners and so on. So, inferring from the explicit context that A has won the race makes it possible to infer from the implicit context that A is likely to be pleased, that he may be given a medal or a prize and so on.

In a third experiment, we manipulated relevance by evoking different implicit contexts. In general, when a context is explicitly provided, participants may expect the premises of a problem to be relevant in this explicit context or, at least, in the wider context implicitly evoked by this explicit context. If the explicit and implicit contexts are related in content to the premises, this should strengthen the expectation of relevance and encourage participants to derive positive conclusions from the premises rather than answer that nothing follows. Inversely, if the explicit and implicit contexts are unrelated in content to the premises, this should lower participant's expectations of relevance and encourage them to say that nothing follows. Here is how we tested this prediction.

Consider Problems 7 and 8:

Problem 7

Context: *A, B, and C, who were measured during a medical examination, are not of the same height.*
Premises: A is taller than B
A is taller than C

Problem 8

Context: *A, B, and C did not win the same amount of money at the last lottery.*
Premises: A is taller than B
A is taller than C

The explicit context of Problem 7, by mentioning measurements of height as part of a medical examination, evokes an implicit context of common knowledge where differences in height may have implications for health, performance, accessibility to certain jobs and so on. This should encourage

Table 7.3 Percentage of conclusion types for Problems 7 and 8

	Problem 7 N = 162	Problem 8 N = 168
Determinate conclusions	76.5	42.9
Nothing follows	19.8	50.6
Errors and weird answers	3.7	6.5

participants to see the conclusion 'A is taller than B and C' as potentially relevant in this implicit context. The explicit context of Problem 8, mentioning the winning of money in a lottery, evokes an implicit context of common knowledge where individual height plays no role at all. Hence, we should observe a much lower rate of determinate conclusions for Problem 8 than for Problem 7. Our results (see Table 7.3) confirmed that there were many more determinate conclusions for Problem 7 (76.5 %) than for Problem 8 (42.9 %, χ^2 (1) = 38.8, p < 0.0001).

The experiments presented in this section give support to the Cognitive Principle of Relevance, that is, the claim that human cognition tends to be geared to the maximization of relevance, by corroborating some of its consequences in the area of psychology of reasoning. More specifically, the choice to draw or not to draw conclusions from a given set of premises, and the choice of which particular conclusion to draw, if any, are guided by considerations of relevance. People are inclined to draw a specific conclusion from a set of premises to the extent that this conclusion seems potentially relevant. This is a non-trivial consequence of the Cognitive Principle of Relevance. It has, in turn, non-trivial consequences for the study of reasoning in general. In particular, people's failure to derive some specific conclusion in a reasoning task may be due, not to poor logical capacities or to pragmatic problems concerning the comprehension of the task but to the failure to see as relevant either the conclusion they were intended to draw, or, more subtly, to the failure to see the relevance of some intermediary inferential step necessary for deriving the intended conclusion. In spontaneous inference, being guided by consideration of relevance should contribute to the overall efficiency of inferential processes, but it may also, on occasion, prevent one from reaching some highly relevant conclusion because crucial intermediary steps didn't seem relevant at all.

4 Testing the Communicative Principle of Relevance with the Wason Selection Task

Wason's Selection Task (Wason, 1966) has been the most commonly used tool in the psychology of reasoning (see Manktelow, 1999). Genuine versions of Wason's Selection Task share the same basic four-component structure:

1. An introduction (sometimes in a narrative form).
2. A conditional statement known as the 'rule', with the linguistic form 'If P, then Q', and either a descriptive content stating how things are, or a deontic content stating how they should be.
3. Four cards: one representing a case where P is satisfied, one where P is not satisfied, one where Q is satisfied, and one where Q is not satisfied (known respectively as the P, the not-P, the Q, and the not-Q cards). When the card displays information about P, information about Q is hidden, and conversely.
4. The instruction to select all and only those cards where the hidden information must be made visible in order to judge whether the rule is true (in descriptive versions) or is being obeyed (in deontic versions).

For example, the text of an 'abstract' descriptive selection task might be: 'Here are four cards. Each has a number on one side and a letter on the other side. Two of these cards are here with the letter-side up, and two with the number-side up. Indicate which of these cards you need to turn over in order to judge whether or not the following rule is true: "If there is a 6 on one side, there is an E on the other side."'

| 6 | 7 | E | G |

With such an abstract version of the task, typically only about 10 per cent of participants make the correct selection of the 6 and G cards, that is, the cards that represent the P case and the not-Q case.

In a typical example of a deontic version of the task (Griggs and Cox, 1982), participants are presented with a rule such as 'If a person is drinking beer, then that person must be over 18 years of age', with cards representing four individuals in a bar, with what they are drinking indicated on one side of the cards and their age indicated on the other side. The four cards represent respectively a person drinking beer, a person drinking soda (with the age hidden for these first two persons), a person aged 29, and a person aged 16 (with the drink hidden for these two other persons). Participants are instructed to select the cards that must be turned over to see whether any of these four people is breaking the rule. Typically, the correct card combination (i.e., the P-card 'This person is drinking beer' and the not-Q card 'This person is 16 years old') is selected by well over 50 per cent of the participants.

Work on the selection task has been the basis of a variety of claims about human reasoning and rationality. In particular, it has been taken to show that most individuals do not, in general, reason in accordance with the rules of logic, not even the elementary rules of propositional calculus, as evidenced by their failure to select the P and the not-Q cards in descriptive versions of the task (e.g., Cheng and Holyoak, 1985; Griggs and Cox, 1982). Does the Selection Task really provide a tool to test general claims about human reasoning? Evans (1989) maintained that participants understand the task as one of identifying the *relevant* cards, and use, for this, heuristic cues of relevance rather than deductive reasoning. Extending this insight, Sperber, Cara and Girotto (1995) put forward a general explanation of the selection task based on Relevance Theory. They argued that participants' performance on the selection task is best explained by considering that: (i) the very process of linguistic comprehension provides participants with intuitions of relevance; (ii) these intuitions, just as comprehension generally, are highly content and context dependent; and (iii) participants trust their intuitions of relevance and select cards accordingly. In standard versions of the task, these intuitions are misleading. In other versions, many deontic versions in particular, people's intuitions of relevance point towards the correct selection of cards. If, in the selection task, pragmatic comprehension mechanisms determine participants' response and thus pre-empt the use of whatever domain-general or domain-specific reasoning mechanisms people are endowed with, the task cannot be a good tool for the study of these reasoning mechanisms. On the other hand, it may be of some use in studying people's intuitions of relevance.

Participants presented with a Wason Selection Task approach the text of the problem, and in particular the conditional rule, in the same way in which they approach all utterances in conversation or in reading. They make use of their standard comprehension abilities. The very fact that a text is presented to them raises expectations of relevance, and they search for an interpretation that satisfies these expectations (which, given the artificiality of the task, may be quite modest). In doing so, they follow the relevance-guided comprehension procedure explained above in (8), that is, they follow a path of least effort in constructing interpretive hypotheses and stop when their expectations of relevance are satisfied. This is, in particular, what participants do with the conditional rule of the selection task: guided by expectations of relevance, they derive from it consequences that might justify these expectations.

The rule itself, being a conditional statement, is not directly testable. Merely by looking at the two sides of a card, you can check the truth or falsity of a plain atomic statement or of a conjunction of atomic statements such as 'there is a 6 on one side of this card and an E on the other side'. It is true if it matches your observations, and false otherwise. You cannot however confirm a conditional statement such as 'if there is 6 on one side, then there is an E on the other side' by matching it to your observations. The truth of a conditional statement is tested indirectly, by deriving from it consequences that are directly testable and testing these. Participants have therefore two reasons to derive consequences from the rule. The first reason is to interpret it in a way that satisfies their expectation of relevance. The second reason is to find directly testable consequences of the rule in order to give a sensible response to the experimenter. What they do in practice is give a response that is based on the consequences they spontaneously derived in interpreting the rule, without looking for other consequences that might provide a better test of the rule. What they should do, in principle, is make sure that not only the consequences they derive are entailed by the rule but also that, conversely, the rule is entailed by these consequences. Otherwise, the consequences might be true and the rule false. This would involve more than just reasoning in accordance with the rules of propositional calculus. It would also require higher-order reasoning about the structure of the problem. People's failure to do so shows not that, presented with such a problem, they are illogical, but that they are unreflective or, at least, insufficiently reflective, and overconfident in their intuitions of relevance.

In the case of the abstract task described above, participants may infer from the rule 'If there is a 6 on one side, there is an E on the other side' that the card with 6 must have an E on the other side. They may also infer from the rule the consequence that there are cards with a 6 and an E (otherwise the rule would be irrelevant). Making either or both of these consequences part of the interpretation of the rule contributes to its relevance by indicating what one might expect to see when turning over the cards. If participants use the first of these two consequences to decide which cards must be turned over in order to see whether the rule is true or false, they will select just the card with a 6 (the P card). If they use just the second consequence, or if they use both, they will turn over the card with a 6 and that with an E (the P card and the Q card). These are indeed the most frequent selections with standard selection tasks. In a deontic case such as that of the drinking-age problem, participants might, in order to satisfy their expectations of relevance, derive from the rule ('If a person is drinking beer, then that person must be over 18 years of age') the consequence that there should be no beer drinker under 18. They would then select the card representing a beer drinker (the P card) and that representing a person under 18 (the not-Q card), thus, as it happens, providing the correct selection.

Why should the consequences derived in the two problems be different? Because they are derived in their order of accessibility until expectations of relevance are reached, and both order of accessibility and expected level of relevance are context dependent. In both problems – and in general with conditional statements – the most accessible consequence is the *modus ponens* one: in the abstract problem, it is that the card with a 6 should have an E on the other side, and in the drinking-age problem, it is that the beer drinker should be 18 or above. In both cases, this implication determines the selection of the P card, which is indeed selected by most participants in both experiments. Why, then, do many participants select also the Q card in the abstract version, and the majority of participants select the not-Q card in the drinking-age problem (as in most deontic versions of the task)?

In the abstract problem above, the implication 'there are cards with a vowel and an even number' is much more easily accessed than the implication 'there are no cards with a vowel and without an even number', and satisfies the low expectations of relevance raised by this artificial problem. In the drinking-age problem, on the other hand, the implication that there should not be underage beer drinkers is the most accessible and the only one that satisfies expectation of relevance: commonsensically, the point of a normative rule such as 'If a person is drinking beer, then that person must be over 18' is not to make adult beer drinkers more common, but to make underage beer drinkers less common.

By pairing rules and contexts more approriately, the order of accessibility of consequences and expectations of relevance can be manipulated and it should be possible to elicit different patterns of selection, including logically correct selections. Sperber and colleagues (1995) produced several *descriptive* versions of the task that elicited a higher percentage of correct responses than had ever been found before with such versions. They showed that – contrary to what was generally believed at the time – good performance is not restricted to deontic versions.[6] Girotto, Kemmelmeir, Sperber and Van der Henst (2001) provided further evidence for the relevance approach by demonstrating how it can be used to manipulate *deontic* versions of the task and obtain at will either the common correct P and not-Q selections or incorrect P and Q selections (more commonly found in descriptive versions). Further experiments and comparisons with the approach of Leda Cosmides and her collaborators (Cosmides, 1989; Fiddick, Cosmides and Tooby, 2000) can be found in Sperber and Girotto (2002). Here, by way of illustration, we give just two examples of these experiments, one succinctly, the other in greater detail.

Girotto and colleagues (2001) used the following problem (adapting a problem from Cheng and Holyoak, 1985): 'Imagine that you work in a travel agency and that the boss asks you to check that the clients of the agency had obeyed the rule "If a person travels to any East African country, then that person must be immunized against cholera", by examining cards representing these clients, their destinations and their immunizations.' The four cards indicated 'Mr Neri. Destination: Ethiopia', 'Mr Verdi. Destination: Canada', 'Immunizations done: Cholera' and 'Immunizations done: None', respectively, and as usual, participants were asked which card had to be examined in order to find out whether the rule had been obeyed by the clients of the agency. In this context, the relevance of the rule is to prevent people without cholera immunization from travelling to East African countries. We predicted therefore that participants would choose the P card (a traveller to

---

[6] Other studies have confirmed this: e.g., Green and Larking, 1995; Hardman, 1998; Johnson-Laird and Byrne, 1995; Liberman and Klar, 1996; Love and Kessler, 1995.

an East African country) and the not-Q card (a person without cholera immunization). Such a prediction is not specific to Relevance Theory. It would be shared by all researchers in the area, whatever their theoretical viewpoint. It reiterates, after all, common findings, that have been explained, for instance, by proposing that people have pragmatic reasoning schemas for reasoning about obligations and permission (Cheng and Holyoak, 1985), or that they have an evolved 'Darwinian algorithm' for reasoning about social contracts (Cosmides, 1989).

According to the relevance-theoretic approach, what causes the selection of the P and not-Q cards in this deontic scenario is that the presence of individuals violating the cholera rule among the people represented by the cards would be more relevant than the presence of individuals obeying the rule. Could this relative relevance of cases of violation versus cases of conformity be reversed by altering the context, which, if the relevance approach is correct, should cause participants to choose the P and the Q cards? To do this, we used the same scenario, with a twist. The narrative stated that contrary to what the boss of the agency had thought, cholera immunization is not required anymore when travelling to East Africa. The boss is now worried that she may have misinformed clients and caused them to follow a rule that is no longer in force. She then asks the employee to see whether or not clients have obeyed the rule 'If a person travels to any East African country, then that person must be immunized against cholera' by looking at cards similar to those used in the previous condition. In this context, what is relevant is that some clients may have followed the false rule and that they may have been immunized unnecessarily (and might, for instance, sue the agency). On the other hand, the case of clients who have ignored the rule is no longer relevant. We predicted therefore that participants would select the P card (a traveller to an East African country) and the Q card (a person with cholera immunization). Note that this prediction is non-standard but follows from the relevance-based explanation of the selection task. This prediction was confirmed. Table 7.4 shows the results we obtained in a within-participants design. (We also obtained practically the same results with a between-subjects design.)

This cholera-rule experiment gives, we hope, an intuitively clear illustration of the role of relevance in participants' response to selection task problems.

Table 7.4 Percentage of the main selection patterns in the true and false cholera rule selection task

Pattern	True Rule	False Rule
P and not-Q	62	15
P and Q	26	71
Other	12	14

However, it remains too intuitive to give a truly specific confirmation to the Communicative Principle of Relevance. In particular, it throws no light on the respective role of effect and effort in guiding participants' intuitions of relevance and selection of cards.

In their Experiment 4, Sperber and colleagues (1995) aimed at taking apart the two factors of relevance, effect and effort, testing their respective roles, and ascertaining whether relevance, which combines the effort and the effect factors in a principled manner, is more explanatory than effort or effect taken alone. For this, they created four scenarios, varying the effect and the effort factors separately in four conditions, namely **effect–/effort+**, **effect–/effort–**, **effect+/effort+**, and **effect+/effort–** (see Figure 7.1). All four scenarios involved a machine that manufactures cards with a number on one side and a letter on the other side. A character, Mr Bianchi, asserts: 'If a card has a 6 on the front, it has an E on the back.' In all conditions, the four cards had respectively a 6, a 4, an E and an A on the visible side, and participants were asked which card or cards had to be turned over to check whether what Mr Bianchi says is true.

From the conditional 'If a card has a 6 on the front, it has an E on the back' participants are sure to derive consequence (12). They may also derive either or both of (13) and (14):

(12) The card with a 6 has an E on the other side
(13) There are cards with a 6 and an E
(14) There are no cards with a 6 and without an E

In the two **effort+** conditions, (13) is easier to derive than (14), which involves two negations. Moreover (14) does not carry any obvious effect worth the extra effort. So we should expect participants to base their selections either on (12) and to select just the E, or on (12) and (13), and to select both the E and the 6.

To increase the probability that participants would derive consequence (14) before (13), we could act on the effort side or on the effect side. To act on the effort side, we had, in the two **effort–** conditions, the machine print only 6s and 4s on one side and Es and As on the other side. Instead of an indefinite number of possible number–letters combinations (e.g. $9 \times 26 = 234$ if only numbers from 1 to 9 are used), we have now four possible combinations; 6 and E, 6 and A, 4 and E, and 4 and A, which are all equally easy to represent. This makes it possible to simplify (14) and replace it with (14′)

(14′) There are no cards with a 6 and an A

We predicted that (14′) being easier to represent than (14), more participants would derive it and would, accordingly, select the card with an

Testing Principles of Relevance   163

Effect−/Effort+	Effect−/Effort−	Effect+/Effort+	Effect+/Effort−
\multicolumn{4}{c}{A machine manufactures cards. It is programmed to print at random, on the front of each card,}
A number	A 4 or a 6	A number	A 4 or a 6
On the back of each card, it prints a letter at random.	On the back of each card, it prints either an E or an A at random.	On the back of each card, it prints a letter: – When there is a 6, it prints an E. – When there is not a 6, it prints a letter at random.	On the back of each card, it prints a letter: – When there is a 6, it prints an E. – When there is a 4, it prints an E or an A at random.
		\multicolumn{2}{l	}{One day, Mr Bianchi, the person in charge, realises that the machine has produced some cards it should not have printed. On the back of the cards with a 6, the machine has not always printed an E:}
		sometimes it has printed any letter at random.	sometimes it has printed an A instead of an E.
The person in charge, Mr Bianchi, examines the cards and has the strong impression that the machine does not really print letters and numbers at random. I think, he says, that		Mr Bianchi fixes the machine, examines the newly printed cards and says: don't worry, the machine works fine,	
\multicolumn{4}{c}{**if a card has a 6 on the front, it has an E on the back**}

Figure 7.1   The four conditions of the machine experiment (Sperber, Cara and Girotto, 1995)

A rather than the card with an E in the effort− conditions than in the effort+ conditions.

To increase the probability that participants' expectations of effect would be satisfied with an interpretation of the rule as implying (14) rather than (13), we developed the scenario, in the two **effect+** conditions, as follows: the machine was supposed to print an E on the back of cards with a 6; however, the machine ceased to function properly and printed cards with a 6 and a letter other than an E; after having repaired it, Mr Bianchi, asserted: 'If a card has a 6 on the front, it has an E on the back.' In such a context, the

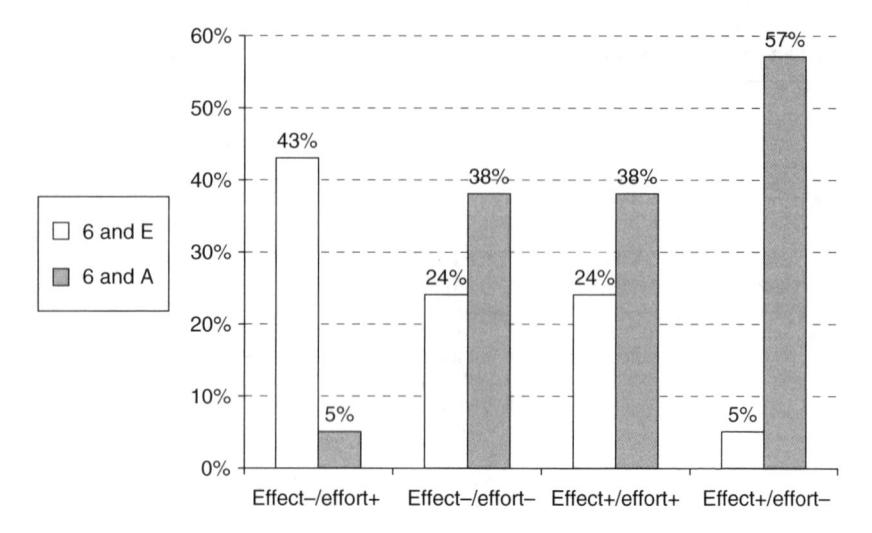

Figure 7.2 Percentage of 6 and E (incorrect) and 6 and A (correct) responses in the four versions of the machine problem

relevance of Mr Bianchi's assertion went through the implication that there were no cards with a 6 and a letter other than an E (in other terms, consequence (14)). On the other hand, in such a context, consequence (13) does not contribute to the relevance of the conditional. We predicted therefore that in the two **effect+** conditions, participants would more often infer (14) and select the 6 and the A card than in the **effect−** conditions.

The two **effect+** conditions on the one hand, and the two **effect−** conditions on the other hand, differ from one another only on the effort side, while the two **effort+** and the two **effort−** conditions differ from one another only on the effect side. Given this, the predictions that follow from the relevance-theoretic account of the task are self-evident: the best performance should be with the **effect+/effort−** condition, and the worse one with the **effect−/effort+** condition. The performance on the **effect+/effort+** and on the **effect−/effort−** condition should be at an intermediary level between the two other conditions. Moreover, the two factors, effect and effort, should, each on its own, contribute to good performance. The results are summarized in Figure 7.2.

These results confirm our prediction. Both factors of relevance, effect and effort, were shown to play a role in performance. These results show how effort and effect factors can be manipulated independently or jointly so as to favour one interpretation of a conditional statement over another. The advantage of the selection task paradigm in this context is that participants' interpretations of the rule are rendered manifest by their selection of cards.

5 Testing the Communicative Principle of Relevance with a speech production task

According to the Communicative Principle of Relevance, utterances convey a presumption of their own optimal relevance, and do so whether or not they actually are optimally relevant. Speakers may fail to achieve relevance, or they may not even try, and, in such cases, the presumption of optimal relevance is unjustified. Justified or not, it is automatically conveyed by every utterance used in communication, and it guides the process of comprehension. Most research exploring the consequences of the Communicative Principle of Relevance have, accordingly, focused on the comprehension process. Still, the communicative principle could not be right – and relevance could not guide comprehension – if speakers were not, often enough, trying to be optimally relevant, and successful at it. In the study that we report in this section, we investigate the degree to which speakers actually aim at being relevant, even when talking to perfect strangers from whom they have little to expect in return.

Imagine the following exchange between two strangers in the street:

(15) *Mr X*: Hello, do you have the time, please?
    Mrs Y: Oh yes, it is 4:30

In fact, Mrs Y's watch does not indicate 4:30 but 4:28. She has chosen to round her answer even though she could have been more accurate. Rounding numbers is quite common. People round when talking about money, distance, time, weight and so on. What explains this behaviour? We recently proposed that rounding is in part explained by considerations of relevance (Van der Henst, Carles and Sperber, 2002). A rounded answer is generally more relevant than an accurate one, and speakers round in order to be relevant to their hearer.

In a few situations, when taking a train for instance, a person asking for the time is better off with an answer precise to the minute. If your train leaves at 4:29, and you are told that it is 4:30 while it is in fact 4:28, you may believe that you've missed it when in fact you could still catch it. On the other hand, if you were told that it is 4:25, you might end up missing your train by considering that you still had four minutes to board it. In most situations, however, the consequences you would draw from a time rounded to the nearest multiple of five minutes are the same as those you would draw from a time accurate to the minute. So, in general, rounding does no harm. Does it do any good? Rounded numbering requires less *processing effort*; 4:30 is easier to manipulate than 4:28. Communicating rounded numbers may thus be a way to provide an optimally relevant answer to addressees by reducing their processing effort without compromising any cognitive effect likely to be derived.

In most situations, then, a speaker who is asked for the time and wishing to be as relevant as possible would round her answer. She might, however, be rounding for other reasons. In particular, if she wears an analogue watch indicating only numbers that are multiples of five, it may be easier for her to round than not to round. She might then round to minimize, not her audience's effort, but her own. In fact, a sceptic might argue, the goal of minimizing one's audience's effort might not play any role in the tendency of people asked for the time to give a rounder answer.

In order to find out whether a tendency to optimize relevance was a factor in rounding the time, we approached people on the campus of the University of Paris VII and just asked them: *'Hello, do you have the time please?'* (Van der Henst, Carles and Sperber, 2002). We took note of their response and of the type of watch they were wearing: analogue or digital, and distinguished two groups, the 'analogue' and the 'digital' group. For people with a digital watch, it requires less effort to just read aloud the exact time indicated by their watch than to round it to the closest multiple of five. If people asked for the time were just trying to minimize their own effort, then they should always round when their watch is analogue, and never do so when it is digital. On the other hand, if people are also motivated by the goal of reducing their audience's effort, then, not only people with analogue watches, but also a significant percentage of people with digital watches should round.

What we found is that people rounded in both conditions. The percentage of rounders is calculated on the basis of the percentages of responses which indicate the time in a multiple of five minutes. If people never rounded there should be 20 per cent of such responses (this is the theoretical distribution of numbers which are multiples of 5). However, the percentages we observed in the two conditions were much higher: 98 per cent of answers were a multiple of 5 in the analogue group, and 65.8 per cent in the digital group. This means that 97 per cent of people rounded in the analogue condition and 57 per cent in the digital one (see Figure 7.3).[7] Hence, even though participants of the digital group rounded less than participants of the analogue group, a majority of them did, remarkably, make an extra effort in order to diminish the effort of their audience.

Some people with analogue watches may round just in order to save their own effort, but the case of people with digital watch shows that a majority of people are disposed to round, even when this means making an extra effort. We attributed this disposition to a more general disposition, that of

---

[7] To calculate the percentage of rounders we used the following formula: *Percentage of rounders* = (M – 20)/80, where M is equal to the percentage of answers given in a multiple of five. When M is equal to 20, the percentage of rounders is equal to 0, when it is equal to 100, so is the percentage of rounders.

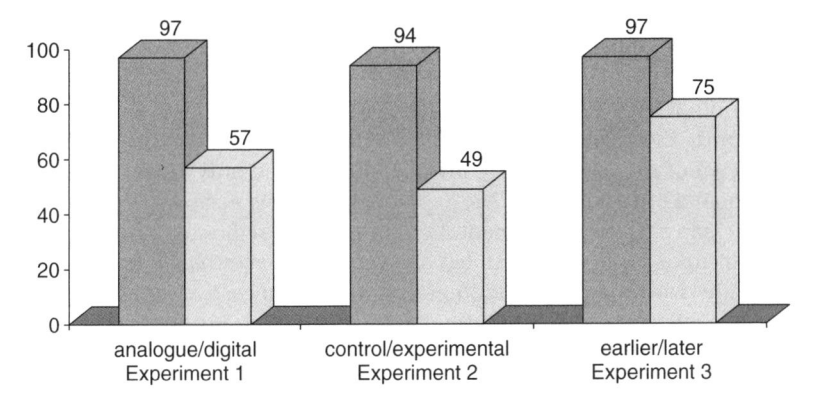

Figure 7.3 Percentages of rounders in the three experiments

Note to Table: In Experiment 1, participants wore analogue watches in the 'analogue' group, and digital watches in the 'digital' group; in Experiment 2, participants were just asked for the time in the 'control' group, and were asked for the time by an experimenter setting his watch in the 'experimental' group; in Experiment 3, participants were asked for the time more than 15 minutes before the time at which the experimenter said he or she had an appointement in the 'earlier' group, and less than 15 minutes before the appointment in the 'later' group.

trying to produce optimally relevant utterances. Still, an alternative explanation could be that people round in order to minimize their commitment: they may not be sure that their watch is precise to the minute, and be more confident that it is accurate within a five-minute interval. Indeed, this desire to minimize commitment may account for some of the rounding we observed, but could it be enough to make the relevance-based explanation superfluous? To investigate this possibility, we created a situation where accuracy manifestly contributed to relevance.

Although rounded answers are easier to process than non-rounded ones, there are some situations, such as that of the train evoked above, where optimal relevance depends upon cognitive effects that are carried only by a more accurate answer. Speakers guided by the goal of producing an optimally relevant answer should, in this condition, provide, if they can, a more precise answer than in the ordinary kind of situation in which our first experiment took place.

We tested this prediction in Experiment 2 with two groups of people. In the control group, participants were approached in the same way as in the previous experiment and were just asked for the time. In the experimental group, the request for the time was framed in a context in which an accurate answer was obviously more relevant. The experimenter approached the participant with a watch held in his hand and said: *'Hello! My watch isn't working properly. Do you have the time please?'* In this context, it was clear that the experimenter was asking for the time in order to set his own watch and that, for this purpose, an answer precise to the minute would be more relevant. Only the answers of participants with an analogue watch were recorded. Participants had therefore to make an extra effort in order to provide an accurate answer. We found that participants were much more accurate in the experimental than in the control condition: there were 94 per cent of rounders in the control condition and only 49 per cent in the experimental one (see Figure 7.3, Experiment 2). This means that 51 per cent of participants of the experimental group gave the requester a time accurate to the minute. Note that rounded answers may nevertheless have been in conformity with the presumption of optimal relevance: even if approximate, they were relevant enough to be worth the hearer's attention, as required by the first clause of the presumption, and, as required by the second clause, they may have been the most relevant ones compatible with the speakers' abilities (if they had doubts about the accuracy of their watch), or preferences (if they were reluctant to work out a more precise answer). Our results show anyhow that a majority of the people not only understood that accuracy was more relevant in this condition, but also were able and willing to make the effort of giving an accurate answer.

That accuracy to the minute is relevant to someone setting his watch is easy enough to understand. It need not involve the kind of refined concern for relevance that Relevance Theory presupposes. In a third experiment, we manipulated the relationship between relevance and accuracy in a much subtler way.

Suppose you want to know how much time you have left before an appointment at 4:00pm. The closer you get to the time of the appointment the more accuracy is likely to be relevant. At 3:32, being told that it is 3:30 is likely to have practically the same effect as being told, more accurately, that it is 3:32. On the other hand, being told at 3:58 that it is 4:00, is likely to be misleading. Two minutes may, for instance, be the time you need to reach the place of your appointment. In other words, the closer you are to the time of the appointment, the more accuracy becomes relevant.

In the third experiment, all participants were approached in the same way and told *'Hello, do you have the time please? I have an appointment at T'*. We then divided participants into two groups: the 'earlier' group who gave a time between 30 to 16 minutes before the time of the appointment and a 'later' group who answered with a time between 14 minutes before the time of the appointment and the time of the appointment itself. As we had predicted, the results show that participants rounded less in the 'later' group (75 per cent of participants) than in the 'earlier' group (97 per cent): 22 per cent difference may not seem so impressive until you realize that those people in the later group who did give an accurate answer not only were willing to make the effort of reading their analogue watch more carefully and had enough confidence in its accuracy, but also made the extra effort of taking the perspective of the stranger who was addressing them and of inferring that accuracy, at this point in time, would contribute to the relevance of their utterance.

The experiments described in this section show how subtle aspects of people's spontaneous speech behaviour can be predicted on the basis of the Communicative Principle of Relevance: speakers tend to produce utterances that justify the presumption of optimal relevance these utterances automatically convey.

6 Conclusion

The studies reported in this chapter tested and confirmed predictions directly inspired by central tenets of Relevance Theory and, in particular, by the Cognitive and the Communicative Principles of Relevance. Of course, it would take many more successful experiments involving a variety of aspects of cognition and communication to come anywhere near a compelling experimental corroboration of Relevance Theory itself. Still, from a pragmatic point of view, the few experiments we have presented here, together with others we have mentioned, show, we hope, how imagining, designing and carrying out experiments helps expand and sharpen pragmatic theory. From an experimental psychology point of view, these experiments illustrate how a pragmatic theory that is precise enough to have testable consequences can put previous experimental research in a novel perspective and can suggest new experimental paradigms.

References

Braine, M. D. S., and O'Brien, D. P. (1998). *Mental Logic*. Mahwah, NJ: Lawrence Erlbaum Associates.

Byrne, R. M. J., and Johnson-Laird, P. N. (1989). Spatial reasoning. *Journal of Memory and Language* 28: 564–75.

Cheng, P. N., and Holyoak, K. J. (1985). Pragmatic reasoning schemas. *Cognitive Psychology* 17: 391–416.

Cosmides, L. (1989). The logic of social exchange: Has natural selection shaped how humans reason? Studies with the Wason selection task. *Cognition* 31: 187–276.

Evans, J. St B. T. (1989). *Bias in Human Reasoning: Causes and Consequences*. Hove: Lawrence Erlbaum Associates.

Evans, J. St B. T., Newstead, S. E., and Byrne, R. M. J. (1993). *Human Reasoning: The Psychology of Deduction*. Hove: Lawrence Erlbaum Associates.

Fiddick, L., Cosmides, L., and Tooby, J. (2000). No interpretation without representation: The role of domain-specific representations in the Wason selection task. *Cognition* 77: 1–79.

Gibbs, R. W., and J. F. Moise (1997). Pragmatics in understanding what is said. *Cognition* 62: 51–74

Girotto, V., Kemmelmeir, M., Sperber, D., and Van der Henst, J. B. (2001). Inept reasoners or pragmatic virtuosos? Relevance and the deontic selection task. *Cognition* 81: 69–76.

Green, D. W., and Larking, R. (1995). The locus of facilitation in the abstract selection task. *Thinking and Reasoning* 1: 183–99.

Griggs, R. A., and Cox, J. R. (1982). The elusive thematic-materials effect in Wason's selection task. *British Journal of Psychology* 73: 407–20.

Happé, F. (1993) Communicative competence and theory of mind in autism: A test of relevance theory. *Cognition* 48: 101–19.

Hardman, D. (1998). Does reasoning occur in the selection task? A comparison of relevance-based theories. *Thinking and Reasoning* 4: 353–76.

Harman, G. (1995). Rationality. In E. E. Smith, and D. N. Osherson (eds), *Thinking: An Invitation to Cognitive Science*, vol. 3 (2nd edn). Cambridge, MA: MIT Press.

Johnson-Laird, P. N., and Byrne, R. J. M. (1991). *Deduction*. Hove: Lawrence Erlbaum Associates.

Johnson-Laird, P. N., and Byrne, R. M. J. (1995). A model point of view. *Thinking and Reasoning* 1: 339–50.

Jorgensen, J., Miller, G., and Sperber D. (1984) Test of the mention theory of irony. *Journal of Experimental Psychology: General* 113: 112–20.

Liberman, N., and Klar, Y. (1996). Hypothesis testing in Wason's selection task: Social exchange, cheating detection or task understanding. *Cognition* 58: 127–56.

Love, R., and Kessler, C. (1995). Focussing in Wason's selection task: Content and instruction effects. *Thinking and Reasoning* 1: 153–82.

Manktelow, K. I. (1999). *Reasoning and Thinking*. Hove: Psychology Press.

Matsui, T. (2000). *Bridging and Relevance*. Amsterdam: John Benjamins.

Matsui, T. (2001). Experimental pragmatics: Towards testing relevance-based predictions about anaphoric bridging inferences. In V. Akman et al. (eds), *Context 2001*: 248–60. Berlin: Springer-Verlag.

Medin, D. L., Coley, J. D., Storms, G., and Hayes, B. K. (2003). A Relevance Theory of induction. *Psychonomic Bulletin and Review*. 10(3): 517–32.

Nicolle, S., and Clark, B. (1999): Experimental pragmatics and what is said: A response to Gibbs and Moise. *Cognition* 66: 337–54.

Noveck, I. A. (2001). When children are more logical than adults: Investigations of scalar implicature. *Cognition* 78: 165–88.

Noveck, I. A., and Posada, A. (2003). Characterizing the time course of an implicature: An evoked potentials study. *Brain and Language* 85: 203–10.

Noveck, I. A., Bianco, M., and Castry, A. (2001). The costs and benefits of metaphor. *Metaphor and Symbol* 16: 109–21.

Politzer, G. (1996). A pragmatic account of a presuppositional effect. *Journal of Psycholinguistic Research* 25: 543–51.

Rips, L. J. (1994). *The Psychology of Proof*. London: MIT Press.

Ryder, N., and Leinonen, E. (2003). Use of context in question answering by 3-, 4- and 5-year-old children. *Journal of Psycholinguistic Research* 32: 397–415.

Sperber, D. (1994). The modularity of thought and the epidemiology of representations. In L. A. Hirschfeld and S. A. Gelman (eds), *Mapping the Mind: Domain Specificity in Cognition and Culture*: 39–67. New York : Cambridge University Press.

Sperber, D., Cara, F., and Girotto, V. (1995). Relevance theory explains the selection task. *Cognition* 52: 3–39.

Sperber, D., and Girotto, V. (2002). Use or misuse of the selection task? Rejoinder to Fiddick, Cosmides, and Tooby. *Cognition* 85: 277–90.

Sperber, D., and Girotto, V. (2003). Does the selection task detect cheater-detection? In J. Fitness and K. Sterelny (eds), *From Mating to Mentality: Evaluating Evolutionary Psychology*. Monographs in Cognitive Science. Hove: Psychology Press.

Sperber, D., and Wilson, D. (1981). Irony and the use-mention distinction. In Peter Cole (ed.), *Radical Pragmatics*: 295–318. New York: Academic Press
Sperber, D., and Wilson, D. (1995). *Relevance: Communication and Cognition* (2nd edn). Oxford: Blackwell.
Sperber, D., and Wilson, D (2002). Pragmatics, modularity and mind-reading. *Mind and Language* 17: 3–23.
Van der Henst, J. -B., Carles, L., and Sperber, D. (2002). Truthfulness and relevance in telling the time. *Mind and Language* 17: 457–66.
Van der Henst, J. -B., Sperber, D., and Politzer, G. (2002). When is a conclusion worth deriving? A relevance-based analysis of indeterminate relational problems. *Thinking and Reasoning* 8: 1–20.
Wason, P. C. (1966). Reasoning. In B. M. Foss (ed.), *New Horizons in Psychology*. Harmondsworth: Penguin.

8
Contextual Strength: the Whens and Hows of Context Effects

Orna Peleg, Rachel Giora* and Ofer Fein

1 Introduction

Highlighting the role context plays in shaping our linguistic behaviour is the major contribution of pragmatics to language research. Indeed, pragmatics has shifted the focus of research from the code to contextual inference (Carston, 2002; Sperber & Wilson, 1986/1995). It is widely agreed now that contextual information is a crucial factor determining how we make sense of utterances. The role of context is even more pronounced within a framework that assumes that the code is underspecified allowing for top-down inferential processes to narrow meanings down and adjust them to the specific context.

There is, however, ample evidence suggesting that the acknowledged supremacy of context should be qualified. Findings show that, at times, even a strong context does not filter out incompatible meanings and therefore does not allow frictionless processing. The conversation in (1), which took place between M and W who is interested in biology and genetics, is a case in point:

(1) *M*: I wanted to talk with you about something, but I can't remember what.
W: [NOTES SEWING THREADS ON THE TABLE] It must have to do with thread (Joking).
M: Yea, I wanted you to do Maya's jeans (Joking).
W: You know, the first interpretation I got was genes with a g.
(15 June 2001, reconstructed from memory, Ariel, in press.)

In spite of strong contextual evidence to the contrary – the cognitive environment manifest to the hearer strongly supports the 'jeans' meaning of the homophone – coupled with the implausibility of 'doing Maya's genes' as

---
*Corresponding author

opposed to 'doing Maya's jeans', the interpreter came up with the less likely interpretation first. Context did not inhibit what was foremost on his (genetics-oriented) mind – the 'genes' meaning of the linguistic code.

How come the hearer did not activate the relevant interpretation first? In what follows we will lay out the whens and hows of context effects: we will specify the conditions under which context may be more or less powerful and question the hypothesis that a strong context may affect comprehension entirely. Specifically, we will focus on the distinction between lexical processes involving coded but contextually inappropriate meanings versus contextual processes involving appropriate interpretations.

2 Effects of contextual strength on initial processing

Though no theory which accounts for comprehension denies the effect of context on how we make sense of utterances, various theories have different views on the speed and locus of these effects. Particularly, they diverge with regard to the very early moments of comprehension.

2.1 The direct access view

Proponents of the direct access view assume that context affects comprehension entirely. According to this view, top-down (contextual) processes interact with bottom-up (lexical) processes rather early on. If context is sufficiently rich and specific, it penetrates lexical processes and selects the appropriate meaning exclusively so that initial comprehension is effortless and seamless, involving no incompatible phase at all (e.g., Marslen-Wilson and Tyler, 1980; Martin, Vu, Kellas and Metcalf, 1999; McClelland and Rumelhart, 1981; Vu, Kellas and Paul, 1998; Vu, Kellas, Metcalf and Herman, 2000). Thus, upon processing

(2) The gardener dug a hole. She inserted the *bulb*.

comprehenders activate only the compatible 'flower' meaning of *bulb*, since this is the only interpretation of *bulb* that would be relevant in the given context. In contrast, the 'light' sense of *bulb*, though salient, should not be activated, since, in the set of accessible assumptions, it is irrelevant (Vu et al., 1998; Vu et al., 2000).

A more moderate version of the direct access view, while assuming that context affects comprehension significantly, also acknowledges the influence of meaning salience on comprehension. In this view, contextual processes are of primary effect: They interact with lexical processes and select the contextually appropriate meaning instantly. However, they do not inhibit irrelevant meanings, which get activated upon encounter of the lexical stimulus (Rayner, Pacht and Duffy, 1994; Kawamoto, 1993). Importantly, however, though lexical processes operate regardless of contextual processes and allow activation of various meanings, the appropriate interpretation always reaches sufficient levels of activation first (Bates, 1999, personal communication, July 2001; Gibbs, 1994; McRae, Spivey-Knowlton and Tanenhaus, 1998).

2.2 The modular view

Unlike the direct access view, the modular view assumes independent – modular and non-modular – systems that do not interact initially (Fodor, 1983). A modular system (lexical access) is sensitive only to its domain-specific (lexical) information. It is encapsulated and does not have access to information outside the module. Rather, initial input analyses are stimulus driven. They are automatic, rapid and on some traditional interpretations, exhaustive: all the responses (meanings) to a stimulus (word) are activated upon its encounter. In contrast, non-modular systems (contextual processes) are sensitive to all kinds of information (linguistic and nonlinguistic) and integrate various outputs into a coherent representation. Non-modular, contextual processes thus affect comprehension post-lexically: they operate after all the meanings of a linguistic stimulus have been activated. Within this framework, context effects are limited. They are slower than lexical processes and either integrate contextually appropriate outputs or suppress them as irrelevant and interfering with comprehension (Swinney, 1979). As a result, initial processes are not always smooth and may involve contextually inappropriate responses that would trigger sequential processes.

The Gricean model (1975) can be viewed as compatible with this view. For Grice, linguistic processes are primary. Context affects comprehension only after the initial (literal) interpretation of the (sentence) unit has been accomplished. If this interpretation reaches contextual fit, no more processes are required. If, however, it fails, further inferential processes follow, involving suppression of irrelevant meanings and derivation of contextually appropriate interpretations (implicatures). In this view, then, comprehension may initially go astray, with a later revision and adjustment stage.

2.3 The graded salience hypothesis

The graded salience hypothesis (Giora, 1997, 2003; Peleg, 2002, Peleg, Giora and Fein, 2001) shares a number of assumptions with the modular view. It too assumes distinct mechanisms: one bottom-up, sensitive only to domain-specific (linguistic) information; and another, top-down, sensitive to all kinds of (linguistic and extra-linguistic) knowledge. Unlike the traditional modular assumption, however, it assumes that the modular (lexical access) mechanism is itself ordered:[1] more salient responses (meanings) are accessed faster than and reach sufficient levels of activation before less salient ones.

---

[1] For a similar view see Duffy, Morris and Rayner, 1988; Rayner and Frazier, 1989; Rayner and Morris, 1991; Sereno, Pacht and Rayner, 1992, among others.

A response is salient to the extent that it is coded. The relative salience of the coded meaning is a function of its prototypicality, or amount of experiential familiarity induced by exposure (frequency). Uncoded responses (implicatures) are nonsalient. According to the graded salience hypothesis, then, salient meanings would be activated automatically upon encounter of the lexical stimulus, regardless of contextual information.[2]

In this framework, contextual information may also affect comprehension immediately. Particularly, a highly informative context may be predictive enough to avail meanings on its own accord very early on without even penetrating lexical access. Indeed, strong contextual information may be faster than lexical processes, so much so, that it may avail meanings before the relevant stimulus is even encountered (fostering an impression of direct access). This may be particularly true when the stimulus is placed at the end of a strong sentential context, after most information has been accumulated and integrated, allowing for effective guessing based on inferential processes. Importantly, however, context does not interact with lexical processes but runs in parallel (Giora, Peleg and Fein, 2004; Peleg et al., 2001). According to the graded salience hypothesis, then, even a strong context has limited effects initially. It may be predictive but it cannot block salient meanings.

Assuming a simultaneous operation of the encapsulated, linguistic mechanism on the one hand and the integrative, central system mechanism on the other allows the graded salience hypothesis to predict when contextual information may be faster than, coincidental with, or slower than linguistic processes. Unlike the modular view, then, the graded salience hypothesis does not always predict slower contextual effects that result in sequential processes.[3] Neither does it assume that activation of a whole linguistic unit should be accomplished before contextual information comes into play (as assumed by Grice, 1975). Rather, along the communication path, context and linguistic effects run in parallel, with contextual information availing meanings on its own accord, affecting only the end product of the linguistic process.

2.4 Predictions

The various theories have different predictions with regard to the whens and hows of context effects. According to the direct access view, a strong context will always win over initially even if lexical effects are strong (as when it biases a polar ambiguity toward the less-salient meaning, having thus to inhibit or be faster than a highly accessible response). According to

---

[2] Coded meanings of low salience, however, may not reach sufficient levels of activation and may not be visible in a context biased toward the more salient meaning of the word (but see Hillert and Swinney, 2001, for a different view).
[3] Note, however, that Fodor (1983: 75) did not exclude predictive effects.

the modular view, lexical processes will always be faster, since they are automatic and encapsulated. Salience imbalance would not affect processing either, since response is exhaustive and unordered. The graded salience hypothesis takes both strength of context and salience effects into consideration. While lexical effects are constant across sentence position, being sensitive only to degree of (coded) salience, contextual effects may vary with respect to predictability and sentential position. Given these variables, the graded salience hypothesis predicts that:

a. Context effects might precede lexical effects when the stimulus is placed in sentence final position, provided the preceding context is highly predictive. Under this condition, guessing the compatible concept(s) would be fast and often occur before the lexical stimulus is encountered.
b. Contextual effects would not precede lexical effects in sentence initial position. In this position, even a strong prior context will not have speedy enough effects to enable it to predict oncoming concepts long before lexical accessing occurs. The assumption is that in initial position, predictive effects are less pronounced than in final position, since beginnings are less constrained than ends.[4]
c. Under all conditions, the incompatible coded meanings will be activated upon encounter of the stimulus, albeit at different levels of activation, determined by their relative salience (see note 2).

3 Findings

Findings in Giora and colleagues (2004), and Peleg and colleagues (2001) support the graded salience hypothesis. In all, our studies demonstrate that lexical and contextual processes make up independent mechanisms that do not interact initially. Specifically, we showed that, as predicted (see (a) above) when placed in final position, constraining contexts can predict the appropriate meaning of a lexical stimulus even before that stimulus is encountered, thus availing appropriate concepts without interacting with lexical processes (Experiment 1 below). We then compared access of coded meanings in sentence initial versus final position. As predicted (see (b) above), we showed that while final position favours context effects, initial position does not (Experiment 2 below). This is true even when the preceding context is highly predictive (as when the target sentence features the previous sentence topic; Experiment 4 below). However, even in final position, coded meanings get activated, despite contextual information to the contrary, as predicted (see (c) above; Experiments 3 and 4 below).

---

[4] See also Gernsbacher (1990) on the processes involved in building a new substructure that are initially insensitive to information in prior context.

3.1 Experiment 1

Review of the literature reveals that experimental data suggestive of selective access induced by prior context was based on materials whose targets were placed at the end of strong sentential contexts. For instance, Vu, Kellas and Paul (1998) and Vu, Kellas, Metcalf and Herman (2000) showed that homonyms such as *bat* activated contextually appropriate meanings exclusively when placed at the end of a highly constraining context such as (3)–(4):

(3) The slugger splintered the *bat*.*

(Probes displayed at *: salient-wooden; unrelated-safe; less-salient-fly; unrelated-station.)

(4) The biologist wounded the *bat*.*

(Probes displayed at *: salient-wooden; unrelated-safe; less-salient-fly; unrelated-station.)

In their studies, subjects read such sentences and named one of four probes (presented in (3)–(4). Findings demonstrated that they always named the contextually compatible probe faster than the unrelated one. On the face of it, then, such findings support the direct access view. They show that only contextually appropriate meanings were tapped initially, irrespective of meaning salience. Indeed, if these findings were a result of context penetrating lexical accessing, they would question the graded salience hypothesis.

To support the alternative view proposed by the graded salience hypothesis that lexical and contextual processes do not interact initially, one should be able to show that results, accounted for by an interactive system, can also be accounted for by non-interactive machinery. Thus, if Vu and collaborators' findings are replicated in the absence of the relevant lexical stimulus, this would support the view that these results are the end-product of contextual processes alone.

To do that, we used Vu and colleagues' materials, but presented the probes in sentence pre-final position in order to see whether contextual processes could induce the appropriate meaning even *before* the target word is encountered. In our study (Peleg et al., 2001), 60 native speakers of English read the sentences off a computer screen and had to make lexical decisions as to whether a probe presented before the final (target) word was a word or a non-word:

(5) The slugger splintered the* *bat*.

(Probes displayed at *: salient-wooden; less-salient-fly; unrelated-station.)

(6) The biologist wounded the* *bat*.

(Probes displayed at *: salient-wooden; less-salient-fly; unrelated-station.)
(Manipulated items taken from Vu et al., 1998.)

3.1.1 Results and discussion

Results replicated those by Vu and his colleagues, indicating that contextually compatible responses were always faster than incompatible responses, regardless of whether the context was biased in favour of the less or more salient meaning of the target (see Table 8.1). This was true of both the participant and item analyses. Replication of Vu and colleagues' findings under conditions that disallow lexical accessing is consistent with our assumption that contextual processes are speedy toward the end of sentences and can predict the appropriate meaning on their own accord very early on, without interacting with lexical processes.[5]

Notwithstanding, it still remains to show that when the lexical stimulus is eventually encountered, lexical accessing proceeds automatically, irrespective of contextual information. Experiment 2 was therefore designed to show that even a strong and speedy context does not penetrate lexical access when this is triggered.

3.2 Experiment 2

In order to show that lexical processes are encapsulated with respect to contextual information, we attempted to replicate Vu and colleagues' (2000) results, manipulating targets' position in the sentence (Giora et al., 2004). We predicted that, at the beginning of sentences, their (Vu et al., 2000) results will not be replicated, since at this position, effects of a strong prior context would neither inhibit nor precede salient meanings. These predictions do not fall out of interactive models, which assume that, in a rich and supportive context, the appropriate meaning is tapped initially, directly and exclusively, or at least more rapidly than the inappropriate meaning. However, as before, probing targets in sentence final position would yield results similar to those obtained by Vu and colleagues (and by our first experiment). Unlike initial position, we argue, sentence final position allows contextual processes to be fast and obscure but not inhibit lexical processes.

Table 8.1 Mean response times (in milliseconds) to probes by context type

	Salient Probe		Less-Salient Probe		Unrelated Probe	
Context	M	SD	M	SD	M	SD
Salient	951	252	1003	243	1005	255
Less-salient	1057	275	927	237	994	231

---

[5] For an alternative critique of Vu et al.'s findings, suggesting that it is the choice of items that is responsible for their results, see Binder and Rayner (1999).

To test our hypotheses we used Vu and colleagues' (2000) materials. For example, in (7) the context is suggestive of the salient/dominant ('electricity') sense of *bulb*; in (8) it is strongly suggestive of the less-salient/subordinate ('plant') sense of the word:

(7) The custodian found the solution. She inserted the *bulb*.*

   (Probes displayed at *: salient-light; less-salient-flower; unrelated- cliff.)

(8) The gardener dug a hole. She inserted the *bulb*.*

   (Probes displayed at *: salient-light; less-salient-flower; unrelated- cliff.)

To manipulate sentence initial versus final position, we subjected the second sentence of their (2000) materials (*She inserted the bulb*) to passivization. In this experiment, we tested only the less-salient condition, because it involves lexical access of salient meanings that conflict with contextual processes inducing compatible but less-salient meanings:

(9) The gardener dug a hole. The *bulb** was inserted.

   (Probe displayed at *.)

Sixty native speakers of English read the original and the passivized versions of Vu and colleagues' (2000) discourses off a computer screen and were administered lexical decision tasks. Relative salience of target meanings had been established by a pre-test.

3.2.1 Results

Results support the graded salience hypothesis. They show that, as predicted, in sentence initial position, responses were faster to the salient (incompatible) probes than to the less-salient (compatible) probes. In sentence final position, however, the picture was different. Responses to the less-salient (compatible) probes were faster than responses to the salient (incompatible) probes (see Figure 8.1).

3.2.2 Discussion

The above results support our view concerning the whens and hows of context effects. In sentence initial position, where only constraints from a previous discourse can be operative, context effects are slow and do not precede lexical processes. Their slow effects in initial position, then, do not allow it to conceal the effects of the lexical mechanism, thus attesting to the involvement of different, non-interactive mechanisms in discourse comprehension. In contrast, this expectation-driven mechanism is fast toward the end of

[Bar chart showing Mean RT for Salient Probe, Less-Salient Probe, and Unrelated Probe at Initial Position and Final Position]

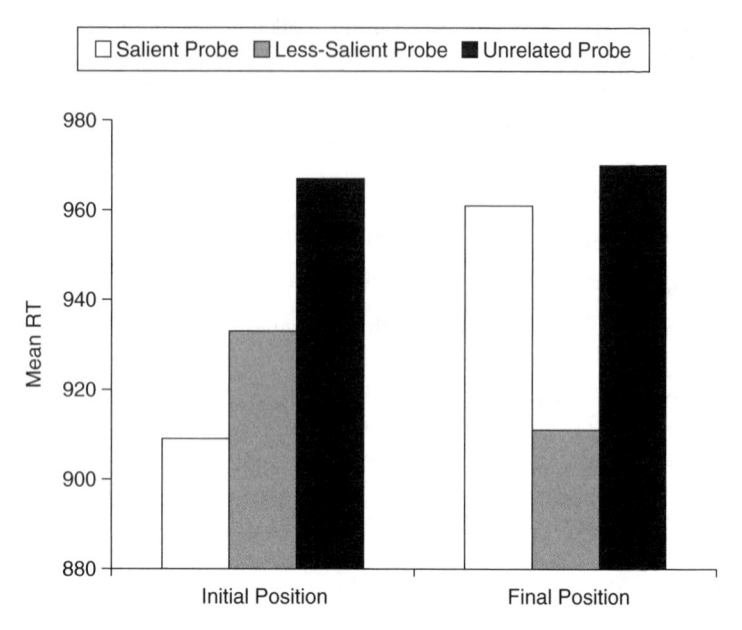

Figure 8.1 Mean response times (in milliseconds) to probes related to the salient (contextually incompatible) and less-salient (contextually compatible) meanings of the target words, and unrelated probes

sentences. At this point, different types of information enable it to predict an upcoming concept swiftly and obscure lexical processes.

Taken together, these findings cannot be accounted for by the context-sensitive, interactive models, which predict that, given enough constraints, either compatible meanings will be activated exclusively or they will be accessed first. These predictions do not hold for sentence initial position, in spite of a prior strong context. We want to further argue that the first prediction – regarding exclusive access of compatible meanings – does not in addition hold for either sentence position.

3.3 Experiment 3

To further demonstrate that, even in sentence final position, salient meanings are not blocked when incompatible, an additional experiment was designed (see Giora et al., 2004). The purpose of this study was to show that salient, but incompatible meanings are not inhibited even when context favours contextual effects. To do that, we compared sentences containing an ambiguous word whose less-salient meaning is contextually compatible (*The gardener dug a hole. She inserted the bulb*) with control sentences ending in a compatible but non-ambiguous word (*The gardener dug a hole. She inserted the flower*). We predicted that, following the ambiguous word (*bulb*), the salient but

Table 8.2 Mean response times (in milliseconds) to probes related to the salient (contextually incompatible) meaning

	Salient Probe	
Target word	M	SD
Ambiguous	993	285
Control	1070	283

incompatible meaning ('light') would be activated compared to the control condition. As in previous experiments, native speakers read the sentences and had to make a lexical decision as to whether related and unrelated probes were a word or a non-word in English.

3.3.1 Results and discussion

As predicted, inappropriate but salient meanings ('light') were activated following the ambiguous condition only (see Table 8.2). This was true for both the subject and item analyses. Such results demonstrate that salient though inappropriate meanings are activated even in a sentential position that benefits contextual processes. Placed in sentence final position, the ambiguous word (*bulb*) facilitated the activation of the probe related to the salient, but contextually inappropriate meaning ('light') compared to the control (*flower*). This finding is inconsistent with the predictions of the radical version of direct access view according to which interactive mechanism should have tapped the contextually compatible meaning exclusively, as allegedly shown by Vu and his colleagues (2000). However, as shown here, this was not the case. Their (Vu et al., 2000), findings are, therefore, more compatible with the assumption that, under conditions that favour contextual processes, a strongly biasing context can avail the appropriate meaning very early on without penetrating lexical accessing that might occur independently somewhat later.

Though our findings so far demonstrate that even in sentence final position, salient but incompatible meanings get activated, they nevertheless show that, in that position, context effects may be faster. The appropriate though less-salient meanings reach sufficient levels of activation faster than salient but incompatible meanings. Only in initial position, is this not the case. Would findings in sentence initial position be subverted if information in sentence initial position is exceedingly predictive? To test this possibility, we designed Experiment 4.

3.4 Experiment 4

Experiment 4 aimed to show that in initial position, contextual effects will not supercede lexical effects even when prior context is highly predictive of oncoming concepts. One kind of high-predictability concepts is topical referents. We therefore compared activation levels of salient but irrelevant meanings with nonsalient but topically compatible interpretation of targets placed in sentence initial position preceded by a context substantiating this topical information. Indeed, initial position is known to be the preferred position for topics (see Giora, 1985a, 1985b, and Reinhart, 1980, and references therein).

Sixty native speakers read Hebrew sentences in which the target word (*delinquent*) appeared either in initial (10) or final (11) position. A prior context strongly biased these sentences toward a nonsalient (metaphorical) meaning which, in all cases, was the topic of the previous context as well as the topic of the target sentence. We took advantage of the relative free word order in Hebrew:

(10) Sarit's sons and mine went on fighting continuously. Sarit said to me: These delinquents* won't let us have a moment of peace.

(Probes displayed at *: salient-criminals; contextually compatible-kids; unrelated-painters.)

(11) Sarit's sons and mine went on fighting continuously. Sarit said to me: A moment of peace won't let us have these delinquents*.[6]

(Probes displayed at *: salient-criminals; contextually compatible-kids; unrelated-painters.)

Readers had to make a lexical decision as to whether the probe was a word or a non-word in Hebrew.

3.4.1 Results and discussion

Results show that context effects were not faster than lexical effects in sentence initial position. Though contextually compatible nonsalient meanings were made available immediately, these effects were not strong enough to supercede lexical effects. However, in final position, results replicated those of Vu and colleagues and of our own (Experiments 1 and 2). In sentence final position, contextual effects were somewhat faster than salience effects, emerging probably before the target word was encountered and processed (cf. Experiment 1). As before, these effects did not inhibit salient though inappropriate meanings in either position (see Figure 8.2 and Peleg et al., 2001).

Such findings support our view that language comprehension involves two distinct mechanisms that run in parallel: one sensitive to contextual

---

[6] The word order in Hebrew is such that the target NP occupies initial position, preceding the demonstrative.

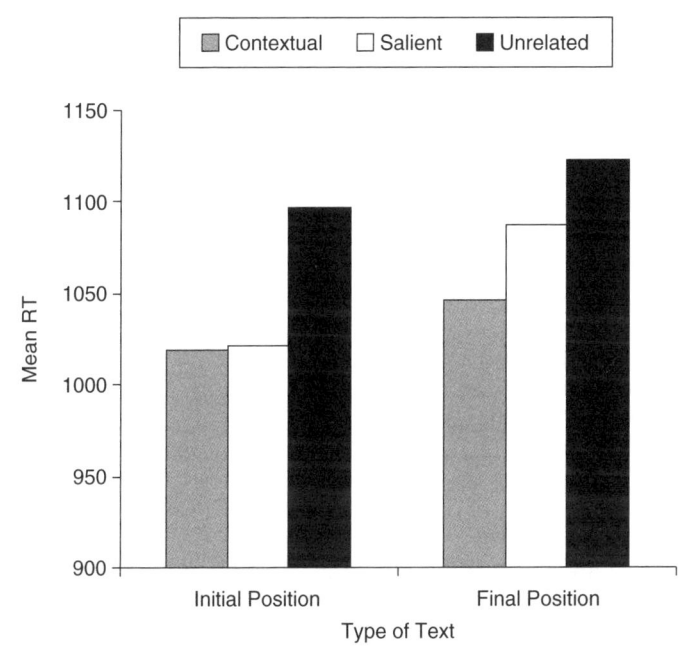

Figure 8.2 Mean response times (in milliseconds) to probes related to the salient (contextually incompatible) and less-salient (contextually compatible) meanings of the target words, and unrelated probes

information and one sensitive to coded, salient information. Thus while contextual information may have fast effects, they do not filter out salience effects. Salient meanings are activated upon encounter of the verbal stimulus, irrespective of context predictiveness. While salience effects are constant across position, speed of context effects varies as a function of the targets' location. They are faster toward the end of sentences, and less pronounced at the beginning of sentences.

4 General discussion

The involvement in comprehension of distinct mechanisms that do not interact initially enables comprehenders to resist exclusive conformity with contextual information. Contextual information, though effective, is limited. This intelligent, integrative mechanism is very powerful, particularly toward the end of discourse units. Still, it does not control other processes entirely. Contrary to appearances, it does not penetrate lexical accessing and it does not activate meanings selectively. Experiment 1 suggests that previous findings supporting selective access (Vu et al., 1998; Vu et al., 2000) might have been affected by contextual processes which did not interact with lexical processes. Experiment 2 indeed demonstrated that contextual processes did not interact with lexical processes, which are automatic and sensitive to lexical stimuli only. Though sentence final position favours contextual processes and allows them to occur even before lexical access is initiated, this is not true of sentence initial position in which contextually compatible meanings are not faster than salient but incompatible meanings. This has been further demonstrated with information that is highly predictive pragmatically. Sentence initial position did not favour contextual information over lexical accessing of salient but incompatible meanings even when such contextual information was highly accessible and useful. Lexical processes, then, are uninterrupted initially, even when context is highly powerful location-wise (Experiment 3) and content-wise (Experiment 4).

The impenetrability of the lexical mechanism allows humans to have access to meanings not invited by information accumulated outside the module. This multiplicity of sources of meanings (originating in the context and the lexicon) allows for non-standard choices. Indeed, findings in Giora (2003) attest that comprehenders do not always suppress salient but contextually incompatible information (as assumed by Fodor, 1983, and Grice, 1975), but occasionally utilize it for various purposes such as humour, pleasure, innovativeness and subversion. The existence of a mechanism and a set of privileged meanings that resist immediate compliance with contextual information even when it is very strong provide for 'a variety of situations' which allow the individual an insight into different alternatives and a second (critical?) thought.

References

Ariel, M. (in press). *Pragmatics and Language*. Cambridge: Cambridge University Press.
Bates, E. (1999). On the nature and nurture of language. In E. Bizzi, P. Calissano and V. Volterra (eds), *Frontiere della biologia* [Frontiers of Biology]. *The Brain of Homo Sapiens*. Rome: Giovanni Trecanni.
Binder, K. S., and Rayner, K. (1999). Does contextual strength modulate the subordinate bias effect? A reply to Kellas and Vu. *Psychonomic Bulleting and Review* 6: 518–22.
Carston, R. (2002). *Thoughts and Utterances: The Pragmatics of Explicit Communication*. Oxford: Blackwell.
Duffy, S. A., Morris, R. K., and Rayner, K. (1988). Lexical ambiguity and fixations times in reading. *Journal of Memory and Language* 27: 429–46.
Fodor, J. (1983). *The Modularity of Mind*. Cambridge, MA: MIT Press.
Gernsbacher, M. A. (1990). *Language Comprehension as Structure Building*. Hillsdale NJ: Lawrence Erlbaum Associates.
Gibbs, R. W. Jr. (1994). *The Poetics of Mind*. Cambridge: Cambridge University Press.
Giora, R. (1985a). Towards a theory of coherence. *Poetics Today* 6: 699–716.

Giora, R. (1985b). A text-based analysis of nonnarrative texts. *Theoretical Linguistics* 12: 115–35.

Giora, R. (1997). Understanding figurative and literal language: The graded salience hypothesis. *Cognitive Linguistics* 7: 183–206.

Giora, R. (2003). *On Our Mind: Salience, Context, and Figurative Language*. New York: Oxford University Press.

Giora, R., Peleg, O., and Fein, O. (2004). Resisting contextual information: You can't put a salient meaning down. Paper submitted for publication.

Grice, P. H. (1975). Logic and Conversation. In P. Cole and J. Morgan (eds), *Speech Acts: Syntax and Semantics* vol. 3: 41–58. New York: Academic Press.

Hillert, D., & Swinney, D. (2001). The processing of fixed expressions during sentence comprehension. In A. Cienki, B. Luka and M. Smith (eds), *Conceptual and Discourse Factors in Linguistic Structure*: 107–22. Stanford, CA: CSLI Publications.

Kawamoto, A. H. (1993). Nonlinear dynamics in the resolution of lexical ambiguity: A parallel distributed processing account. *Journal of Memory and Language* 32: 474–516.

Marslen-Wilson, W. D., and Tyler, L. K. (1980). The temporal structure of spoken language understanding. *Cognition* 8: 1–71.

Martin, C., Vu, H., Kellas, G., and Metcalf, K. (1999). Strength of discourse context as a determinant of the subordinate bias effect. *The Quarterly Journal of Experimental Psychology* 52A: 813–39.

McClelland, J. L., and Rumelhart, D. E. (1981). An interactive activation model of context effects in letter perception. Part 1: An account of basic findings. *Psychological Review* 88: 375–407.

McRae, K., Spivey-Knowlton, M. J., and Tanenhaus, M. K. (1998). Modeling the influence of thematic fit (and other constraints) in on line sentence comprehension. *Journal of Memory and Language* 38: 283–312.

Peleg, O. (2002). Linguistic and nonlinguistic mechanisms in language comprehension. Unpublished Ph.D. dissertation. Tel Aviv University.

Peleg, O., Giora, R. and Fein, O. (2001). Salience and context effects: Two are better than one. *Metaphor and Symbol* 16: 173–92.

Rayner, K., and Frazier, L. (1989). Selection mechanisms in reading lexically ambiguous words. *Journal of Experimental Psychology: Learning, Memory, and Cognition* 15: 779–90.

Rayner, K., and Morris, R. K. (1991). Comprehension processes in reading ambiguous sentences: Reflections from eye movements. In G. B. Simpson (ed.), *Understanding Word and Sentence*: 175–98. Amsterdam: North Holland.

Rayner, K., Pacht J. M., and Duffy, S. A. (1994). Effects of prior encounter and global discourse bias on the processing of lexically ambiguous words: Evidence from eye fixations. *Journal of Memory and Language* 33: 527–44.

Reinhart, T. (1980). Conditions for text coherence. *Poetics Today* 1: 161–80

Sereno, C. S., Pacht, J. M., and Rayner, K. (1992). The effect of meaning frequency on processing lexically ambiguous words: Evidence from eye fixations. *Psychological Science* 3: 269–300.

Sperber, D., and Wilson, D. (1986/1995). *Relevance: Communication and Cognition*. Oxford: Blackwell.

Swinney, D. A. (1979). Lexical access during sentence comprehension: (Re)consideration of context effects. *Journal of Verbal Learning and Verbal Behavior*, 18: 645–59.

Vu, H., Kellas, G., Metcalf. K., and Herman, R. (2000). The influence of global discourse on lexical ambiguity resolution. *Memory and Cognition* 28: 236–52.

Vu, H., Kellas, G., and Paul, S. T. (1998). Sources of sentence constraint in lexical ambiguity resolution. *Memory and Cognition* 26: 979–1001.

9
Electrophysiology and Pragmatic Language Comprehension
Seana Coulson

1 Introduction

At the outset of their book *Relevance*, Sperber and Wilson (1986/1995) remind the reader: 'In writing this book, we have not literally put our thoughts down on paper. What we have put down on paper are little dark marks, a copy of which you are now looking at. As for our thoughts, they remain where they always were, inside our brains.' With these witty remarks, they note that while we often think and speak as if language were a conduit for thought, this is only a metaphor, and a deceptive one at that. It is deceptive because it implies that language comprehension can be reduced to a decoding process (Reddy, 1979). Another, perhaps more appropriate, metaphor involves a portrait of the language user as a paleontologist who constructs theories about extinct animals based on linguistic fossil input. But regardless of one's favourite metaphor for the relative import of coded and inferential aspects of language comprehension, recent advances in the study of language suggest that the many-headed beast we call meaning depends importantly on electrical activity in the brains of the speakers and hearers who construct it.

It might seem odd to suggest that pragmatics, as the study of language in context, should be investigated in the controlled conditions of the laboratory. Perhaps even more bizarre is the suggestion that pragmatics – the aspect of language comprehension that requires the appreciation of cultural conventions, that describes the expression of social relationships, and that frequently appeals to explanatory frameworks which transcend the individual – might profitably be studied with physiological methods. In answer to the first worry, we point to the other contributions in this volume. In answer to the second, we note pragmaticists' increasing appeal to inference in their accounts of how the listeners construct the speakers' intended message (Bach, 1994; Barwise, 1983; Carston, 2002; Fauconnier, 1997; Recanati, 1989; Sperber and

Wilson, 1986/1995), and point to the importance of the brain for cognitive activity.

In cognitive neuroscience, language can be treated in three different ways: first, as an overt behaviour; second, as an activity subserved by mental computation; and third, as neural activity (Zigmond, Bloom, Landis, Roberts and Squire, 1999). Ultimately, the goal is to build a three-level account of pragmatic language competence that involves a description of the phenomena, a description of the cognitive processes or mental computations that underlie those phenomena, and a description of the neural activity that implements the cognitive processes (as in Marr, 1982). Due both to advances in technology and an increasing appreciation for the utility of establishing the link between cognitive processes and their neural implementation, recent years have seen a dramatic increase in the number of investigators using measures of neural activity to elucidate cognitive processes. One such measure is the event-related brain potential (ERP) derived from the electroencephalogram (EEG). A non-invasive measure of electrical brain activity, the ERP has proven to be a useful tool for studying cognitive and language processes. It provides a link to the neurobiology of behaviour, it has a high temporal resolution, and it allows the investigator to draw inferences about qualitative and quantitative processing differences. Indeed, even investigators whose primary concern is in mental computations and who have little interest in the relationship between cognitive and neural processes can find this methodology useful as a multi-dimensional index of on-line language comprehension.

Below we review how electrophysiological methods and data can inform the study of pragmatic language comprehension. We begin with a general description of the EEG and ERPs, and give an overview of language sensitive ERP components. Findings from the cognitive ERP literature are discussed in a way intended to highlight how ERPs can be used to address questions about the representation and timing of cognitive processes, and how electrophysiological data can complement experimental findings using behavioural paradigms. Finally, we suggest how ERPs might be used to experimentally address issues pertaining to pragmatics such as the comprehension of direct versus indirect speech acts; the computation of entailments, explicatures and implicatures; and the importance of non-linguistic cues for language comprehension.

2 EEG and ERPs

Work on the cognitive neuroscience of language has attempted to monitor how the brain changes with manipulations of particular linguistic representations. The assumption is that language sub-processes are subserved by different anatomical and physiological substrates that will generate distinct patterns of biological activity. These patterns can then be detected by methods sensitive to electromagnetic activity in the brain, such as the electroencephalograph, or EEG. An EEG is a non-invasive measure of physiological activity in the brain made by hooking up electrodes to the subject's scalp. These electrodes pick up electrical signals naturally produced by the brain and transmit them to bioamplifiers. Early versions of the EEG used galvanometers to move pens on a rolling piece of paper. In more modern EEG systems, the bioamplifiers convert information about voltage changes on the scalp to a digital signal that can be stored on a computer.

2.1 EEG

Because the brain constantly generates electrical activity, electrodes placed on the scalp can be used to record the electrical activity of the cortex. The EEG amplifies tiny electrical potentials and records them in patterns called brain waves. Brain waves vary according to a person's state, as different patterns can be observed when a person is alert and mentally active than when she is relaxed and calm, or than when she is sleeping. The pattern of electrical activity in the fully awake person is a mixture of many frequencies but is dominated by waves of relatively fast frequencies between 15 and 20 cycles per second (or Herz), referred to as beta activity. If the subject relaxes and closes her eyes, a distinctive pattern known as the alpha rhythm appears. The alpha rhythm consists of brain waves oscillating at a frequency that ranges from 9–12 Herz. As the subject falls asleep, her brainwaves will begin to include large-amplitude delta waves at a frequency of 1 Herz.

Although the EEG provides overall information about a person's mental state, it can tell us little about the brain's responses to specific stimuli. This is because there is so much background activity in the form of spontaneous brain waves it is difficult to identify which brain wave changes are related to the brain's processing of a specific stimulus and which are related to the many ongoing neural processes occurring at any given time. In order to better isolate the information in the EEG that is associated with specific processing events, cognitive electrophysiologists average EEG that is time-locked to the onset of particular sorts of stimuli, or to the initiation of a motor response. The average EEG signal obtained in this way is known as the event-related potential, or ERP.

2.2 ERPs

ERPs are patterned voltage changes in the ongoing EEG that are time-locked to classes of specific processing events. Most commonly these events involve the onset of stimuli, but they can also include the execution of a motoric response (Hillyard, 1983; Rugg and Coles, 1995). As noted above, we obtain ERPs by recording subjects' EEG and averaging the brain response to stimulus events. For example, in early work on language processing Kutas and Hillyard (1980) recorded ERPs to the last word of sentences that either ended congruously (as in (1)), or incongruously (as in (2)):

(1) I take my coffee with cream and sugar.
(2) I take my coffee with cream and dog.

Although the EEG associated with the presentation of a single event is relatively inscrutable, cognitive neuroscientists have detected certain regularities in EEG elicited by sensible sentence completions (like 'sugar' in (1)) that differ from those in EEG elicited by bizarre sentence completions (like 'dog' in (2)). To date, the best method for highlighting these regularities in the EEG is to average the signal associated with a given category of stimulus. The logic behind averaging, of course, is to extract from the EEG only that information which is time-locked to the processing of the event. Cognitive neuroscientists refer to the averaged signal as the event-related potential (ERP) because it represents electrical activity in the brain associated with the processing of a given class of events.

For example, in their landmark study Kutas and Hillyard constructed 70 sentences, half of which ended congruously, half incongruously (Kutas and Hillyard, 1980). By averaging the signal elicited by congruous and incongruous sentence completions, respectively, these investigators were able to reveal systematic differences in the brain's electrical response to these stimulus categories in a particular portion of the ERP that they referred to as the N400 component. Subsequent research has shown that N400 components are generated whenever stimulus events involve meaningful processing of the stimuli, and that its size is sensitive to fairly subtle differences in the processing difficulty of the words that elicit it. As such, many investigators have used the N400 component of the brain waves as a dependent variable in psycholinguistic experiments (see Kutas, Federmeier, Coulson, King and Muente, 2000a, for review).

2.3 ERP components

While EEG measures spontaneous activity of the brain and is primarily characterized by rhythmic electrical activity, the ERP is a waveform containing a series of deflections that appear to the eye as positive and negative peaks. Such peaks are often referred to as *components*, and much of cognitive electrophysiology has been directed at establishing their functional significance. ERP components are characterized by their *polarity*, that is, whether they are positive- or negative-going; their *latency*, the time point where the component reaches its largest amplitude; and their *scalp distribution*, or the pattern of relative amplitudes the component has across all recording sites. The N400, for instance, is a negative-going wave that peaks approximately 400 msec after the onset of the stimulus, and has a centro-parietal distribution which is slightly larger over the right side of the head.

The ERP approach seeks correlations between the dimensions of ERP components elicited by different stimuli and putatively relevant dimensions of the stimuli themselves. ERP components with latencies under 100 msec are highly sensitive to systematic variations in the physical parameters of the evoking stimulus. Because their amplitudes and latencies seem to be determined by factors outside the subject, they are referred to as *exogenous* components. In contrast, *endogenous* ERP components are less sensitive to physical aspects of the stimulus, reflecting instead the psychological state of the subject. While exogenous components are modulated by the intensity, frequency and duration of the stimulus events, endogenous components are modulated by task demands and other manipulations that affect the subjects' expectancies, strategies and mental set.

The P300 component of the ERP is a paradigmatic example of an endogenous component because its amplitude (or size) is modulated by subjective aspects of experimental stimuli, such as their salience, their task relevance, and their probability. Actually a whole family of positive-going components of varying latency, P300s are elicited by any stimulus that requires the participant to make a binary decision. The amplitude of this response is proportional to the rarity of the target stimulus, as well as how confident the participant is in her classification judgement. The latency of the P300 (i.e., the point in time at which it peaks) varies with the difficulty of the categorization task, and ranges from 300 to over 1000 msec after the onset of the stimulus (Donchin, Ritter and McCallum, 1978; Kutas, McCarthy and Donchin, 1977; Magliero, Bashore, Coles and Donchin, 1984; McCarthy, 1981; Ritter, Simpson and Vaughan, 1983).

While not specifically sensitive to language, P300s will be elicited in any psycholinguistic paradigm that requires a binary decision. As long as the experimenter is aware of the conditions known to modulate the P300, this component can serve as a useful dependent measure of language-relevant decision making. For example, participants might be presented with one statement, and then asked to signal whether another statement was entailed or implied by the first. The amplitude of the P300 in such a case varies with the participant's confidence in her decision, and its latency indexes when the decision is made. However, because P300 amplitude is very sensitive to stimulus probability, the number of critical stimuli in each experimental condition must be held constant. Another thing to be aware of is the fact that task-induced P300s may overlap in time with more specifically language-sensitive ERP effects such as the N400.

3 Language-sensitive ERPs

Since the discovery of the N400, cognitive neuroscientists interested in language have frequently appealed to ERPs as a dependent measure in psycholinguistic experiments. As a result, a number of language-sensitive ERP components have been reported (see Kutas, Federmeier, Coulson, King and Muente, 2000b, for review). Although most of this research has been motivated by issues in sentence processing, these findings may prove valuable to researchers interested in pragmatic aspects of language comprehension. Below we review ERP components known as the N400, the lexical processing negativity (LPN), the left anterior negativity (LAN), the P600, as well as slow cortical potentials, and briefly discuss the utility of each for studies of pragmatic language comprehension.

3.1 N400

The N400 is a negative-going wave evident between 200 and 700 msec after the presentation of a word. Though this effect is observed all over the scalp, it is largest over centroparietal areas and is usually slightly larger on the right side of the head than the left (Kutas, VanPetten and Besson, 1988). The N400 is elicited by words in all modalities, whether written, spoken or signed (Holcomb, 1990). Moreover, the size, or amplitude, of the N400 is affected in a way that is analogous in many respects to popular measures of priming in psycholinguistics, such as naming and lexical decision latencies.

For instance, in both word lists and in sentences, high-frequency words elicit smaller N400s than low-frequency words (Smith and Halgren, 1989). The N400 also evidences semantic priming effects, in that the N400 to a word is smaller when it is preceded by a related word than when it is preceded by an unrelated word (Bentin, 1987; Holcomb, 1988). Third, the N400 is sensitive to repetition – smaller to subsequent occurrences of a word than to the first (Rugg, 1985; VanPetten, Kutas, Kluender, Mitchiner and McIsaac, 1991a). Further, while pseudowords (orthographically legal letter strings) elicit even larger N400s than do real words, orthographically illegal non-words elicit no N400 at all (Kutas and Hillyard, 1980).

In addition to its sensitivity to lexical factors, the N400 is sensitive to contextual factors related to meaning. For example, one of the best predictors of N400 amplitude for a word in a given sentence is that word's cloze probability (Kutas, Lindamood and Hillyard, 1984). Cloze probability is the probability that a given word will be produced in a given context on a sentence completion task. The word 'month' has a high cloze probability in 'The bill was due at the end of the...', a low cloze probability in 'The skater had trained for many years to achieve this...', and an intermediate cloze probability in 'Because it was such an important exam, he studied for an entire...'. N400 amplitudes are large for unexpected items, smaller for words of intermediate cloze probability, and are barely detectable for contextually congruous words with high cloze probabilities. In general, N400 amplitude varies inversely with the predictability of the target word in the preceding context.

Because initial reports of the N400 component involved the last word of a sentence, many people have the misconception that N400 is an ERP component elicited by sentence final words. However, N400 is elicited by all words in a sentence. Interestingly, the size of the N400 declines across the course of a congruent sentence, starting large and becoming smaller with each additional open-class word. This effect has been interpreted as reflecting the buildup of contextual constraints as a sentence proceeds because it does not occur in grammatical but meaningless word strings (VanPetten, 1991b). In general, the amplitude of the N400 can be used as an index of processing difficulty: the more demands a word poses on lexical integration processes, the larger the N400 component will be. This feature of the N400 makes it an excellent dependent measure in language comprehension experiments. As long as words in different conditions are controlled for length, frequency in the language, ordinal position in a sentence and cloze probability, N400 amplitude can be used as a measure of processing effort.

3.2 LPN

The lexical processing negativity (LPN) is a brain potential to written words that is most evident over left anterior regions of the scalp. Its association with lexical processing derives from the fact that its latency is highly correlated with word frequency, peaking earlier for more frequent words (King, 1998). This component was originally thought to be an electrophysiological index of the brain's distinction between open-class content words and closed-class function words as the so-called N280 component was elicited by closed- but not open-class words (Neville, Mills and Lawson, 1992). However, subsequent testing indicated that word class effects are attributable to quantitative differences in word length and frequency (Osterhout, Bersick and McKinnon, 1997). That is, two words with the same frequency in the language elicit LPNs with the same latency even if one is an open-class word and the other a closed-class word (King, 1998). Because its latency is sensitive to word frequency, this component is useful as an index that the initial stages of lexical processing have been completed.

3.3 LAN

Researchers have also identified ERP components that seem to be sensitive to syntactic manipulations. The first is a negativity that occurs in approximately the same time window as the N400 (i.e., 300–700 msec post-word onset) and is known as the LAN (left anterior negativity) because it is most evident over left frontal regions of the head. Kluender and Kutas described this component in a study of sentences containing long distance dependencies that required the maintenance of information in working memory during parsing (Kluender, 1993). Similarly, King and Kutas (1995) described this component as being larger for words in object relative sentences like (3) that induce a greater working memory load than subject relative sentences like (4):

(3) The reporter who the senator attacked admitted the error.
(4) The reporter who attacked the senator admitted the error.

As an ERP component sensitive to working memory load, the LAN can be used to index differences in the processing difficulty of appropriately controlled stimuli.

3.4 P600

Another ERP component sensitive to syntactic and morphosyntactic processing is the P600, sometimes called the syntactic positive shift (SPS). This slow positive shift has been elicited by violations of agreement, phrase structure and subcategorization in English, German and Dutch (Coulson, King and Kutas, 1998b; Hagoort, Brown and Groothusen, 1993; Mecklinger, Schnefers, Steinhauer and Friederici, 1995; Neville, Nicol, Barss, Forster and Garrett, 1991; Osterhout, 1992). This component is typically described as beginning around 500 ms post-stimulus, and peaking at approximately 600 ms. Its scalp distribution tends to be posterior, although anterior effects have also been reported (see Coulson et al., 1998b, for review). Because the broad positivity is elicited by syntactic errors, it has been hypothesized to reflect a re-analysis of sentence structure triggered by such errors (Hahne, 1999).

However, Coulson and colleagues (Coulson et al., 1998b) found that the amplitude of the P600 varied with the probability of ungrammatical trials within an experimental block. In fact, the P600 to all improbable trials (collapsed across grammaticality) was indistinguishable from that to all grammatical violations. Thus Coulson and colleagues (Coulson et al., 1998a; Coulson et al., 1998b) argue that the P600 is not a syntax-specific component, but rather a variant of a domain-general component in the P300 family which has been hypothesized to reflect 'context updating', a process in which the subject recalibrates her expectations about the environment (Donchin, 1988). Nonetheless, the fact that syntactic violations are associated with the late positive ERP known as the P600 provides a convenient tool for testing hypotheses about grammatical processing and its impact (or, perhaps even dependence) on contextual, pragmatic factors.

3.5 Slow cortical potentials

Besides phasic ERPs, temporally extended tasks such as reading or speech comprehension also elicit electrical changes with a slower time course. In order to examine slow brain potentials, it is necessary to average several seconds worth of data (namely average EEG that begins at the onset of a particular class of language stimuli and ends several seconds later), apply a low-pass filter to the ERP data, and restrict analysis to activity less than 0.7 Hertz. Kutas describes three slow brain potentials which might be elicited in experimental studies of the comprehension of pragmatic aspects of written language (Kutas, 1997). The first is a left lateralized negative shift over occipital sites that is thought to reflect early visual processing. The second is the clause ending negativity (CEN), an asymmetric negativity larger over left hemisphere sites that may be associated with sentence wrap-up operations. The third is an ultra-slow (< 0.2 Hertz) positivity over frontal sites that may be associated with sentential integration. Since inferential aspects of language comprehension might be expected to develop slowly over the course of a sentence, or set of sentences, it is likely that experimental manipulations that promote or inhibit the generation of inferences might be detected as modulations of these slow cortical potentials.

4 ERP studies of pragmatic language comprehension

Because it can provide a continuous on-line index of processing that occurs at the advent of a linguistic stimulus, the ERP is well-suited for addressing questions that have to do with what sorts of information experimental participants are sensitive to and when. One constraint to keep in mind, however, is that (by definition) the ERP is the brain response to numerous stimuli that share some theoretically interesting property such as occuring in a true sentence rather than a false one, or being a prototypical category member as opposed to a non-prototypical one. For language experiments, a minimum of 30 trials in each experimental condition (namely each cell) is recommended to obtain a reasonable ratio of signal to noise. As several components of the ERP are sensitive to stimulus repetition (VanPetten et al., 1991a), most experimenters construct multiple stimulus lists in order to fully counterbalance their designs without requiring individual subjects to read multiple 'versions' of a single stimulus. Finally, because ERPs can vary greatly between individuals, it is advisable to use a within-subjects design whenever possible.

Given these caveats, there are a number of ways to use ERPs to test hypotheses about language comprehension. For instance, given the assumption that qualitative differences in the ERP waveform reflect the operation of qualitatively different cognitive processes, ERPs can be used to identify the operation of different cognitive processes as they occur in the interpretation of langauge. One possible use of the ERP measure, then, would be to identify different components in the waveforms that index different levels of processing. For example, the existence of separate components in the waveforms that index functionally distinct levels of processing could be seen as implicit support for a firm distinction between semantics and pragmatics. Moreover, once identified, electrophysiological indices of semantic and pragmatic processing could be used to test hypotheses about the relative contribution of each to the comprehension of any given linguistic stimulus.

To date, very little ERP language research has concerned pragmatic aspects of language comprehension. Moreover, the little that has been done has not revealed an ERP index specifically sensitive to pragmatic language comprehension. However, extant work has suggested that ERPs are sensitive to experimental modulations of higher level contextual factors. For example, St Georges, Mannes and Hoffman recorded participants' ERPs as they read ambiguous paragraphs that either were or were not preceded by a disambiguating title (St George et al., 1994). Although the local contextual clues provided by the paragraphs were identical in the titled and untitled conditions, words in the untitled paragraphs elicited greater amplitude N400s.

Similarly, Van Berkum, Hagoort and Brown found that words which elicit N400s of approximately equal amplitude in an isolated sentence, do not elicit equivalent N400s when they occur in a context that makes one version more plausible than the other (Van Berkum et al., 1999). For instance, 'quick' and 'slow' elicit similar N400s in 'Jane told her brother that he was exceptionally quick/slow'. However, 'slow' elicits a much larger N400 when this same sentence is preceded by 'By five in the morning, Jane's brother had already showered and had even gotten dressed'. This sensitivity of the N400 component to higher-order aspects of language makes it an excellent measure for testing hypotheses about processing difficulty associated with the comprehension of various sorts of pragmatic language phenomena.

Muente, Schlitz and Kutas have used slow cortical potentials evident in recorded ERPs to reveal processing differences between superficially similar sentences that required readers to differentially exploit their background knowledge (Muente et al., 1998). Muente and colleagues hypothesized that people's conception of time as a sequential order of events determines the way we process statements referring to the temporal order of events. Consequently, they recorded ERPs as participants read sentences such as 'Before/After the psychologist submitted the article, the journal changed its policy'. Because 'Before X, Y' presents information in the reverse chronological order, it was hypothesized that these sentences would be more difficult to process than the 'After' sentences. Indeed, Muente and colleagues found that ERPs recorded at electrode sites on the left frontal scalp were more negative for the more difficult 'Before' sentences. Perhaps more compelling, they found that the size of this effect was correlated with individual participants' working memory spans.

4.1 Considerations in ERP language research

The basic paradigm in ERP language research involves recording ERPs to (minimally) different sorts of langauge stimuli in order to observe modulations in the amplitude and/or latency of particular components. Although there are known limitations to using ERP data to localize neural generators in the brain, it is an excellent measure for determining precisely when the processing of two classes of stimuli begins to diverge. Because brain wave measures are acquired with a high degree of temporal resolution (on the order of milliseconds), ERPs can potentially reveal the exact moment of divergence in the processing of particular categories of events. In any case, the detection at time t of a reliable difference in the ERP waveforms elicited by two categories of events suggests that processing of those categories differs at that instant, and began at least by time t (Coles, Gratton and Fabiani, 1990).

Because the N400 is sensitive to the same processes indirectly assessed in the reaction-time paradigm, we can view its use in investigations of pragmatic language comprehension as an analogous version of behavioural measures. One advantage of ERP measures is that they can be collected in the absence of an explicit task (other than that of language comprehension itself). Moreover, ERP measures can also be collected while the participant performs a behavioural task, thus giving the experimenter a measure of ongoing brain activity before, during and after the performance of the task. Regardless of whether one conducts two experiments – one behavioural and one ERP – or whether the two sorts of measures are collected concomitantly, ERP data can greatly aid in the interpretation of the behavioural results.

In fact, ERP and reaction-time data are often complementary as reaction-time data can provide an estimate of how long a given processing event took, while ERP data can suggest whether distinct processes were used in its generation. An experimental manipulation that produces a reaction-time effect might produce two or more ERP effects, each of which is affected by different sorts of manipulations. By giving the experimenter the means to explore these dissociations, ERPs can help reveal the cognitive processes that underlie the pragmatic phenomenon of interest. In fact, to the extent that ERP effects can be identified with specific cognitive processes, they provide some evidence of how processing differed in the different conditions (King and Kutas, 1995).

4.2 Joke comprehension

Joke comprehension is one area of language comprehension relevant for pragmatics because of the way it highlights the importance of background knowledge for comprehension and the development of expectations. For example, 'I let my accountant do my taxes because it saves time: last spring it saved me ten years', is funny both because the reader or listener initially expects the amount of time saved by the accountant to be on a different order of magnitude than years, and because it is possible to formulate a coherent interpretation of the statement whereby the 'time saved' is jail time. While lexical reinterpretation plays an important part in joke comprehension, to truly appreciate this joke it is necessary to recruit background knowledge about the particular sorts of relationships that can obtain between business people and their accountants so that the initial busy professional interpretation can be mapped into the 'crooked-businessman' frame. Coulson refers to the pragmatic reanalysis needed to understand examples like this one as *frame-shifting* (Coulson, 2000).

Given the impact of frame-shifting on the interpretation of one-line jokes, one might expect the underlying processes to take time, and consequently be reflected in increased reading times in behavioural tests of processing difficulty such as self-paced reading. In this paradigm, the task is to read sentences one word at a time, pressing a button to advance to the next word. As each word appears, the preceding word disappears, so that the experimenter gets a record of how long the participant spent reading each word in the sentence. Accordingly, Coulson and Kutas measured how long it took people to read sentences that ended with jokes that required frame-shifting than with non-funny 'straight' endings consistent with the contextually evoked frame (Coulson, 1998). Two types of jokes were tested, high constraint jokes like (5) which elicited at least one response on a sentence completion task with a cloze probability of greater than 40 per cent, and low constraint jokes like (6) which elicited responses with cloze probabilities of less than 40 per cent. (For both (5) and (6) the word in parentheses is the most popular response on the cloze task.)

(5) I asked the woman at the party if she remembered me from last year and she said she never forgets a (face 81 per cent).
(6) My husband took the money we were saving to buy a new car and blew it all at the (casino 18 per cent).

To control for the fact that the joke endings are (by definition) unexpected, the straight controls were chosen so that they matched the joke endings for cloze probability, but were consistent with the frame evoked by the context. For example, the straight ending for (5) was *name* (the joke ending was *dress*); while the straight ending for (6) was *tables* (the joke ending was *movies*). The cloze probability of all four ending types (high and low constraint joke and straight endings) was equal, and ranged from zero per cent to 5 per cent. Coulson and Kutas found that readers spent longer on the joke than the straight endings, and that this difference in reading times was larger and more robust in the high constraint sentences (Coulson and Kutas, 1998). This finding suggests there was a processing cost associated with frame-shifting reflected in increased reading-times for the joke endings, especially in high constraint sentences that allow readers to commit to a particular interpretation of the sentence.

In a very similar ERP study of the brain response to jokes, Coulson and Kutas found that ERPs to joke endings differed in several respects from those to the straight endings, depending on contextual constraint as well as participants' ability to get the jokes (Coulson and Kutas, 2001). In poor joke comprehenders, jokes elicited a negativity in the ERPs between 300 and 700 milliseconds after the onset of the sentence-final word. In good joke comprehenders, high but not low constraint endings elicited a larger N400 (300–500 msec post-onset) than the straights. Also, in this group, both sorts of jokes (high and low constraint) elicited a positivity in the ERP (500–900 msec post-onset) as well as a slow, sustained negativity over left frontal sites. Multiple ERP effects of frame-shifting suggest the processing difficulty associated with joke comprehension involves multiple neural generators operating with slightly different time courses.

Taken together, these studies of frame-shifting in jokes are far more informative than either study alone. The self-paced reading-time studies suggested that frame-shifting needed for joke comprehension exerts a processing cost that was especially evident in high constraint sentence contexts

(Coulson and Kutas, 1998). ERP results suggested the processing cost associated with frame-shifting is related to higher-level processing (Coulson and Kutas, 2001). In the case of the high constraint jokes, the difficulty includes the lexical integration process indexed by the N400, as well as the processes indexed by the late-developing ERP effects. In the case of the low constraint jokes, the difficulty was confined to the processes indexed by the late-developing ERP effects. The added difference in lexical integration indexed by the N400 may explain why joke effects on reading times were more pronounced for high constraint sentences than for low. Because the late developing ERP effects were only evident for good joke comprehenders who successfully frame-shifted, they are more likely to be direct indices of the semantic and pragmatic re-analysis processes involved in joke comprehension.

As a general methodological point, the demonstration of individual differences in memory, vocabulary, language ability or, in this case, on-line comprehension, and their relationship to various pragmatic phenomena has a great deal of potential. Experimental approaches to pragmatics, especially when the topic concerns whether readers generate inferences in response to certain sorts of contextual cues, would do well to consider how individual differences in cognitive abilities affect these phenomena. Moreover, work on joke comprehension by Coulson and Kutas demonstrates how ERPs and reaction-time data for the same stimuli can provide complementary information about the underlying cognitive processes.

4.3 Metaphor comprehension

In fact, ERPs can reveal reliable differences even when no reaction-time differences are evident. This is important because reaction times are typically interpreted as reflecting processing difficulty, yet it is quite possible for two processes to take the same amount of time, but for one to recruit more neural processing resources. One issue in pragmatics where this has been an important issue is the study of metaphor comprehension. Because classical accounts of metaphor comprehension (Grice, 1975; Searle, 1979) depict a two-stage model in which literal processing is followed by metaphorical processing, many empirical studies have compared reading times for literal and non-literal utterances and found that when the metaphorical meaning was contextually supported, reading times were roughly similar. However, as Gibbs notes, parity in reading times need not entail parity in the underlying comprehension processes (Gibbs, 1994). It is possible, for example, that literal and metaphorical meaning might take the same amount of time to comprehend, but that the latter required more effort or processing resources. Alternatively, comprehension processes for literal versus metaphoric utterances might take the same amount of time to complete, and yet involve quite different computations (Gibbs, 1989).

Because they involve a direct and continuous measure of brain activity, ERPs can potentially distinguish between qualitatively different sorts of processing, even if their corresponding behavioural manifestations require the same amount of time. Taking advantage of the known relationship between N400 amplitude and processing difficulty, Pynte and colleagues contrasted ERPs to familiar and unfamiliar metaphors in relevant versus irrelevant contexts (Pynte, Besson, Robichon and Poli, 1996). They found that regardless of the familiarity of the metaphors, N400 amplitude was a function of the relevance of the context. Moreover, by using ERPs, Pynte and colleagues employed a measure which is in principle capable of revealing the qualitative processing differences by the standard (Gricean) pragmatic model. In fact, they observed no evidence of a qualitative difference in brain activity associated with the comprehension of literal and metaphoric language.

Reports that literal and non-literal language comprehension display a similar time course and recruit a similar set of neural generators are consistent with a number of modern models of metaphor comprehension (Coulson and Matlock, 2001; Gibbs, 1994; Giora, 1997; Glucksberg, 1998). Coulson's (2000) model also makes predictions for comprehension difficulty, predicting a gradient of processing difficulty related to the extent to which comprehension requires the participant to align and integrate conceptual structure from different domains. This prediction was tested by Coulson and Van Petten (2002) when they compared ERPs elicited by words in three different sentence contexts on a continuum from literal to figurative, as suggested by conceptual blending theory (Fauconnier and Turner, 2002). For the literal end of the continuum, Coulson and Van Petten used sentences that promoted a literal reading of the last term, as in 'He knows that whiskey is a strong INTOXICANT'. At the metaphoric end of the continuum, they used sentences which promoted a metaphoric reading of the last term, as in 'He knows that power is a strong INTOXICANT'. Coulson and Van Petten also posited a *literal mapping* condition, hypothesized to fall somewhere between the literal and the metaphoric uses, such as 'He has used cough syrup as an INTOXICANT'.

Literal mapping stimuli employed fully literal uses of words in ways that were hypothesized to include some of the same conceptual operations as in metaphor comprehension. These sentences described cases where one object was substituted for another, one object was mistaken for another, or one object was used to represent another – all contexts that require the comprehender to set up mappings between the two objects in question, and the domains in which they typically occur. In positing a continuum from literal to metaphorical based on the difficulty of the conceptual integration needed to comprehend the statement, Coulson and Van Petten (2002) predicted a graded difference in N400 amplitude for the three sorts of stimuli.

Data reported by Coulson and Van Petten were largely consistent with these predictions. In the early time window, 300–500 msec post-onset and before, ERPs in all three conditions were qualitatively similar, displaying similar waveshape and scalp topography. This suggests that during the initial stages, processing was similar for all three sorts of contexts. Moreover, as predicted, N400 amplitude differed as a function of metaphoricity, with literals eliciting the least N400, literal mappings the next-most, and metaphors eliciting the most N400, suggesting a concomitant gradient of processing difficulty. The graded N400 difference argues against the literal/figurative dichotomy inherent in the standard model, and is consistent with the suggestion that processing difficulty associated with figurative language is related to the complexity of mapping and conceptual integration.

5 Future directions

While a few ERP studies of figurative language comprehension have been conducted, serious investigation of the pragmatic aspects of language comprehension has barely begun. This is no easy task as understanding language as an integrated, goal-directed process will require elucidating the relationships that hold among language subcomponents and between language and other cognitive abilities. For instance, understanding language utterances necessarily requires that relevant linguistic, contextual and background knowledge be integrated. However, very little is known about the relative importance of local context and background knowledge, or how these factors interact. Because different components of the brain waves are modulated by different factors, ERPs are a potentially powerful tool for teasing apart the different contributions of linguistic and non-linguistic sources of information.

5.1 Direct versus indirect speech acts

For example, ERPs could be recorded while a subject read or listened to sentences that constituted either direct or indirect speech acts as in (7) and (8). The effect of posing the speech act as a question might be assessed by including a contrast between a direct interrogative speech act posed as a question (as in (9)) and an indirect speech act posed in a declarative sentence (as in (10)):

(7) Give me the mustard.
(8) Can you give me the mustard?
(9) Did Harry marry Sally?
(10) I'd like to know if Harry married Sally.

By recording ERPs to words in direct and indirect speech acts it is possible to observe whether one form of these speech acts is easier to process and whether the relative processing difficulty can be altered by varying the contextual conditions in which they occur.

5.2 Entailments, explicatures and implicatures

One issue in pragmatics to which ERP research might be productively directed is the distinction between explicatures, implicatures, and entailments. In the framework of Relevance Theory (Sperber and Wilson, 1986/1995), the *explicature* is a fully specified linguistic meaning of an utterance, an *implicature* is an implicit inferred meaning of an utterance, and an *entailment* is a proposition that is logically implied by the sentence. For example, in a context in which the speaker has been asked about how many people went to the Cognitive Science Holiday Party, the explicature of (11) is akin to 'Not all of the people invited to the Cognitive Science Holiday Party went to the Cognitive Science Holiday Party':

(11) All of the boys went to the party.
(12) Some of the boys went to the party.
(13) Not all of the boys went to the party.

In this context, (12) is entailed by (11), because it is true in all situations in which (11) is true. Interestingly, the speaker who asserts (12) implicates (13) but does not entail it, as (truth-functionally) (12) is true when (11) is, and (11) and (13) are mutually incompatible.

By recording ERPs to explicatures, implicatures and entailments it might be possible to detect whether these categories of language-induced inferences elicit the same pattern of brain waves. ERPs could also be used to evaluate the adequacy of pragmatic theories by testing whether or not the same processes underlie the derivation of explicatures and implicatures, or whether certain sorts of information are derived automatically (see Chapter 14). One might, as Gibbs and Moise (1997) did with behavioural measures, ask readers to judge what a speaker says when he asserts 'Jane has three children,' and compare ERPs elicited by the minimal meaning 'Jane has at least three children but may have more than three', and the enriched meaning 'Jane has exactly three children and no more than three', to see which category elicits more signs of surprise and/or processing difficulty.

5.4 The importance of non-linguistic cues for language comprehension

Although cognitive neuroscientists have learned a great deal about language comprehension, it remains the case that most studies have employed experimenter-constructed stimuli in the controlled and artificial setting of the laboratory. However, language 'in the wild' occurs in a much richer context. Not only are the units of processing larger than those typically studied – that is, texts and discourses rather than the words and sentences so dear to the hearts of psycholinguists – but there are social and physical cues to guide the language user. In the future, we must exploit technological advances to bring more of the world into the laboratory. For instance, using MP3 technology it is possible to present subjects with auditory stimuli, such as naturally occurring conversation, or more controlled, scripted versions of the same phenomena. EEG could be collected while subjects listened to these stimuli, and ERPs to theoretically interesting events could be examined.

The ongoing nature of the EEG signal makes it a good measure for assessing the on-line comprehension of linguistic materials. However, recording ERPs to auditorally presented stimuli comes with its own set of challenges. Perhaps the main problem is that words in continuous speech do not generally elicit distinct ERPs because word boundaries are often absent from the speech signal. Fortunately, it is possible nonetheless to observe measurable differences in N400 amplitude to the last word of congruous and incongruous sentence completions (e.g., Holcomb, 1991; Van Petten, Coulson, Plante, Rubin and Parks, 1999). Moreover, Mueller and colleagues point to the utility of examining slow brain potentials when investigating the comprehension of spoken language (Mueller, King and Kutas, 1997). In a sentence processing study that compared ERPs elicited by subject-relative sentences with the more demanding object-relative sentences, Mueller and colleagues identified an ultra-slow frontal positivity whose amplitude varied as a function of comprehension difficulty in both written and spoken materials.

Similarly, Steinhauer, Alter and Friederici have used ERPs to study how intonational phrasing guides the initial analysis of sentence structure (Steinhauer et al., 1999). They recorded ERPs as subjects listened to syntactically ambiguous sentences with appropriate and inappropriate prosodic cues. In naturalistic stimuli, Steinhauer and colleagues found that participants' ERPs showed a positive-going waveform at prosodic phrase boundaries, this they call the Closure Positive Shift. In cases where the prosodic cues conflicted with syntactic ones, the mismatch elicited an N400–P600 pattern of ERP components suggesting participants used prosodic features to determine their intial (incorrect) parse of the sentence. These results show that the ERP is a good measure for revealing the time course and neural basis of prosodic information processing.

It seems possible that the ERP effects that proved useful for sentence processing studies by Mueller and colleagues (1997) and Steinhauer and colleagues (1999) might also prove useful in elucidating the pragmatic import of prosodic intonation. For example, ERPs could be recorded as subjects listen to sentences such as (14) intoned as promises or as threats:

(14) I'll be there.

Moreover, such sentences could be embedded in contexts which are either appropriate or inappropriate in order to compare the time course of semantic and prosodic information on the ERP. Similarly, our intonational promises and threats could be embedded in contexts that vary in the degree to which they disambiguate the speech act in order to explore the interaction of semantic and prosodic variables on ERP indices of on-line language comprehension.

Another facet of normal language comprehension typically absent from laboratory studies is the presence of visual information. This visual information includes both the local context as well as visual information about the speaker, such as her facial expressions and her gestures. As EEG can in principle be time-locked to the onset of visual events in MP3 videos, it is possible to record ERPs as subjects watch videos of speakers interacting in real contexts. Although the continuous nature of videographic stimuli present some of the same problems as continuous speech, it seems plausible that large differences in processing difficulty would be evident in ERP effects to visual stimuli, just as they are to speech.

6 Conclusions

As noted in the introduction to this chapter, the study of pragmatics can never be totally divorced from the observation of language use in naturally arising communicative contexts. However, if we hope to develop a science that extends beyond the scope of the ethnographic site, it is important to test the generalizability of hypotheses in pragmatics with other tools in the cognitive scientist's toolbox, including native-speaker intuitions, reaction times and accuracies on various judgement tasks, eye-movement registration, and even scalp recorded ERPs. This chapter has reviewed how ERPs are recorded, outlined the strengths and weaknesses of the technique, discussed what sorts of linguistic manipulations are known to give rise to ERP effects, and offered suggestions as to how the technique could be applied to address issues in pragmatic language comprehension. Just as the meaning of an utterance cannot be simply decoded, the significance of an ERP effect requires consideration of the motivating hypotheses, the experimental design, and knowledge of the sorts of manipulations that have led to similar effects in the past. Hopefully the reader who has made it this far has gleaned enough about the use of the ERP technique to see its power and utility for studying the inferential comprehension processes so essential for pragmatics.

References

Bach, K. (1994). *Mind and Language* 9: 124–62.
Barwise, J. P., John (1983). *Situations and Attitudes*, Cambridge, MA: MIT Press.
Bentin, S. (1987). *Brain and Language* 31: 308–27.
Carston, R. (2002). In *Thoughts and Utterances: The Pragmatics of Explicit Communication*. Oxford: Blackwell.
Coles, M., Gratton, G., and Fabiani, M. (1990). In J. C. a L. Tassinary (ed.), *Principles of Psychophysiology: Physical, Social and Inferential elements*: 413–55. Cambridge: Cambridge University Press.
Coulson, S. (2000). *Semantic Leaps: Frame-Shifting and Conceptual Blending in Meaning Construction*. Cambridge: Cambridge University Press.
Coulson, S., and Van Petten, C. (2002). *Memory and Cognition* 30: 958–68.
Coulson, S., King, J. W., and Kutas, M. (1998a). *Language and Cognitive Processes* 13: 653–72.
Coulson, S., King, J. W., and Kutas, M. (1998b). *Language and Cognitive Processes* 13: 21–58.

Coulson, S., and Kutas, M. (1998). UCSD Cognitive Science Technical Report: 98–103.
Coulson, S., and Kutas, M. (2001). *Neuroscience Letters* 316: 71–4.
Coulson, S., and Matlock, T. (2001). *Metaphor and Symbol* 16: 295–316.
Donchin, E. C. M. (1988). *Behavioral and Brain Sciences* 11: 357–74.
Donchin, E., Ritter, W., and McCallum, C. (1978). In E. Callaway, P. T., and S. H. Koslow (eds), *Brain Event-Related Potentials in Man*. New York: Academic.
Fauconnier, G. (1997). *Mappings in Thought and Language*. Cambridge: Cambridge University Press.
Fauconnier, G., and Turner, M. (2002). *The Way we Think*. New York: Basic Books.
Gibbs, R. (1994). *The Poetics of Mind: Figurative Thought, Language, and Understanding*. Cambridge: Cambridge University Press.
Gibbs, R., and Gibbs, R. J. (1989). *Metaphor and Symbolic Activity* 4: 145–58.
Gibbs, R., and Moise, J. F. (1997). *Cognition* 62: 51–74.
Giora, R. (1997). *Cognitive Linguistics* 7: 183–206.
Glucksberg, S. (1998). *Current Directions in Psychological Science* 7: 39–43.
Grice, H. (1975). In P. C. J. Morgan (ed.), *Syntax and Semantics: Volume 3, Speech Acts*. New York: Academic Press.
Hagoort, P., Brown, C., and Groothusen, J. (1993). *Language and Cognitive Processes* 8: 439–83.
Hahne, A. and Frederici, A. D. (1999). *Journal of Cognitive Neuroscience* 11: 194–205.
Hillyard, S., and Kutas, M. (1983). *Annual Review of Psychology* 34: 33–61.
Holcomb, P. J. (1988). *Brain and Language* 35: 66–85.
Holcomb, P., and Neville, H. (1990). *Language and Cognitive Processes* 5: 281–312.
Holcomb, P., and Neville, H. (1991). *Psychobiology* 19: 286–300.
King, J. W., and Kutas, M. (1998). *Neuroscience Letters* 244: 61–4.
King, J. W., and Kutas, M. (1995). *Journal of Cognitive Neuroscience* 7: 376–395.
Kluender, R., and Kutas, M. (1993). *Journal of Cognitive Neuroscience* 5: 196–214.
Kutas, M. (1997). *Psychophysiology* 34: 383–98.
Kutas, M., Federmeier, K., Coulson, S., King, J. W., and Muente, T. F. (2000a). In J. T. Caccioppo, L. G. Tassinary and G. G. Berntson (eds), *Handbook of Psychophysiology*: 576–601. Cambridge: Cambridge University Press.
Kutas, M., Federmeier, K. D., Coulson, S., King, J. W., and Muente, T. F. (2000b). In John T. Cacioppo, Louis G. Tassinary et al. (eds), *Handbook of Psychophysiology* (2nd edn): xiii, 1039. Cambridge: Cambridge University Press.
Kutas, M., and Hillyard, S. (1980). *Science* 207: 203–5.
Kutas, M., Lindamood, T. E., and Hillyard, S. (1984). In S. K. J. Requin (eds), *Preparatory States and Processes*: 217–37. Hillsdale, NJ: Erlbaum.
Kutas, M., McCarthy, G., and Donchin, E. (1977). *Science* 197: 792–7.
Kutas, M., Van Petten, C., and Besson, M. (1988). *Electroencephalography and Clinical Neurophysiology* 69: 218–33.
Magliero, A., Bashore, T. R., Coles, M. G. H., and Donchin, E. (1984). *Psychophysiology* 21: 171–86.
Marr, D. (1982). *Vision*. San Francisco: W. H. Freeman.
McCarthy, G. D. (1981). *Science* 211: 77–80.
Mecklinger, A., Schriefers, H., Steinhauer, K., and Friederici, A. D. (1995). *Memory and Cognition* 23: 477–94.
Mueller, H., King, J. W., and Kutas, M. (1997). *Brain Research: Cognitive Brain Research* 5: 192–203.
Muente, T., Schiltz, K., and Kutas, M. (1998). *Nature* 395: 71–3.
Neville, H., Mills, D., and Lawson, D. (1992). *Cerebral Cortex* 2: 244–58.

Neville, H., Nicol, J., Barss, A., Forster, K. I., and Garrett, M. F. (1991). *Journal of Cognitive Neuroscience* 3: 151–65.

Osterhout, L., and Holcomb, P. (1992). *Journal of Memory and Language* 31: 785–804.

Osterhout, L., Bersick, M., and McKinnon, R. (1997). *Biological Psychology* 46: 143–68.

Pynte, J., Besson, M., Robichon, F., and Poli, J. (1996). *Brain and Language* 55: 293–316.

Recanati, F. (1989). *Mind and Language* 4: 295–329.

Reddy, M. (1979). In A. Ortony (ed.), *Metaphor and Thought*: 284–324. Cambridge: Cambridge University Press.

Ritter, W., Simpson, R., and Vaughan, H. G. (1983). *Psychophysiology* 20: 168–79.

Rugg, M. (1985) *Psychophysiology* 22: 642–7.

Rugg, M. D., and Coles, M. G. H. (eds). (1995). *Electrophysiology of Mind: Event-Related Brain Potentials and Cognition*. Oxford: Oxford University Press.

Searle, J. (1979). *Expression and Meaning: Studies in the Theory of Speech Acts*. Cambridge: Cambridge University Press.

Smith, M. E., and Halgren, E. (1989). *Journal of Experimental Psychology: Learning, Memory, and Cognition* 15: 50–60.

Sperber, D., and Wilson, D. (1986/1995) *Relevance: Communication and Cognition*. Oxford: Blackwell.

St George, M., Mannes, S., and Hoffman, J. E. (1994). *Journal of Cognitive Neuroscience* 6: 70–83.

Steinhauer, K., Alter, K., and Friederici, A. D. (1999). *Nature Neuroscience* 2: 438–53.

Van Berkum, J., Hagoort, P., and Brown, C. (1999). *Journal of Cognitive Neuroscience* 11: 657–71.

Van Petten, C., and Kutas, M. (1991) *Memory and Cognition* 19: 95–112.

Van Petten, C., Coulson, S., Plante, E., Rubin, S., and Parks, M. (1999). *Journal of Experimental Psychology: Learning, Memory, and Cognition* 25: 394–417.

Van Petten, C., Kutas, M., Kluender, R., Mitchiner, M., and McIsaac, H. (1991a). *Journal of Cognitive Neuroscience* 3: 131–50.

Zigmond, M., Bloom, F. E., Landis, S. E., Roberts, J. L., and Squire, L. R. (1999). *Fundamental Neuroscience*. San Diego, CA: Academic Press.

10
Speech Acts in Children: the Example of Promises

Josie Bernicot and Virginie Laval

1 Introduction

Promises are central to human exchanges, especially in adult–child interactions. They consist of a commitment on the part of the speaker to perform a future act, as in *'je promets de ranger ma chambre'* ('I promise to clean my room'). For the past ten years, we have been investigating promise comprehension among children from the point of view that language is a communication system and that language competence is the acquisition and use of that system. The emphasis is therefore placed on the functional aspects of language (Bates, 1976; Bruner, 1983; Ervin-Tripp and Mitchell-Kernan, 1977; Halliday, 1985 ; Ninio and Snow, 1996; Tomasello, 2000). It has been shown in this perspective that interaction formats or routines (prototypical exemplars of social relations) are very important for young children (Bernicot, 1994; Marcos and Bernicot, 1994, 1997).

Some of the questions that we have been addressing are the following: How do children understand utterances that express a promise? How does their comprehension evolve with age? What cues do children use to interpret utterances expressing promises? Do they consider contextual cues, such as the listener's wishes about the accomplishment of an action (Bernicot and Laval, 1996) or do they rely on textual cues such as the utterance's linguistic form or its temporal markers (Laval and Bernicot, 1999)? To answer these questions we use both the theoretical perspectives offered by speech act theory (Austin, 1962; Searle, 1969, 1979; Searle and Vanderveken, 1985; Vanderveken, 1990a, 1990b) and the methodology of experimental psychology. The goal of the present chapter is to examine the role of one aspect of the interlocutors' intentions (listener's wishes about the accomplishment of an action) and one of the textual characteristics of utterances (verb tense) in promise comprehension among children aged 3 to 10.

According to speech act theory, a promise is an illocutionary act known as *commissive* because it is the speaker's goal to indicate that the speaker is committed to some course of action.[1] Four fulfilment conditions accompany this definition. To illustrate these, consider the following promise – '*je promets de ranger ma chambre*' ('I promise to clean my room').

1. *Propositional content condition.* The utterance says something about a future act to be performed by the speaker (the speaker says he is going to clean his room).
2. *Preparatory condition.* The listener would rather have the speaker accomplish the future act than not accomplish it (the listener wants the speaker to clean his room).
3. *Sincerity condition.* The speaker intends to carry out the future act (the speaker intends to clean his room).
4. *Essential condition.* It becomes the speaker's obligation to carry out the future act (the speaker is obligated to clean his room).[2]

The interlocutor's intentions are defined by the preparatory and the sincerity conditions (see Gibbs and Delaney, 1987, on the importance of these conditions among adults).

Prior work (Astington, 1988b) has shown that the comprehension of promises by children evolves with age. For children between the ages of 5 and 9, a promise appears to correspond to a true statement that can refer to a past or future action. What is important at these ages is that the action corresponding to the propositional content of the statement be accomplished; the fact that the speaker has (or does not have) control over the action is not considered. Starting at age 9, children make the distinction between a promise and a prediction, based on whether or not the speaker has control over the occurrence of the action. The distinction between a promise and an assertion begins to appear between the ages of 11 and 13.

Astington (1990) related the production of commissive speech acts by children to the 'metapragmatic' knowledge they have about such acts. At age 5, children know how to make promises in the appropriate situations; at the age of 6, they correctly use the verb *promise*. According to this author, metapragmatic knowledge about promises – assessed by having children judge speech acts produced by other individuals – appears at about the age of 10. The metapragmatic knowledge possessed at that age pertains to the speaker's responsibility to perform the action corresponding to the propositional content of the promise-making statement.

---

[1] This is opposed to Assertives, Directives, Expressives and Declarations.
[2] Mey (2001) highlighted that for anthropologists the speech act of promising is not only defined by these four fulfilment conditions, but that its success also depends on the ways it is supposed to sustain and confirm the existing order of things (see Duranti (1996) and Keating (1998) about Philippino people). In this case, 'promises' are similar to 'assertives' (in the meaning of 'predictions').

Understanding a promise implies being able to process linguistic cues concerning the future, that is, the listener must be capable of processing verb tense. A promise is very often an ordinary utterance produced during everyday events, as in the example above. From a linguistic standpoint, the expression 'I promise' is not a necessary part of a promise utterance. There are other linguistic forms (in an interaction context) capable of achieving the same function, as in *'je vais ranger ma chambre'* ('I am going to clean my room') or *'je rangerai ma chambre'* ('I will clean my room'). The utterance *'je promets de ranger ma chambre'* ('I promise to clean my room') is the surface form of *je promets+je rangerai ma chambre* (I promise+I will clean my room). This analysis leads us to the conclusion that all linguistic forms specific to promises must express a future action, whether in their deep structure, as in *'je promets de ranger ma chambre'* ('I promise to clean my room') or in their surface structure, as in *'je vais ranger ma chambre'* or *'je rangerai ma chambre'* ('I am going to clean my room' or 'I will clean my room'). In other words, the future as a temporal marker is a textual characteristic specific to promise utterances.

In French, as in other languages, future markers are used to express the aspect of an action (e.g., the desiderative future) and to situate it in time. Several studies have shown that until the age of 6, children rely primarily on adverbs and time prepositions to place an action in time; it is not until after that age that they begin to use verb tense. For promises in particular, we are interested in the future tense as a temporal marker: it specifies that the action described in the propositional content of the utterance will take place at some time after the utterance is produced.

From a morphological standpoint, the future in French can be expressed in four ways (Fleischman, 1982):

- Via the present tense or *praesens pro futuro*, as in *'Paul joue au tennis demain'* ('Paul is playing tennis tomorrow')
- Via sentences combining a modal auxiliary and an infinitive verb, as in *'Paul doit jouer au tennis demain'* ('Paul must be playing tennis tomorrow')
- Via the immediate future tense, as in *'Paul va jouer au tennis'* ('Paul is going to play tennis')
- Via the simple future tense, as in *'Paul jouera au tennis'* ('Paul will play tennis')

Only the immediate future and the simple future, which provide interpretable temporal markers in cases where an adverb or time preposition is lacking, are relevant to the study of promises. Although both of these ways of expressing the future are used in promise utterances, they are not interchangeable. A number of studies on this subject (e.g., Harner, 1981a, 1981b), have shown that the immediate future commits the speaker to the accomplishment of the action in the very near future, whereas the simple future is less suitable for talking about an action that will follow immediately. For example, if a little boy asks his mother to buy a toy and the mother says *'Oui, je vais te l'acheter'* ('Yes, I am going to buy it for you'), he is usually satisfied with that answer; but if she answers *'Oui, je te l'acheterai'* ('Yes, I will buy it for you'), the boy will usually say, 'Yes, but when?'. In everyday situations, the immediate future does not necessarily correspond to an immediate action, but it serves as a sort of guarantee that the action will be executed right away. In contrast, the simple future may look like a threat that things will not happen as expected.

In this chapter, we present two experiments. The first was designed to determine how children's comprehension of promises is affected by: (a) the presence or absence of the preparatory condition; and by b) the linguistic form of the statement (i.e., does it contain or does it not contain the verb *promise*). The second experiment was designed to gain insight into one of the textual characteristics of promises: the future tense as a temporal marker of utterances. More specifically, the aim was to determine the role of the future tense in the comprehension of promises by children.

2 Experiment 1

Two main objectives guided this study. The first was to gain an accurate understanding of the role of the preparatory condition in the comprehension of promises. Astington's (1988b, 1990) findings demonstrated the importance of the accomplishment of the action (an essential component of the sincerity condition), which showed that the children mastered the sincerity condition from the age of 5. Here, we sought experimental data that will be revealing of the role of an interlocutor's desires (in other words, data concerning the preparatory condition).

The second objective was to test linguistic forms which do not contain the verb *promise* but which, according to the speech acts classification (Searle and Vanderveken, 1985; Vanderveken, 1990a, 1990b), are specifically commissive, that is, they contain verbs in the future tense (active or passive voice).

To meet these objectives in Experiment 1, the variables manipulated were the satisfaction/non-satisfaction of the preparatory condition, the linguistic form of the commissive statement, and the children's age. The sincerity condition was always met.

2.1 Method

Subjects. Seventy-two, native French-speaking children participated in the experiment (42 girls and 30 boys). They were divided into three groups of 24 subjects on the basis of age. The three groups will hereafter be called the *3-year-olds* (mean age: 3;10, range: 3;3 to 4;1); the *6-year-olds* (mean age: 6;10, range: 6;2 to 7;0); and the *10-year-olds* (mean age: 10;10, range: 10;2 to 11;1).

Materials. Eighteen stories about the adventures of a young boy named Bill were constructed. In each story, made up of four frames consisting of a picture with a caption, Bill makes a promise. The linguistic context and situational context combined created realistic, everyday-life situations. The material was designed to keep the child's attention focused on the task. Two examples are given in Table 10.1. All of the stories had the same 4-frame structure, as follows:

Picture 1: Theme. The picture showed two characters, the speaker and the listener, in the story's setting. The caption was used to manipulate the preparatory condition (PC). The preparatory condition was clearly satisfied in half of the stories (PC+) and was not satisfied in the other half (PC–), that is, either the listener wanted the speaker to keep his promise or the listener did not want the speaker to keep his promise.

Picture 2: Promise. The picture showed the speaker up close, talking to the listener. The caption contained the statement made by the speaker (Bill). The promise was being made to a different listener in each story (a friend or one of Bill's parents) using one of the following three statement forms, which varied in illocutionary force:

- *Promise-to-act* statements explicitly contained the verb *promettre* (promise) followed by a verb in the infinitive form. The grammatical subject of the

Table 10.1 Examples of stories from Experiment 1

Condition	Preparatory Condition Satisfied (PC+)/promise-to-act statement	Preparatory Condition Not Satisfied (PC–)/future-action statement
Picture 1: story theme	Bill is supposed to go to bed at 8:30. He's allowed to look at a book before going to sleep. One night, Bill's father thinks Bill is very tired. He wants Bill to turn off the lights very soon.	Bill's best friend is called Bungo: it's his dog. They played together all afternoon in the woods and Bungo is dirty. Bungo really needs to be washed, but Bill's father doesn't want Bill to use the hose alone.
Picture 2: promise-making statement	Bill says to his father: 'I promise I'll turn out the lights right away.'	Bill says to his father: 'I'll wash Bungo tomorrow.'
Picture 3: fulfilment of promise	Five minutes later, Bill's father sees that the lights are out.	The next day, Bill's father sees Bill washing his dog with the hose.
End of Story	Picture 4.1 Bill's father is happy. Picture 4.2 Bill's father is unhappy.	Picture 4.1 Bill's father is happy. Picture 4.2 Bill's father is unhappy.

sentence was the person making the promise. The social act intentionally posed by the speaker was a firm commitment (e.g., *'Je te promets de laver mon vélo'*, 'I promise to wash my bike').

- *Future-action* statements employed a verb that was conjugated in the future tense. The verb *promettre* (promise) did not appear and the grammatical subject of the sentence was the person making the promise. The social act intentionally posed by the speaker was a commitment, but not a firm one (e.g., *'Je laverai mon vélo'*, 'I'll wash my bike').
- *Predictive-assertion* statements employed a verb that was in the passive voice and future tense. The verb *promettre* (promise) did not appear and the grammatical subject of the sentence was not the person making the promise. In this case, there was no commitment on the part of the speaker (e.g., *'Mon vélo sera lavé'*, 'My bike will be washed').

Picture 3: Promise fulfilment. The picture showed Bill accomplishing the action corresponding to the propositional content of the commissive statement. The caption described the fulfilment of the promise made in the second frame of the story. The sincerity condition was satisfied in all stories.

Pictures 4: End of story. Two different pictures were constructed for Frame 4, each depicting a possible ending to the story. In one, the listener was shown with a clearly contented expression on his/her face and the caption described him/her as *happy* (Picture 4.1). In the other, the listener was shown with a clearly discontented expression on his/her face and the caption described him/her as *unhappy* (Picture 4.2). These two endings reflected the listener's reactions to the fulfilment of the promise, depending on his/her desire for the promise to be kept or not kept.

Procedure. The children were tested individually using a story completion task. The experimenter told the beginning of the story, that is, the first three frames. The child was to complete the story by choosing one of the two pictures proposed for Frame 4 (happy or unhappy listener).

A total of nine stories were presented to each child (the preparatory condition being an intergroup variable): three stories with a promise-to-act statement (PA), three with a future-action statement (FAC), and three with a predictive-assertion statement (PAS).

Four story-presentation orders were used, each of which was randomly assigned to three children. The presentation order of the two endings was also varied randomly across stories.

Experimental design. The experimental design included three independent variables: (i) subject age (3, 6, 10; independent samples); (ii) preparatory condition (PC+: satisfied; PC: not-satisfied; independent samples); and (iii) linguistic form of the commissive statement (PA: promise-to-act; FAC: future-action; PAS: predictive-assertion; related samples).

Coding. The procedure was designed to determine the extent to which children distinguish statements that are promises (defined by speech act theory) from ones which are not. Following the speech act theory proposal (Searle and Vanderveken, 1985; Vanderveken, 1990a; 1990b), two types of right answers are defined, depending on the satisfaction/non-satisfaction of the preparatory condition and the happiness/unhappiness of the listener at the end of the story.

The first type of right answer occurs when the preparatory condition is met and the children appropriately choose the picture of the *happy* listener. In the sense proposed by speech act theory, this is a promise. The second type of right answer occurs when the preparatory condition is not met and appropriate choice is the picture of the *unhappy* listener. In the sense proposed by speech act theory, this is not a promise.

2.2 Results: correct answers

For each subject, the rate of correct answers was obtained by taking the ratio of the number of right answers to the total number of responses (9). This ratio was multiplied by 100 and then treated in a 3-factor analysis of variance with the following design: Age (3) x Preparatory Condition (2) x Linguistic form of the promise (3). Figure 10.1 indicates the mean per-subject right answer rate as a function of the three factors.

The analysis yielded a significant effect of age ($F(2, 66) = 7.71$, $p < 0.001$), preparatory condition ($F(1, 66) = 28.64$, $p < 0.0001$), and linguistic form of the promise ($F(2, 132) = 7.84$, $p < 0.001$), and an interaction between the preparatory condition and the linguistic form ($F(2, 132) = 7.19$, $p < 0.001$). Below we summarize our results:

1. Three-year-olds and 6-year-olds give fewer right answers (61.9 per cent and 70.7 per cent respectively) than 10-year-olds (91.6 per cent).
2. Children give a greater number of right answers when the preparatory condition is met (91.6 per cent) than when it is not met (57.9 per cent). Satisfaction of the preparatory condition appears to promote correct responding, regardless of age. In other words, children seem to have difficulty functioning in a context that is not prototypical of a promise situation; this appears to be especially the case at the ages of 3 and 6.
3. The linguistic form of the commissive statement has no effect when the preparatory condition is met. In contrast, when it is not met, the promise-to-act form results in fewer right answers (PA = 47.1 per cent, FAC = 66.6 per cent, PAS = 60.1 per cent). A strong contradiction between the linguistic form of the statement (Promise-to-act) and the statement in the Preparatory Condition Not Satisfied (PC−) seems to induce the highest number of wrong answers. Indeed, in promise-to-act statements, the speaker's intentions are explicitly expressed as a firm commitment, at the same time as the non-satisfaction of the preparatory condition generates

Three-year-olds' Mean RA

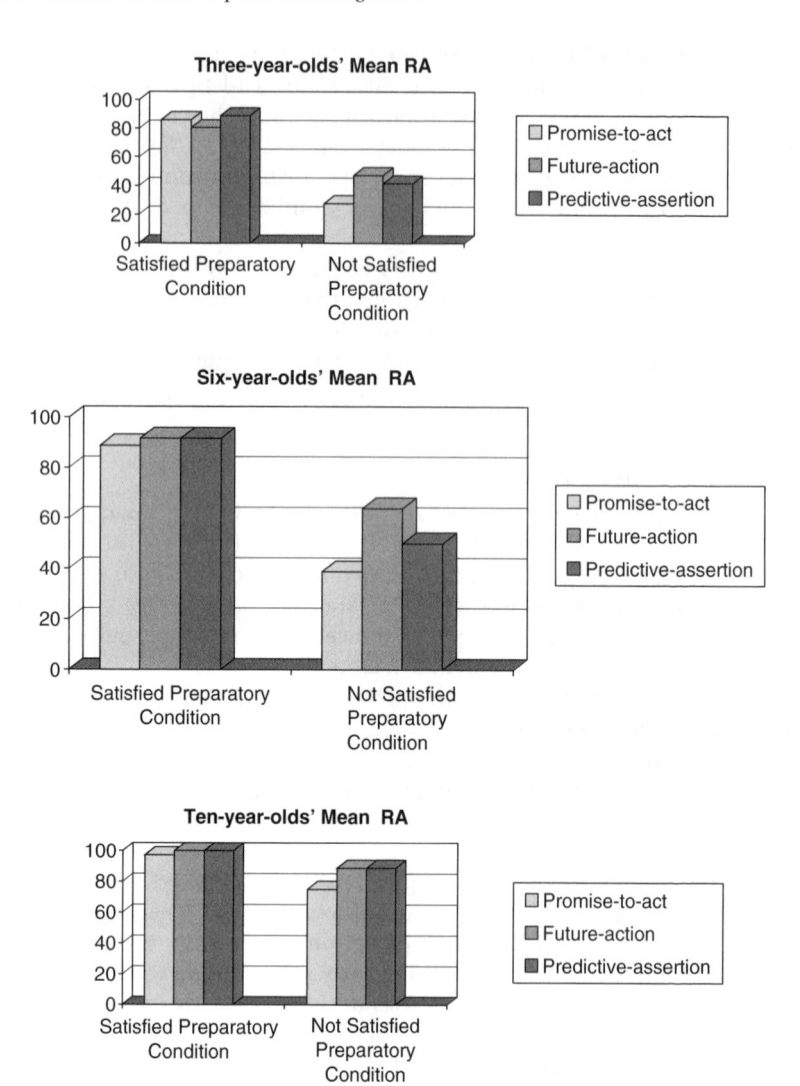

Six-year-olds' Mean RA

Ten-year-olds' Mean RA

Figure 10.1 Mean per-subject percentage of right answers (RA), by subject age (3, 6 and 10), preparatory condition (satisfied, not-satisfied), and linguistic form of the promise (Promise-to-act, Future-action, and Predictive-assertion)

a context in which the listener does not want the speaker to accomplish the promised action. In other words, the listener's desires (contextual cue) radically oppose the speaker's intentions (linguistic cue). The large number of 'The listener is happy' responses observed here shows that in cases of strong conflict between contextual cues and linguistic cues, children tend to base their interpretation on linguistic cues, that is, on the promise-to-act statement.

2.3 Discussion of Experiment 1

This study deals with the comprehension of promises by children. Two factors contributed to generating the experimental 'promises': the communication situation in which the statement was made and a verbal statement. The preparatory condition is a cue used to comprehend promises by children as early as age 3. When the preparatory condition was satisfied, that is to say in prototypical situations, understanding of the promise statement was facilitated for the three ages: 3, 6 and 10. This result extends Astington's (1988b, 1990) data for children between the ages of 5 and 9, where action accomplishment, an essential component of the sincerity condition, was found to be critical to success on the task. Thus, sincerity condition cues are understood before preparatory condition cues. In line with Bruner (1983) and other authors (Bernicot, 1994; Marcos and Bernicot, 1994, 1997), comprehension of prototypical situations with interaction formats, that is, situations in which the preparatory condition is met, was found to be superior to comprehension of non-prototypical situations. Prototypical situations continue to facilitate the correct interpretation of promise statements until the age of 10.

The linguistic form of the statement (promise-to-act, future-action and predictive-assertion) appears to play a minor role in children's comprehension of promises. However, our results point out that promise-to-act statements, which explicitly contain the verb *promise*, are generally not interpreted any better than future-action (future tense, active voice) and predictive-assertion statements (future tense, passive voice). For children between the ages of 3 and 10, future-action and predictive-assertion statements are just as specific to promising as statements containing the verb *promise* itself. It is therefore not necessary to systematically use this verb to test promise comprehension (for a similar conclusion, see Astington, 1988b; and Chapter 3).

The interaction between the situation in which the statement was made and the verbal statement revealed that a large number of wrong answers arise when the preparatory condition is not met. When the contextual cues and the linguistic cues are highly contradictory, children tend to base their interpretation on linguistic information.

3 Experiment 2

Although to our knowledge there are no studies dealing specifically with the comprehension of temporal markers in promise utterances, studies on the comprehension of the future have a few interesting points to offer. In French as in other languages, future markers are used to express the aspect of an action (e.g., the desiderative future) and to situate it in time. Several studies (e.g., Bronckart, 1976) have shown that until the age of 6, children rely primarily on adverbs and time prepositions to place an action in time; it is not until after that age that they begin to use verb tense. For promises in particular, we are only interested here in the future tense as a temporal marker: it specifies that the action described in the propositional content of the utterance will take place at some time after the utterance is produced. These studies have shown that for both English-speaking children (Harner, 1981a, 1981b) and Spanish-speaking children (Van Naerssen, 1979, 1980), the immediate future is understood earlier (at about the age of 3 or 4) than the simple future. They confirm the results on the production of time and aspect obtained by Bronckart (1976), who showed that until the age of 6, children are better at using time adverbs than verb tense to locate an action in time. Note, however, that none of these studies used a task involving a communication situation.

In the light of the above findings, three major objectives were set up for Experiment 2. Our first goal was to determine the role of the future tense in the comprehension of promises. In accordance with Searle's (1979) analysis, only linguistic forms that express a future action would be considered to be specific to promises. We are interested here in two temporal markers of the future which unambiguously express the future in French without adverbs or time prepositions: the immediate future and the simple future (Fleischman, 1982). These two tenses are not equivalent: the degree of certainty about whether the upcoming action will be accomplished is higher with the immediate future than it is with the simple future. Accordingly, if the future (immediate and/or simple) is the tense specific to promises, then understanding an utterance expressing a promise means being able to process textual markers that place actions in the future, that is, linguistic forms that indicate verb tense. In this perspective, it was hypothesized that the comprehension of promises by children would vary with the temporal characteristics of the utterance, and that the use of a future tense would promote promise comprehension. In addition, in line with the results of studies on future-tense comprehension by children (see Harner, 1981a, 1981b), it was predicted that the immediate future (*'je vais te donner la pelle'*, 'I am going to give you the shovel') would facilitate promise comprehension more than the simple future (*'je te donnerai la pelle'*, 'I will give you the shovel'), especially for the youngest children.

Our second goal was to determine the role of the preparatory condition as one of the contextual parameters of promise comprehension. The study by Bernicot and Laval (1996) pointed out that children under 10 have trouble taking the preparatory condition into account (when this condition was not satisfied). This finding was obtained by comparing the comprehension of promise utterances in communication situations where the preparatory condition was satisfied (the listener wanted the speaker to accomplish the promised action), with situations where the preparatory condition was contravened (the listener did not want the speaker to accomplish the promised action). It was very difficult, particularly for the youngest children, to interpret the subjects' responses in this study because the utterance and the preparatory condition were radically opposed. In order to better understand these results, here we propose considering utterance production contexts involving a lesser degree of variation in the preparatory condition. To this end, contexts in which the preparatory condition was fulfilled were compared to 'neutral' situations where the preparatory condition was neither explicitly fulfilled nor explicitly unfulfilled. It was hypothesized that, in this case, even the youngest children would take the preparatory condition into account, and that explicit fulfilment of the preparatory condition would promote interpretation as a promise, while the neutral condition would promote interpretation as a non-promise.

Our third objective was to determine the potential links between the text and the context in language functioning, particularly during language acquisition. How does the future tense promote promise comprehension in children? To what extent do promises promote the comprehension of future tense markers? Given the importance of interaction formats and the results already obtained on the impact of context on request comprehension in young children (Bernicot, 1991), it was predicted here that fulfillment of the preparatory condition would facilitate the comprehension of future tense markers by the youngest children, and that future tense markers would promote the comprehension of promises in the oldest children.

3.1 Method

Subjects. Fifty-four native French-speaking children participated in the experiment (26 girls and 28 boys). They were divided into three groups of 18. The mean ages of the three groups were 3 years 4 months (range: 2; 11 to 3; 10); 6 years 3 months (range: 5; 11 to 6; 10 months); and 9 years 4 months (range: 8; 11 to 9; 10). Hereafter, these three groups will be called the 3-year-old group, the 6-year-old group, and the 9-year-old group.

Materials. Eighteen stories about the adventures of a character were devised. In all 18, a little boy named Bill was speaking to a same-age peer named Loulou (a nickname for a boy in French). Each story was composed of six pictures (10 × 10cm) with short captions. The pictures, which provided a situational context for the linguistic expressions, showed real-life situations taken from children's everyday experiences, and helped keep the subject's attention focused on the task. Two examples are given in Table 10.2. Each story had four pictures:

Picture 1. The caption stated the general theme of the story, presented the two interlocutors (Bill and Loulou), and stated that the listener (Loulou) had

Table 10.2 Examples of stories from Experiment 2

Condition	Preparatory Condition Context with Immediate Future	Neutral Context with Simple Future
Picture 1: preparatory condition (PC) context	Bill and Loulou are on vacation at the seaside. Building sand castles is one of their favourite activities. Since the beginning of the vacation, Bill and Loulou have had only one shovel. Loulou is building a castle with the shovel.	It's a nice day. After school, Bill and Loulou are going to play on the swings. Loulou is on the swings. The dog is rolling over in the grass next to the swings.
Picture 2: (PC) continued	Now Bill has the shovel. But Loulou really needs it to finish his castle or else it will fall down.	Now Bill is on the swings. Loulou is playing with the dog. He's throwing a stick for the dog to fetch.
Picture 3: utterance	Bill says to Loulou: 'I'm going to give you the shovel.'	Bill says to Loulou: 'I will give you the swings.'
Picture 4: End of Story	**Picture 4.1** Bill gives the shovel to Loulou.	**Picture 4.2** Bill is still playing with the shovel.	**Picture 4.3** Bill and Loulou are playing in the water.	**Picture 4.1** Bill gives the swings to Loulou	**Picture 4.2** Bill is still playing with the swings.	**Picture 4.3** Bill and Loulou are playing with the dog.

the focal object. The corresponding picture showed the two characters in the story setting, and made it very clear that the listener had the focal object.

Picture 2. The caption emphasized the fact that the situation had changed since Picture 1, because the speaker now had the focal object (e.g., a bicycle). Two contexts were manipulated in the caption: a preparatory condition context and a neutral context. In the former, the preparatory condition was fulfilled. In other words, the listener's desire was made plain in the caption: the listener obviously wanted the speaker to accomplish the action described in the propositional content of the utterance. The corresponding picture showed the speaker and listener together, with the speaker in possession of the bicycle and the listener wishing he had it (depicted as a bubble with a drawing of the focal object inside). In the neutral context, the listener's desire about the accomplishment of the action described in the propositional content of the utterance was not clear: nothing was said about whether or not the listener wanted the promised action to be accomplished. The corresponding picture showed the speaker and listener together, with the speaker in possession of the focal object and the listener doing something else.

Picture 3. This picture showed the speaker in the foreground talking to the listener. The caption was used to manipulate the utterance produced by the speaker. The verb in the utterance was in one of three tenses:

- Immediate future (structure: I + am going + infinitive verb + you + direct object, as in *'je vais te donner le vélo'*, 'I am going to give you the bike').
- Simple future (structure: 'I + verb in simple future + you + direct object', as in *'je te donnerai le vélo'*, 'I will give you the bike').
- *Passé composé*, hereafter simply called the past tense (structure: 'I + verb in past tense + you + direct object', as in *'je t'ai donné le vélo'*, 'I gave you the bike').

We included the past tense because the most direct way for us to determine the role of the future in children's comprehension of promises was to oppose utterances in the future to utterances in the past. In other words, the past tense was used here as a control for the verb-tense variable.

Picture 4. Three different pictures, each corresponding to a different ending, were proposed for Picture 4. The subject had to complete the story by choosing one of the three. Picture 4.1 depicts the speaker giving the focal object to the listener. For utterances in the future tense, this picture corresponded to the accomplishment of the propositional content of the utterance, and the selection of this picture was indicative of textual and/or contextual processing based on the experimental conditions. Picture 4.2 showed the speaker keeping the object for himself and the listener in the background. For utterances in the future tense, this picture corresponded to the non-accomplishment of the propositional content. For utterances in the past, the selection of this picture was justified by the fact that the listener had the focal object in Picture 1, giving the speaker every right to keep it for himself. This choice was indicative of textual and/or contextual processing based on the experimental conditions. Picture 4.3 showed the two interlocutors together, with one of the elements of the setting. The element in question had nothing to do with the propositional content of the utterance. In all experimental conditions, the choice of this picture meant that the text had not been processed, and therefore, that processing was purely contextual.

Procedure. Each child performed the story-completion task individually. The experimenter first made sure that the child could distinguish the two characters in the story (Bill and Loulou). The procedure was as follows: the experimenter told the beginning of the story by reading the captions of the first three pictures, and then asked the child to complete the story by choosing a picture from among the three proposed. The children's answers were written down by the experimenter. Each child saw all 18 stories. The story presentation order was varied randomly across subjects. The six presentation orders for the three possible choices were randomly assigned to six children.

Experimental design. There were three independent variables in the experimental design: three levels of age (3, 6 and 9), a between-group variable; two types of utterance production context (preparatory condition, neutral), a within-group variable; and three verb tenses (immediate future, simple future and past), a within-group variable.

Coding. The subjects' answers (pictures chosen) were labelled according to the textual characteristics of the utterance (immediate future, simple future, past). This way of coding the results thus reflected the relationship between the utterance and the accomplishment or non-accomplishment of its propositional content. More precisely, if the speaker agreed to execute the act in the future, the logical ending to the story was accomplishment of the action: this was the case of utterances in the future tense. On the other hand, if the speaker stated that he had already accomplished the action, the logical ending was the non-accomplishment of the action: this was the case for utterances in the past tense.

This defined three response categories (see Table 10.3):

1. Theoretically *right answers* (RA), which included all choices where the action was accomplished when the utterance was in the future tense, and all choices where the action was not accomplished when the utterance was in the past tense (e.g., the right answer for utterances in examples of Table 2 was Picture 4.1).

Table 10.3 Coding of children's answers (Experiment 2)

	Immediate Future or Simple Future	Past
Right Answers (RA)	Choice 1	Choice 2
	Action accomplished	Action not accomplished
Wrong Answers	Choice 2	Choice 1
(WA)	Action not accomplished	Action accomplished
Contextual Answers	Choice 3	Choice 3
(CA)	Interlocutors with one of the elements of the setting	Interlocutors with one of the elements of the setting

2. *Wrong answers* (WA), which included all choices where the action was accomplished when the utterance was in the past tense, and all choices where the action was not accomplished when the utterance was in the future tense (e.g., the wrong answer for utterance in examples of Table 2 was Picture 4.2).
3. *Contextual answers* (CA), which included all choices of the picture showing the two interlocutors with one element of the setting. Choosing this 'neutral' answer was a way of avoiding textual processing and was a direct proof of purely contextual processing. There were two main reasons for including the contextual answer category. The first was methodological, the idea being to increase the number of choices proposed to the subjects. The second was more theoretical and was aimed at finding out whether the youngest children would do essentially contextual processing rather than textual processing (e.g., the contextual answer for utterances in examples of Table 10.2 was Picture 4.3).

Table 10.3 gives the different types of answers for each category of the verb-tense variable.

3.2 Results: right answers

Given the aims of this chapter, only the results for the right answer are presented here. The dependent variable 'number of right answers' was examined using an analysis of variance with three factors: subject age (3) x type of context (2) x verb tense (3). Figure 10.2 shows the mean percentage of right answers, by age, utterance production context and verb tense. The analysis yielded a significant effect of age ($F(2, 51) = 271.26$, $p < 0.0005$), an interaction between age and utterance production context ($F(2, 51) = 15.44$, $p < 0.0005$), and an interaction between age and verb tense ($F(4, 102) = 59.11$, $p < 0.0005$). The main findings can be described as follows.

The 3-year-olds gave fewer right answers (11.72 per cent) than the 6-year-olds (52.46 per cent), who in turn produced fewer right answers than the 9-year-olds (97.22 per cent). The 3- and 6-year-olds gave more right answers in the

222 *Current Issues in Experimental Pragmatics*

Three-year-olds' mean rate of RA

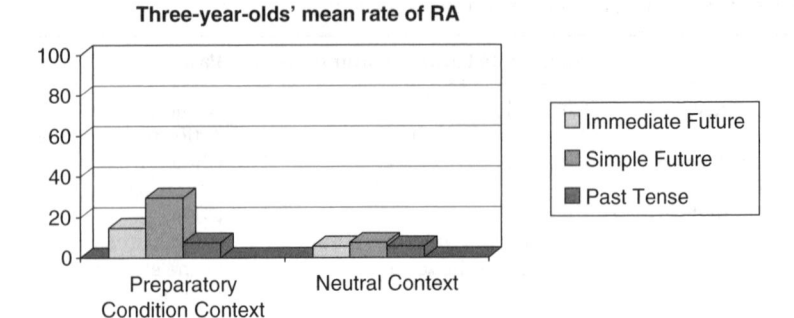

Six-year-olds' mean rate of RA

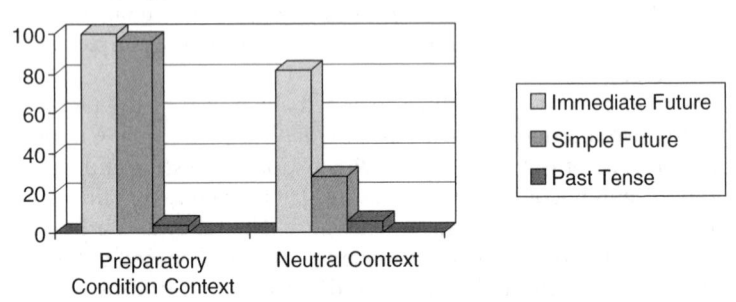

Nine-year-olds' mean rate of RA

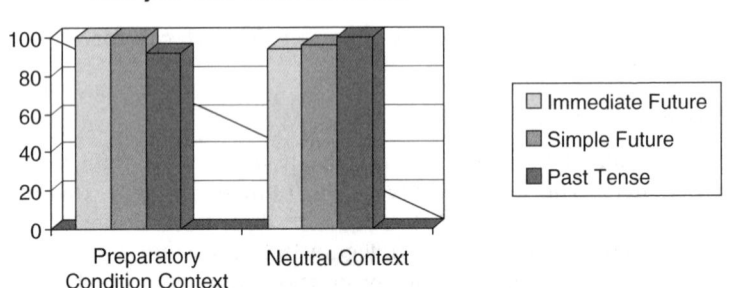

Figure 10.2 Mean percentage of right answers (RA), by age (3, 6 and 9), utterance production context (preparatory condition vs neutral), and verb tense (immediate future, simple future and past)

preparatory condition context (17.28 per cent and 66.66 per cent, respectively) than in the neutral context (6.17 per cent and 38.27 per cent, respectively). This difference did not exist at age-9 (preparatory condition context: 97.53 per cent; neutral context: 96.91 per cent).

At the age of 6, the immediate future led to a greater number of right answers (90.74 per cent) than the simple future (62.03 per cent) ($F(1, 17) = 35.67$, $p < 0.0005$), which in turn triggered more right answers than the past tense (4.62 per cent). For the 9-year-olds, the correct answer rate was nearly 100 per cent for all verb tenses (immediate future: 97.22 per cent, simple future: 98.14 per cent, past tense: 96.29 per cent).

3.3 Discussion of Experiment 2

The results obtained here validate and further refine our hypotheses on the role of the future in the comprehension of promises, both regarding the early processing of the preparatory condition and the link between text and context. As a whole, the 3-year-olds' performance was poor (approximately 12 per cent of their answers were correct). The results for the 3-year-olds point out the importance of the utterance production context in the comprehension of promises. Note that these children were presented situations in which the 'preparatory condition' context was clearly satisfied as well as 'neutral' contexts, which were neither explicitly satisfied nor explicitly contravened. The fact that the children gave more right answers in the preparatory condition context than in the neutral one (along with the very low scores obtained in that context) validates the results already obtained by Bernicot and Laval (1996), and suggests that 3-year-olds have not yet mastered the preparatory condition. This result also confirms the facilitating role of the preparatory condition in the comprehension of promise utterances and, in agreement with other authors (Bruner, 1983; Bernicot, 1994; Marcos and Bernicot, 1994, 1997), points out the importance of prototypical situations for young children.

For the 6-year-olds, the overall performance level was about 53 per cent. Performance varied across contexts and tenses. In the preparatory condition context, the best scores were obtained for utterances in the immediate or simple future. In the neutral context, the immediate future gave rise to higher scores than did the other two tenses. In line with our hypotheses based on Searle's (1979) theory, the future was found to favour the interpretation of the utterances as promises. This result is particularly important for the immediate future in the neutral context, because it suggests that at the young age of 6, children can base their interpretation of utterances on verb tense when contextual cues are lacking. In contrast, 6-year-olds do not yet appear to be capable of using simple future cues in contexts that are not specific to promises. This finding was reinforced by the fact that contextual answers were numerous for utterances in the simple future. An analogous result was obtained for past-tense utterances in the neutral context. The results concerning the earlier acquisition of the immediate future compared to the simple future are compatible with past work.

As a whole for 6-year-olds and 3-year-olds alike, right answers in the preparatory condition context outnumbered those in the neutral context, where the scores were particularly low for both the simple future and the past. These results confirm those obtained by Bernicot and Laval (1996), who compared situations that either obviously fulfilled or obviously contravened the preparatory condition. We can therefore assume that 6-year-olds do not fully master the preparatory condition and acquire it after the sincerity condition. Relating language to a theory of mind (Wellman, 1990), we can conclude that during promise comprehension, children take the speaker's intentions into account before considering the listener's desires. Prototypical situations favour the comprehension of promise utterances in 6-year-olds. The importance of context to utterance comprehension is particularly well illustrated by the difference observed between future-tense utterances and past-tense utterances: when the context was specific, utterances in the past tense were interpreted as promises (in this situation, the percentage of wrong answers was as high as 94.44 per cent). We can regard this finding as an indication that verb tense was completely ignored. In a communication situation where textual cues (past tense) and contextual cues (preparatory condition fulfilled) are contradictory, 6-year-olds consider the production context first. The results obtained here for context with the 3- and 6-year-olds are in line with Fayol, Hickmann, Bonnotte and Gombert's (1993) findings showing that the production of temporal markers by native French speakers (adults and 10-year-olds) is highly dependent upon the narrative context.

At the age of 9, the children's answers were nearly 100 per cent correct. The lack of a variation across tenses and contexts shows that the children systematically based their interpretation on tense markers in the utterance. This result is particularly interesting for the neutral context, since it clearly demonstrates that when contextual cues are lacking, the interpretation of a promise utterance is based on future tense markers, whether it be the immediate future or the simple future. In other words, it is through the processing of future tense markers that utterances are interpreted as promises.

4 Conclusions

The two experiments presented in this chapter highlight the importance of the promise fulfilment preparatory condition in the comprehension of promises: prototypical situations whose preparatory condition is satisfied facilitate the comprehension of promise utterances for the 3-year-olds and the 6-year-olds. This reinforces the idea that context is very important to explain language acquisition (Bates, 1976; Bruner, 1983; Ervin-Tripp and Mitchell-Kernan, 1977; Halliday, 1985 ; Laval and Bernicot, 1999; Ninio and Snow, 1996; Tomasello, 2000). Thus, for promise comprehension tested by means of non-verbal behaviour, it was shown here that in addition to considering the sincerity condition, mastered from the age of 5 (Astington, 1988b), we had to consider the preparatory condition mastered about the age of 9 or 10 (Laval and Bernicot, 1999). Speaker's beliefs and listener's desires are two important elements for the children's comprehension of promises.

The second experiment investigated the precise role of linguistic form in the promise-making statement by comparing statements with verbs in the future tense to statements with other verb forms (see Laval and Bernicot, 1999). It thus appears that context can orient or favour the processing of the textual characteristics of utterances in children aged 3 and 6. It also appears that textual-cue processing can lead to contextual-cue processing. This tendency starts emerging at the age of 6 and becomes general by the age of 9. For 6-year-olds, whether or not the context is processed on the basis of the text depends on the features of the communication situation. In other words, when promise-specific contextual information is lacking, these children correctly process the immediate future but not the simple future and reconstruct the promise from those markers. These results validate the hypothesis that there is a tight link between textual and contextual characteristics during language acquisition and language functioning.

In everyday situations, what is it that 'counts as' a promise? We answer with Mey (2001): 'all depends on the circumstances'. In some cases, we pay attention to the people who promise, rather than to their exact words. While in other contexts, we focus on the social frame in which the promise is given. From a developmental point of view, the youngest children give priority to the social frame when they interpret a 'promise' utterance. By the time they are 9, they begin to give a priority to the linguistic features of the utterance.

References

Astington, J. W. (1988a). Children's production of commissive speech acts. *Journal of Child Language* 15: 411–23.
Astington, J. W. (1988b). Children's understanding of the speech act of promising, *Journal of Child Language* 15: 157–73.
Astington, J. W. (1990). Metapragmatics: Children's conception of promising. In G. Conti-Ramsden and C. Snow (eds), *Children's Language*. Hillsdale, NJ: Lawrence Erlbaum Associates.
Austin, J. L. (1962). *How To Do Things With Words*. Cambridge, MA: Harvard University Press.
Bates, E. (1976). *Language and Context: The Acquisition of Pragmatics*. New York: Academic Press.
Bernicot, J. (1991). French children's conception of requesting: The development of metapragmatic knowledge. *International Journal of Behavioral Development* 14: 285-304.
Bernicot, J. (1994). Speech acts in young children: Vygotsky's contribution. *European Journal of Psychology of Education* 9: 311–19.

Bernicot, J., Comeau, J., and Feider, H. (1994). Dialogues between French-speaking mothers and daughters in two cultures: France and Quebec. *Discourse Processes* 18: 19–34.

Bernicot, J., and Laval, V. (1996). Promises in French children: Comprehension and metapragmatic knowledge. *Journal of Pragmatics* 25: 101–22.

Bronckart, J. P. (1976). *Genèse et organisation des formes verbales chez l'enfant*. Bruxelles: Dessart et Mardaga.

Bruner, J. S. (1983). *Child's Talk: Learning to Use Language*. New York: Norton.

Duranti, A. (1996). *Linguistics Anthropology*. Cambridge: Cambridge University Press.

Ervin-Tripp, S., and Mitchell-Kernan, C. (1977). *Child Discourse*. New-York: Academic Press.

Fayol, M., Hickmann, M., Bonnotte, I., and Gombert, J. E. (1993). The effects of narrative context on French verbal inflections: A developmental perspective. *Journal of Psycholinguistic Research* 22: 453–78.

Fleischman, S. (1982). *The Future in Thought and Language: Diachronic Evidence from Romance*. Cambridge: Cambridge University Press.

Gibbs, R. W., and Delaney, S. M. (1987). Pragmatic factors in making and understanding promises, *Discourse Processes* 10: 107–26.

Halliday, M. A. K. (1985). *An Introduction to Functionnal Grammar*. London: Arnold.

Harner, L. (1981a). Children talk about time and aspect of actions. *Child Development* 52: 498–506.

Harner, L. (1981b). Immediacy and certainty: Factors in understanding future reference. *Journal of Child Language* 9: 115–24.

Keating, E. (1998). *Power Sharing: Rank, Gender, and Social Space in Pohmpei, Micronesia*. Oxford: Oxford University Press.

Laval, V., and Bernicot, J. (1999). How French-speaking children understand promises: The role of future tense, *Journal of Psycholinguistic Research* 28: 179–95.

Levinson, S. C. (1983). *Pragmatics*. Cambridge: Cambridge University Press.

Maas, F., and Abbeduto, L. (1998). Young chidren's understanding of promising: Methodological considerations. *Journal of Child Language* 25: 203–14.

Maas, F., and Abbeduto, L. (2001). Chidren's judgement about intentionally and unintentionally broken promises. *Journal of Child Language* 28: 517–29.

Marcos, H., and Bernicot, J. (1994). Addressee co-operation and request reformulation in young children. *Journal of Child Language* 21: 677–92.

Marcos, H., and Bernicot, J. (1997). How do young children reformulate assertions? A comparison with requests. *Journal of Pragmatics* 27: 781–98.

Mey, J. (2001). *Pragmatics*. Oxford: Blackwell.

Ninio, A., and Snow, C. E. (1996). *Pragmatic Development*. Colorado: Westview Press.

Searle, J. R. (1969). *Speech Acts*. Cambridge: Cambridge University Press.

Searle, J. R. (1979). *Expression and Meaning*. Cambridge: Cambridge University Press.

Searle, J. R., and Vanderveken, D. (1985). *Foundations of Illocutionary Logic*. Cambridge: Cambridge University Press.

Van Naerssen, M. M. (1979). *Proposed Acquisition Order of Grammatical Structures for Spanish as a First Language*. Unpublished Manuscript. Los Angeles: Department of Linguistics, University of Southern California.

Van Naerssen, M. M. (1980). *Ignoring the Reality of Future in Spanish*. Paper presented at the Los Angeles Second Language Research Forum, 1–2 March 1980.

Tomasello, M. (2000). The social-pragmatic theory of word learning. *Pragmatics* 10(4): 401–14.

Vanderveken, D. (1990a). *Meaning and Speech Act: Principles of Language Use*, vol. I. Cambridge: Cambridge University Press.

Vanderveken, D. (1990b). *Meaning and Speech Act: Formal Semantics of Success and Satisfaction*, vol. II. Cambridge: Cambridge University Press.

Verschueren, J. (1999). *Understanding Pragmatics*. London: Arnold.

Wellman, H. M. (1990). *The Child's Theory of Mind*. Cambridge, MA: Bradford Books and MIT Press.

11
Reasoning and Pragmatics: the Case of *Even-If*

Simon J. Handley and Aidan Feeney

1 Introduction

Researchers interested in the psychology of reasoning often regard pragmatics as being somehow less worthy of interest than 'actual reasoning'. Pragmatic factors are often regarded as extraneous variables that interfere with people's ability to compute logical inferences. Another view is that there are separate associative and symbol-manipulating systems for thinking (Evans and Over, 1996; Stanovich, 1999; Sloman, 1996). Under this view, the symbol-manipulating or logical system is often seen as more interesting or more characteristic of higher forms of thought than is the associative or pragmatic system. Although there are notable exceptions to these views (see Sperber, Cara and Girotto, 1995; Hilton, 1995), their preponderance is understandable given the way in which research on the psychology of reasoning has developed since the 1960s when Peter Wason first demonstrated the influence of extra-logical factors on people's thinking. Although he was motivated in this by his disagreement with Piagetian views about thinking, Wason's work had a consequence that he could not have foreseen. By saying that thinking was NOT logical, Wason helped to shape the development of paradigms in the field where performance was measured against the yardstick of logically correct performance.

More recently, an interest in pragmatics has started to emerge in work on reasoning. For example, Rumain, Connell and Braine (1983), Byrne (1989) and others have thrown light on the invited inferences underlying people's performance on conditional reasoning tasks. Similarly, recent work by Barrouillet, Grosset and Lecas (2000) has examined how the ability to resist these pragmatically invited inferences develops. Research has also begun to examine reasoning from everyday conditional assertions and the influence that people's knowledge of the speaker has on the inferences drawn (see, for example, Evans and Twyman-Musgrove, 1998).

Work on logical quantifiers too is very interesting in pragmatic terms. Language researchers (Moxey and Sanford. 1993) have pointed out that the meaning of linguistic quantifiers is, to a large extent, pragmatically determined. Experimental work on syllogistic reasoning (Newstead, 1995) has shown that whilst people do make Gricean errors in their interpretation of logical quantifiers, these errors do not play a large role in deductive reasoning tasks involving quantified premises. Most recently, Noveck (2001) has neatly demonstrated how the tendency to make Gricean errors in the interpretation of logical quantifiers develops over time. The finding that children are more logical in their interpretation of quantifiers than are adults nicely turns most reasoning researchers' conception of the relationship between logic and pragmatics on its head, with pragmatic competence shown to succeed logical competence.

Our interests lie squarely in the study of pragmatic inference. To that end, in this chapter we will describe two experiments that examine the inferences that people draw from everyday conditional utterances and the communicative intentions that they ascribe to utterers of such statements. In so doing we hope to demonstrate how the methodologies and paradigms traditionally used by reasoning researchers, far from being confounded by pragmatic factors, can be used to shed light on wholly pragmatic inferences. Thus, we hope to make the case that these methods can be invaluable to practitioners of the new discipline of Experimental Pragmatics.

The plan of the chapter is as follows. We begin by reviewing the philosophical and linguistic literature pertaining to *even-if*, focusing initially on accounts of *even*. We then present two experiments that examined the inferences that people draw from *even-if* assertions containing arbitrary content (Experiment 1) and realistic content (Experiment 2) as a test of the intuitions derived from the linguistic literature. The findings are discussed in the light of the reviewed literature and recent empirical and theoretical work in the field of human reasoning. Finally, we discuss our methodological approach in a wider context relating to the fields of both human reasoning and Experimental Pragmatics.

2 The case of *even-if*

Any analysis of *even-if* must begin with an understanding of the function that *even* on its own serves in the language. Consider, for example, the following assertion:

(1) Even Hilary distrusts Bill.

There is a general consensus in the literature that *even* serves to pick out an extreme position, that is, 'is less probable, more surprising, contrary to expectation and so forth – in a contextually determined range of alternatives'

(Sanford, 1989, p. 206). This view appeals to the notion of a pragmatic probability scale where the focus of *even* is at the extreme position on the scale (Fauconnier, 1975). Consequently, the assertion cues the listener to access a range of neighbouring statements that are more probable and less surprising, for example that George distrusts Bill, that Tony distrusts Bill, that Monica distrusts Bill etc. In this way it acts to invite the listener to make the inference that Bill is not to be trusted.

Whilst most authors would agree that *even* serves to pick out an extreme position on a scale, with the concomitant activation of propositions that lie on a lower point of that scale, there is a fundamental disagreement in the literature regarding the meaning of the term. Some authors have argued that *even* acts as a universal quantifier and hence changes the truth conditions of a sentence (see, for example, Lycan, 1991). An alternative view is that *even* does not modify truth conditions. Instead the contribution that *even* makes to a sentence is one of conventional implicature, implicature that results in a sense of unexpectedness or surprise (see, for example, Bennett, 1982; Francescotti, 1995).

The quantifier view is most clearly expressed in William Lycan's work. Lycan (1991) claims that an *even* utterance is true in cases where all contextually relevant sentences that lie on the probability scale are also true. Referring back to the example in (1), under this view the sentence is true if and only if it is the case that everybody (who one might reasonably expect to) distrusts Bill, plus Hilary distrusts Bill. Hence the meaning of (1) can be re-phrased as:

(2) Everybody distrusts Bill, and that includes Hilary.

where the fact that Hilary distrusts Bill is presented as an instance of the universal claim that everybody does.

An alternative view is that *even* does not affect the truth conditions of a sentence. That is, the truth of (1) is evaluated against the truth of the constituent proposition, Hilary distrusts Bill. Under this view, most clearly expressed by Bennett and Francescotti, *even* makes a difference in conventional implicature in much the same way as other connectives such as 'but' and 'and' (Carston, 1993). According to Francescotti (1995) *even* represents its focus as unexpected or surprising, and indicates the speaker's belief in the truth of a range of less surprising, less unexpected propositions. The felicitousness of *even* depends upon the constituent proposition being more surprising than at least most of these neighbouring propositions. As such *even* serves to elicit expectation contravening effects. To illustrate with our example concerning Hilary and Bill, even Hilary distrusts Bill is felicitous only if Hilary distrusts Bill and there exist a range of other true sentences, George trusts Bill, Tony distrusts Bill, which are less surprising than Hilary distrusts Bill. This in turn leads to a sense of unexpectedness or surprise.

The distinction between the quantifier and the implicature accounts hinges to a great extent on the question of whether *even* picks out the most extreme point on the scale. If this is not the case, then the quantifier approach clearly fails, since counter-examples to the claim of universal quantification would then be apparent. This issue is discussed in some detail by both Lycan (1991) and Francescotti (1995) and we will return to this point in the discussion of our experimental findings. For now it is worth pointing out that both approaches can and have been extended to the analysis of *even-if* and it is to this construction that we now turn.

Whilst there exist exceptions, most authors have treated *even-if*, not as a distinct logical connective, but as a construction that consists of a combination of the focusing particle *even* and the *if* of a conditional. The analyses presented above can readily be extended to *even-if*. To illustrate imagine that a student makes the following assertion:

(3) Even if Pete studies hard he will fail the History exam.

As Jackson (1987) has argued the basic analyses of *even* outlined above can be readily applied to *even-if*. The assertion denies a commonly held presupposition that if one studies hard, then one will pass an exam.[1] That is, it denies a belief that:

(4) If Pete studies hard he will pass the History exam.

As we have seen *even* picks out an extreme possibility on a scale of related statements that are more probable in a given context. This provides a series of effects related to unexpectedness or surprise. Declerck and Reed (2001) have argued that *even-if* similarly induces a sense of unexpectedness. According to their argument this sense of unexpectedness relates to the conditional as a whole and leads to what they label an expectation understanding, that is, one might expect p to preclude q, and a non-preclusive understanding, p does not (in fact) preclude q. It also calls to mind a range of conditionals with alternative antecedents that serve to make the consequent more probable (i.e., where P(q/p) is higher). The focal conditional lies at the extreme point of a scale, consequently licensing the inference that the conditional relationship holds for all other values of the same scale and hence for a series of antecedents (Konig, 1986). According to Jackson (1987, see also Declerck and Reed, 2001) the range often consists of the conditional with the

---

[1] This account of *even-if* assumes, as does Adams (1975), that the truth of a conditional is its conditional probability. Under material implication 'If p then q' is not incompatible with the assertion 'If p then not-q', but of course under conditional probability the truth of one assertion precludes the truth of the other.

antecedent negated. Hence, given (4) above, from background knowledge people will access information concerning what they commonly understand about the relationship between studying and failing exams, namely:

(5) If Pete doesn't study then he will fail the History exam.

The combination of the representation of the assertion in (3) with the manifest assumption in (5) leads directly to the inference that Pete will fail the History exam, whether or not he studies hard. In formal terms this inference is what logicians would term *constructive dilemma* and it corresponds to the intuition that many *even-if* assertions appear to entail their consequent.

The distinction between the implicature and quantifier accounts in their approach to *even if*, whilst significant from a linguistic perspective, is subtle with regards to the nature of the representation that may be constructed by a listener. In essence the difference hinges on whether the truth of an *even-if* assertion is determined by the consequent holding across all possible antecedents, or simply by the observation that the stated consequent holds in the presence of the stated antecedent. Hence, under a quantifier account (3) is true if and only if:

(6) In all antecedent circumstances Pete will fail the exam, including the circumstance in which he studies hard.

In contrast, under the implicature account the truth of the assertion is guaranteed if there are no circumstances in which Pete studies hard and does not fail the exam. We will consider later whether our experimental evidence can differentiate between these views.

As well as licensing the inference that the consequent will occur, the use of *even-if* also blocks many of the invited inferences that are associated with the everyday understanding of indicative conditionals. For example, given the following conditional assertion:

(7) If Pete studies hard he will pass his exam.

we are invited to infer that the antecedent is not only sufficient, but also necessary for the consequent. That is we infer that if Pete doesn't study hard he will not pass his exam, and if he passes his exam he has studied hard (Geis and Zwicky, 1971). *Even-if* assertions in contrast do not license these inferences, as they cue listeners to assume that the conditional relationship holds for a whole series of antecedents. Hence, given the assertion in (3) we do not infer that if Pete does not study hard, he will not fail the exam. This characteristic of *even-if* leads to clear empirical predictions. It suggests that *even-if* assertions will not support the Affirmation of the Consequent and Denial of the Antcedent inferences that follow from these invited inferences and this is one of the predictions that we test in the studies that follow.

Before describing Experiment 1 it is important to define the scope of our investigations. In the literature, the notion that all *even-if* assertions entail their consequent has been questioned by a number of authors (see, for example, Bennett, 1982), who invariably invoke examples of the following kind to counter the claim:

(8) Even if you drink a little then you will be dismissed.

Clearly the inference that one will be dismissed no matter what is not supported by this assertion. In considering examples of this kind we concur with a number of authors (see for example, Konig, 1986), who argue that the key to determining whether an *even-if* assertion entails the consequent depends upon the focus of *even*. In (8) the focus of *even* is on one part of the antecedent (*even* a little) rather than the whole antecedent as in (3). In the experiments that follow we focus on indicative *even-if* conditionals of this kind, where the focus of *even* is on the whole antecedent clause. Whilst we believe that *even-if* assertions which do not entail their consequent are no less worthy of study than those that do, our key interest here is on the processes that may underlie inferences that the consequent follows no matter what. Hence, we limit our examination to cases of this kind.

3 Experiment 1

Before considering psychological approaches to *even-if*, we present two experiments that examined the inferences that people draw from *even-if* assertions containing relatively unfamiliar (Experiment 1) and realistic content (Experiment 2). These studies provide a test of some of the claims present in the linguistic literature and also provide data pertinent to the development of a psychological account of *even-if*. Our major predictions were simple, but nevertheless provide the first empirical test of the accounts present in the literature. In Experiment 1 we were interested in examining whether people would infer the consequent from *even-if* assertions that contained relatively unfamiliar content. Second we wished to examine whether, given arbitrary content, *even-if* assertions would cue participants to represent alternative antecedents and consequently resist drawing the invited necessity implicatures that are commonly associated with the pragmatics of conditional statements. To this end we presented one group of 30 participants with the following *even-if* assertions:

A Plumber is inspecting the pipe-work at a brewery. The Plumber says: 'Even-if the lever is pressed the water will flow'

An engineer is inspecting a factory that produces widgets for the automotive industry. The Engineer says:
'Even-if the light is on then the conveyor belt will move'

A second group of 30 participants was presented with conditional assertions that were based upon identical scenarios:

A Plumber is inspecting the pipe-work at a brewery. The Plumber says:
'If the lever is pressed the water will flow'

An engineer is inspecting a factory that produces widgets for the automotive industry.
The Engineer says:
'If the light is on then the conveyor belt will move'

The advantage of this approach is that any difference between the inferences drawn by each group must be a result of the pragmatic effects that the addition of *even* has on the representation of the assertion

Participants were asked to complete two tasks, the first of which required them to indicate what the speaker intended to convey by his assertion in relation to the antecedent and consequent clauses. The second task was a standard conditional inference task requiring participants to indicate what followed from the four conditional argument forms, *Modus Ponens* (MP), *Modus Tollens* (MT), *Affirmation of the Consequent* (AC) and *Denial of the Antcedent* (DA). The order of task presentation was counterbalanced within subjects. In the interpretation task participants were asked to rate on a 9-point scale the extent to which they agreed that the speaker intended to convey the following:

(a) That the lever is pressed
(b) That the lever is not pressed
(c) That the water will flow
(d) That the water will not flow

In the conditional inference task, participants were presented with the four conditional argument forms and were asked which of three outcomes was most likely. An example of the MP inference is given below:

Suppose you find out later that the lever is pressed. Given what the Plumber has said, which of the following outcomes is most likely?

(i)   The water is flowing
(ii)   The water is not flowing
(iii) The water may or may not be flowing

We will first present the data for the interpretation task. In all analyses the data was aggregated across the two contents as there were no meaningful differences between them. Considering the consequent data first, a 2 (connective) × 2 (q vs not-q) ANOVA revealed a highly significant interaction between the two factors (F(1, 58) = 10.54, MSE = 9.74, p < 0.01). The means involved in this interaction show that although *if-then* implies that the consequent will occur (5.8) more strongly than that the consequent will not occur (3.4), this effect was much more marked with the *even-if* connective (7.4 vs 2.3). This provides very clear evidence that *even-if* assertions invite the listener to infer that the consequent will occur.

A 2 (If vs even-if) × 2 (ratings for p vs not-p) ANOVA on the ratings for the antecedent clause revealed a main effect of connective (F(1, 58) = 8.47, MSE = 6.50, p < 0.01). Higher ratings were given for both p and not-p for the *even-if* assertions (4.2, 5.4) than for the assertions containing *if-then* (3.2, 4.5). There was also a main effect of antecedent polarity (F(1, 58) = 11.33, MSE = 8.83, p < 0.01). Participants gave higher ratings to not-p (4.9) than to p (3.61). No other effects in the analysis were significant. We made no specific predictions with regard to inferences concerning the antecedent and the findings here are difficult to interpret. On the one hand the main effect of the connective might suggest that *even-if* assertions are interpreted as conveying more information about the antecedent than *If* assertions. On the other hand the results may simply be due to a scaling effect. Perhaps a more intersting finding is that both connectives are interpreted as conveying not-p (i.e., that the light is not on, or the lever is not pressed) more so than p (i.e., that the light is on or the lever is pressed). Conditionals invite one to hypothetically suppose the antecedent condition holds and to imagine the consequent also occurring. Arguably it is only felicitous to assert a conditional if you are uncertain about the antecedent or you know that it has not occurred. If you know that the antecedent has occurred, then your belief in the conditional should lead you to assert the consequent rather than a less informative conditional.

Figure 11.1 shows the rates of inference from the four conditional arguments for both *even-if* and *if-then* connective. The graph shows a very clear biconditional pattern from the *if-then* connective, with high rates of both the valid and fallacious inferences. Rates of drawing MP from *even-if* are similar to *if-then*, but participants do not draw either DA, AC or MT. This provides powerful evidence that *even-if* blocks the invited inferences associated with the necessity implicature. However the pattern of inferences for *even-if* are much more complex than this initial analysis might suggest. *Even-if* does not simply block invited inferences. If this were the case, then we would expect similar rates of MT inferences for *even-if* as for *if-then*.

Figure 11.2 shows a breakdown of proportion of participants who affirmed each of the three response options for the *even-if* assertions. For DA, 1 corresponds to the response that q follows, 2 to the response that not-q

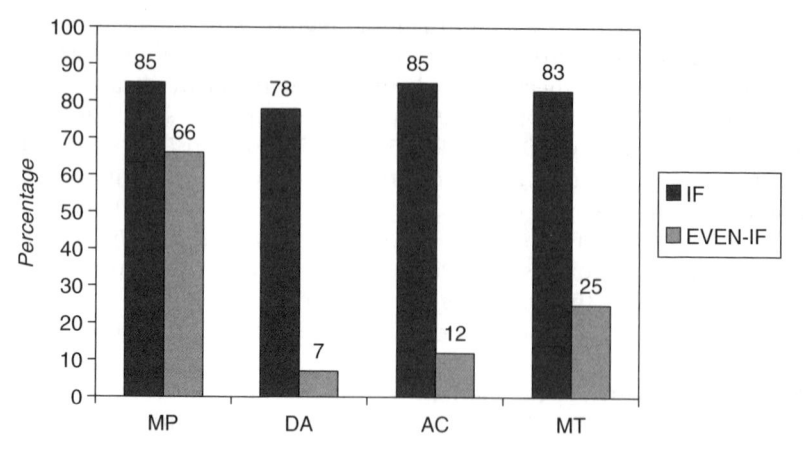

236 *Current Issues in Experimental Pragmatics*

Figure 11.1 The percentage of participants endorsing each of the conditional inferences in Experiment 1

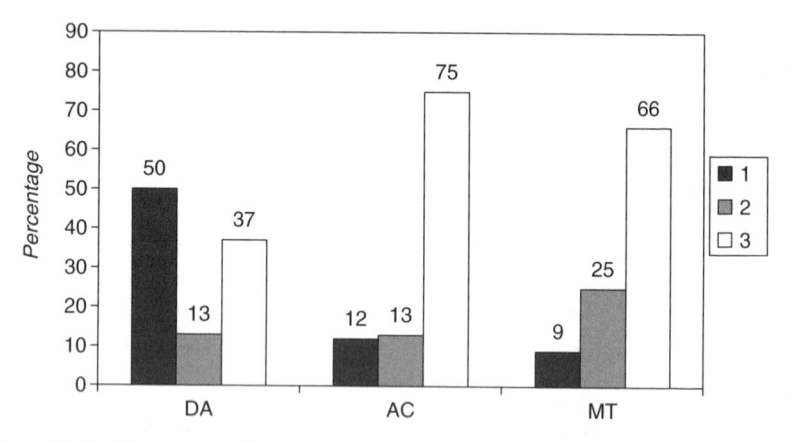

Figure 11.2 The pattern of responses given for the *even-if* assertion in Experiment 1

follows, whilst 3 corresponds to the response that it is not possible to tell what follows. For AC and MT, 1 corresponds to p, 2 to not-p and 3 to impossible to tell. As the graph clearly indicates, for DA the most common response (50 per cent) is to infer that the consequent occurs. That is, people infer that water will flow, or the conveyor belt will move, when told that the lever is not pressed or that the light is not on. Indeed participants are almost as willing to infer the consequent from the denial of the antecedent (50 per cent) as they are from its affirmation (MP – 66 per cent). This provides strong evidence that *even-if* cues people to infer that the consequent will occur no matter what, even in contexts relatively bereft of detail.

The pattern of findings for AC is clear; the majority of participants when told that the consequent occurs infer that the antecedent may or may not have occurred (75 per cent). This finding is consistent with the notion that *even-if* calls up alternative conditionals on the *even-if* scale, which specify antecedents where P(q/p) is higher than the conditional presented. In this case, given the paucity of information in the scenario, it is probable that the alternative consists of a conditional that specifies an antecedent corresponding to the negation of the antecedent in the *even-if* assertion. Hence, participants are likely to imagine two possibilities; one in which, for example, the lever is pressed and the water flows and one in which the lever is not pressed and water flows. Consequently, participants indicate that nothing can be inferred for definite from the knowledge that the antecedent occurred.

We turn now to the rates of inference from MT which are particularly interesting. As the immediate inference data shows, participants tend to infer that *even-if* implies that the consequent will occur. The categorical premise for MT denies this possibility. As Figure 1 shows, the majority of participants make no determinate inference in this case. Instead they indicate that the antecedent may or may not be true (66 per cent). Interestingly, a smaller proportion of participants made a converse inference; that is, from the denial of the consequent they infer the antecedent (25 per cent). We will return to this interesting pattern of inferences in light of the findings in Experiment 2 where the inferences that people draw from everyday examples of *even-if* are examined.

Experiment 1 has established a number of clear findings with regard to the inferences that people make from *even-if* assertions. First, as the interpretation data show, *even-if* assertions convey that the consequent will occur much more readily than standard conditional assertions. Second, as has been argued in the literature, *even-if* blocks the invited inferences that are commonly drawn from conditional assertions. The pattern of inference that is associated with the necessity implicature and involves drawing all four conditional inferences is present with the assertions containing IF, but entirely absent for assertions containing *even-if*. Third, the data for MP and DA demonstrate that participants tend to infer the consequent from both the affirmation and denial of the antecedent. This suggests that the representation constructed for *even-if* encodes both the possibility described by the assertion and alternative possibilities in which the consequent occurs in the absence of the antecedent. The high proportion of uncertainty responses for the AC inference provides additional support for this claim.

4 Experiment 2

The data from Experiment 1 provide support for many of the conjectures in the linguistic and philosophical literatures. In particular they suggest that, in the absence of any knowledge-rich context, *even* makes a dramatic contribution to the meaning of *if*. In this section we present a second experiment that attempts to generalize the results of Experiment 1 to more realistic, everyday instances of *even-if*. As we have seen in the linguistic literature *even-if* is assumed to initiate the activation of alternative antecedents on a scale of unexpectedness, where the given antecedent is least likely to lead to the consequent and the remaining antecedents are more likely. In Experiment 2 we introduced an additional manipulation that concerned the point in the scale that the given antecedent picked out. Consider, for example, the following two assertions:

(9) Even if I read the lecturer's handout I will fail the exam.
(10) Even if I read everything on the reading list I will fail my exam.

In (10) the antecedent picks out a point on the scale of unexpectedness that is higher than the point picked out in (9). That is, one would be less likely to expect to fail an exam given that one read everything on the reading list than to fail an exam given that one only read a lecturer's handout. Consequently, in (10) there are fewer alternative antecedents higher up on the scale than in (9) where the antecedent picks out a less extreme possibility. One way of characterizing this scale is in terms of conditional probabilities, where the probability of the consequent given the antecedent is high in the case of the given antecedent, but higher in the case of alternative antecedents accessed in context. In (9) $P(q/p)$ is higher and hence less extreme than in (10) where $P(q/p)$ is lower.

Our interest in this manipulation was to gauge the extent to which scalar position might influence the inferences that participants would be willing to draw from assertions of this kind. As we have seen under the quantifier account, *even-if* is proposed to lead to an interpretation that the consequent will occur under the stated antecedent condition and all other antecedent conditions. Hence, under this account one might expect that antecedents such as those in (10), that pick out a more extreme, lower probability position will result in more secure inferences than those that pick out a less extreme position. This follows if one assumes that people will be more willing to draw inferences from propositions where background knowledge more readily satisfies the universal requirement. In contrast, the implicature account has no such requirement, only that the given antecedent is more surprising, contrary to expectations, than most of its neighbouring conditionals in context. Hence, an implicature account might not predict any effect of the scalar manipulation.

Another characteristic of the instances that we examine is more specific. In Experiment 2 we consider *even-if* assertions in which the antecedent refers to an action that the speaker may or may not take. Referring back to (9) and (10), an interesting question concerns the extent to which the speaker is not only inviting an inference about the consequent, but also inviting an inference about the possible action described by the antecedent. We will argue that *even-if* assertions of this kind may additionally result in effects for the listener in terms of the information they provide about a speaker's intention to act.

In Experiment 2 we employed a similar method to that employed in Experiment 1, combining an immediate inference task with a conditional arguments task. A group of 107 participants received two scenarios. One group of 53 participants, (the HIGH SCALE group) received scenarios where the given antecedent lay at an extreme point of the probability scale. A second group of 54 participants (the LOW SCALE group) received scenarios where the given antecedent lay at a less extreme point on the probability scale. The scenarios were presented in the following way:

Exam scenario

Pete and Jimmy, students at the University of Durham, are discussing the forthcoming examinations. Jimmy says:

HIGH SCALE
'Even-if I read everything on the reading list I will fail the exam.'
LOW SCALE
'Even-if I read the lecturer's handout I will fail the exam.'

Train scenario

Paul and Joe are sitting in Durham market square, Joe says:

HIGH SCALE
'Even-if I run as fast as I can I will miss my train.'
LOW SCALE
'Even-if I walk quickly I will miss my train.'

Participants were then asked to complete an interpretation task and a conditional inference task. The order of task presentation was counterbalanced within subjects. In the interpretation task the participants were asked to rate on a series of 9-point scales the extent to which they agreed that, for example, (referring to the first scenario) Jimmy intended to convey:

(a) That he will read everything on the reading list
(b) That he will not read everything on the reading list
(c) That he will fail his exam
(d) That he will not fail his exam

In the conditional inference task, participants were presented with the scenario and the *even-if* assertion followed by the four conditional argument forms and were asked which of three outcomes was most likely. An example of the MP inference is given below:

Suppose Jimmy reads everything on the reading list. According to what he has said which of the following outcomes is most likely?

(i) Jimmy failed his exam
(ii) Jimmy passed his exam
(iii) Jimmy may or may not have passed his exam[2]

We will present the interpretation data first. The initial analysis focused upon the consequent inferences. We performed a $2 \times 2 \times 2$ ANOVA, with Group (High Scale vs Low Scale) as the between-participants factor and Content (Exam vs Train) and Polarity (Rating to Q vs rating for Not-Q) as the within participants factors. There was a main effect of polarity ($F(1, 105) = 86.28$, MSE = 11.85, $p < 0.001$), reflecting, as expected, that participants interpreted the speaker's assertion as indicating that they would fail the exam or miss the train (6.53) rather than pass the exam or catch the train (3.44). This again confirms the strong intuition that *even-if* assertions entail their consequent. There was no effect of the scalar manipulation. However, there was a significant interaction between the content of the assertion and polarity ($F(1,105) = 12.07$, MSE = 4.94, $p < 0.001$). This interaction reflected the finding that participants inferred the consequent more readily from the TRAIN content (6.96) than the EXAMS content (6.10) and this difference was significant as revealed by planned comparisons on the interaction ($F(1,105) = 14.46$, MSE = 2.7, $p < 0.001$). Whilst there was no effect of our scalar manipulation, we believe that the effects of content can be explained in scalar terms and we will discuss this interpretation in more detail later.

An identical analysis was carried out on the antecedent inferences. The only notable effect to emerge from this analysis was a significant main effect of polarity ($F(1,105) = 25.59$, MSE = 10.9, $p < 0.001$). Participants inferred that the speaker intended to convey that they did not intend to read the handout/reading list, or walk quickly/run fast (5.64) more so than satisfying the antecedent condition (4.02). This finding suggests that *even-if* results in additional effects over and above entailing the consequent. In the contexts that we presented here participants clearly infer that the speaker is also conveying their intentions with regards to the actions that they will take.

Figure 11.3 presents the data for the antecedent inferences MP and DA. The data is aggregated across the scalar manipulation and the two contents. Again there was no effect of our scalar manipulation, but there were effects

---

[2] Of course the disjunctive alternative will always be at least as likely as either of the other two outcomes. The data suggest, however, that participants did not make their selections on this basis.

Reasoning and Pragmatics: Even-If 241

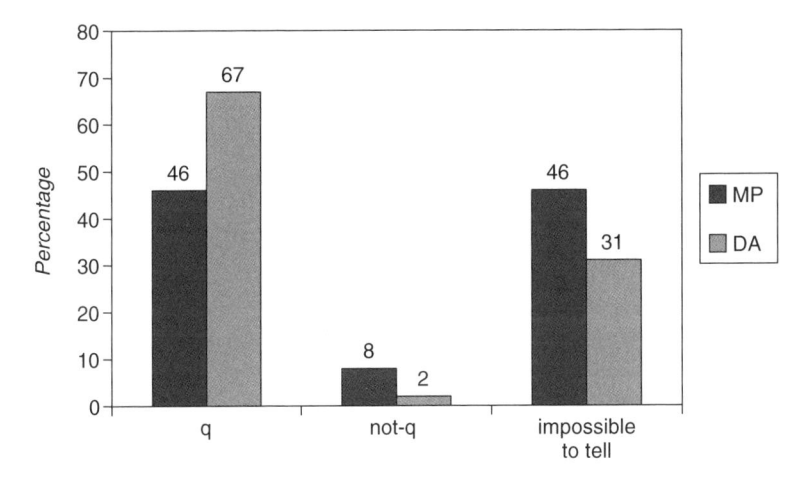

[Bar chart showing percentages for MP and DA across three response categories:
- q: MP 46, DA 67
- not-q: MP 8, DA 2
- impossible to tell: MP 46, DA 31]

Figure 11.3 The pattern of responses to the antecedent inferences on the conditional arguments task in Experiment 2

of content and these will be discussed shortly. Recall in our discussion of the findings of Experiment 1 we suggested that *even-if* elicits a representation consisting of two possibilities, one in which the antecedent and consequent co-occur and one in which the consequent occurs in the absence of the antecedent. Extending this analysis to the present study, we would expect participants to infer that the consequent occurs whether they are told that the antecedent occurs (MP) or that it doesn't occur (DA). Considering DA first, as Figure 11.3 clearly shows, the majority of participants inferred that the protagonist would fail the exam or miss the train on being told that they didn't read the handout/reading list or walk/run to the station (67 per cent). Most of the remaining participants inferred that it was impossible to tell whether the speaker passed the exam or caught the train (31 per cent). Presumably these participants recognize that it is possible to pass exams without studying (whether through luck or through the setting of a generous assessment) or catch trains without rushing to the station (the train may be late or cancelled – a situation many British readers will be familiar with). As the reader will recognize, the majority of inferences here are converse inferences. That is, participants are inferring q when they are told not-p.

Turning now to the rates of inference for MP, participants inferred that the speaker would fail the exam or miss the train less (46 per cent) than they did for DA (67 per cent). Here an equal number of responses indicated that it was impossible to tell what outcome occurred (46 per cent). This interaction suggests that people concurrently represent the assertion that *even-if* serves to deny – that is they represent from background knowledge the normal state of affairs, that one might expect p to preclude q. In Declerck and Reed's (2001) terms this corresponds to the expectation understanding of *even-if* and our data suggests that this understanding affects the inferences drawn. In a sense being told that Jimmy, for example, covers the entire reading list provides a condition that may disable the outcome that has been inferred. Hence, participants are less likely to reason with certainty that Jimmy will fail the exam.

In the analysis of the immediate inference data it was noted that participants inferred the consequent more readily with the TRAIN content than the EXAMS content. An analogous finding emerged here, with participants less likely to infer the consequent in the EXAMS problem for both MP (39 per cent) and DA (59 per cent) than in the TRAIN content (52 per cent and 76 per cent respectively).

Turning now to the consequent inferences, MT and AC, Figure 11.4 shows that, on the AC inference, the majority of participants inferred that it was not possible to tell whether the antecedent occurred (66 per cent). Again this pattern is consistent with the notion that people represent *even-if* assertions through the concurrent representation of two possibilities with common consequents and alternative antecedents. The pattern of responses for MT was more complex. In many ways a *Modus Tollens* inference from an *even-if* assertion is a very peculiar inference to ask people to make. As we have shown, *even-if* serves to cue the immediate inference that the speaker believes that the consequent will occur. The categorical premise in a *Modus Tollens* argument serves to countermand this inference. The question is what should people then infer?

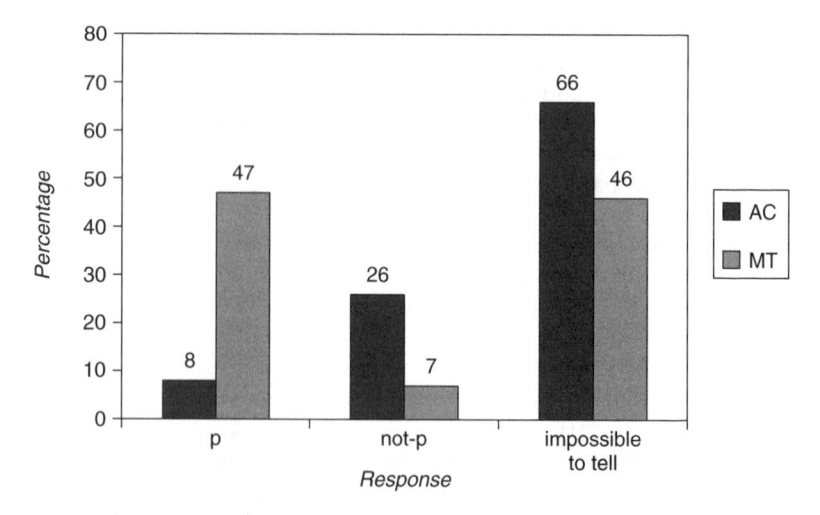

Figure 11.4 The pattern of responses to the consequent inferences on the conditional arguments task in Experiment 2

As Declerck and Reed (2001) have argued, *even-if* makes salient the possibility that is denied, that is, it makes salient the expectation that the antecedent (i.e., Jimmy reading the full reading list) normally leads to the consequent (i.e., that Jimmy will pass the exam). Given that this possibility is denied by *even-if*, providing a categorical premise that is consistent with this possibility, as is the case with MT, may cause participants to doubt the truth of the speaker's assertion. It is not possible to draw any determinate inference from a false premise and indeed a large proportion of our participants responded that it was not possible to tell whether the antecedent condition holds (46 per cent). Most of the remaining participants made the converse inference – that is they inferred that, for example, Jimmy read the handout or Joe ran for the train (47 per cent). We would suggest that these people fall back on their expectation and infer that Joe must have run for the train or Jimmy must have read the handout, given that Joe caught the train or Jimmy passed the exam. As with MP above, this finding suggests that the possibility that the assertion serves to deny is represented concurrently with a representation of the assertion and affects the inferences that people are willing to draw.

5 Discussion

One of the key aims of this chapter was to highlight the way in which the typical methods and approaches employed in research on human reasoning may be applied in examining the rich pragmatic inferences that form part of our understanding of conditional utterances. The findings point to a complex, yet intriguing, picture. In this final section we aim to achieve three things. First, we will examine the extent to which these findings may be reconciled with the contrasting linguistic accounts discussed at the beginning of this chapter. Second, we will briefly consider a psychological account of the phenomena that draws on a contemporary account of human thinking. Finally, we discuss the more general implications of our approach for the development of the field of experimental pragmatics.

5.1 Linguistic accounts

The experimental data presented in this chapter are complex yet a relatively clear picture has emerged. Many of the findings map on to the linguistic intuitions that were described at the outset of the chapter, namely:

1. When an *even-if* assertion is made participants infer that the speaker intended to convey that the consequent will occur. This finding holds across both arbitrary and realistic content.
2. Relative to *if*, *even-if* blocks the necessity implicatures and consequently does not support AC and DA.

3. The inference data suggests, as linguists have argued, that *even-if* is scalar and calls up alternative antecedents on a scale of expectedness or conditional probability. This leads to a pattern of inferences, whereby the affirmation and denial of the antecedent is associated with the consequent outcome and the affirmation of the consequent is associated with an indeterminate outcome.

These findings in themselves might serve to buttress linguistic intuitions, but they do not clearly differentiate between a quantifier and an implicature account. However, there are a number of more subtle effects that may more readily serve this purpose.

The quantifier account suggests that the interpretation of *even-if* is one of universal quantification. One prediction that we derived from this account was that *even-if* assertions would be less amenable to inference if the antecedent picked out a less extreme and consequently a higher point on the probability scale. In these cases a universal interpretation would be less probable and hence less likely to lead to an inference that the consequent would occur. The scalar manipulation in Experiment 2 did not lead to any differences in immediate inferences or on the arguments task. Although null results must be treated with caution, this finding seems not to be predicted by a basic quantifier account. However, as Lycan (1991) has argued, it is possible that the alternatives over which the quantification holds may be highly contextually dependent. Hence, universality may only hold for contextually relevant propositions and the membership of this range may vary as a function of the content of the assertion.

Although the lack of scalar effects may not be decisive in discriminating between accounts, we believe that there are two findings in our data that clearly undermine the quantifier account. The first is based upon the overall analysis of the arguments data of Experiment 2. Here we found very clearly, irrespective of content or the scalar manipulation, that participants were more likely to infer the consequent when they were told that the antecedent did not obtain (DA) than when they were told that the antecedent did obtain (MP). Now under a quantifier account the consequent should be inferred to the same degree irrespective of the antecedent outcome presented, yet participants reasoned with much less certainty when they were told that the antecedent condition was fulfilled. This finding is discussed in more detail below.

A second finding that further argues against the quantifier account is the effect of content demonstrated in Experiment 2. Recall that, for example, of the following two assertions:

(11) 'Even if I read everything on the reading list I will fail the exam.'
(12) 'Even if I run as fast as I can I will miss my train.'

participants were more likely to infer that Joe missed the train than they were to infer that Jimmy failed the exam. We believe that this effect is readily interpretable within a scalar framework, but that this interpretation is completely at odds with the quantifier account. Considering the scale in probabilistic terms the train example is one where the P(q/p) is higher than the equivalent conditional probability in the exams example. That is, the likelihood of missing a train given that you run to the station is higher (in fact running to the station is a condition that suggests you may miss the train anyway) than the likelihood of failing an exam given that you cover a full reading list. On a scale of unexpectedness the train example falls lower on the scale than the exams example. Intuitively this distinction can be understood if one considers the many alternatives higher up on the scale, such as catching a taxi, getting a lift, catching a bus and so on, that would more readily lead to the consequent. Consequently, the quantifier interpretation of

(13) Joe will miss the train in all circumstances, including the circumstances in which he runs.

is less likely to hold. Yet it is this content that leads to more confidence in inferring the consequent and to higher rates of MP and DA. This is directly contrary to what one might expect if the quantifier account is the right one. Our preferred interpretation of this finding is that in the case of our examples *even-if* picks out a point on a scale which may not be the most extreme, but indicates the limit of what the speaker would consider doing to potentially undo an outcome. The speaker is not signalling that the outcome will occur in all circumstances, but simply in the circumstances constrained by the upper bounds of the given antecedent. Now the more likely it is that the outcome would occur anyway given the action, that is, the lower on the scale of unexpectedness the antecedent falls, the more likely participants are to infer the consequent. We would argue that this occurs because people imagine the circumstances in which the action was taken and are less than certain whether such an action would even in normal circumstances undo the outcome.

The quantifier approach claims that the truth of an *even-if* assertion is determined by the consequent holding across all antecedents. The inference data show that this condition does not necessarily hold. Indeed it can be demonstrated quite simply that the truth of *even if* is not undermined by completions that specify an alternative where this condition is not met. Consider, for example, the following:

(14) Even if I run all the way to the station I will miss my train so I will have to take a Taxi.

Here the truth of the *even-if* assertion is not in doubt. The completion simply provides an alternative higher up the scale that would undo the outcome.

It still remains the case that the truth of the assertion can only be judged relative to the likelihood that the train would be missed given that the speaker ran to the station. Indeed, examples of this kind provide strong support for an implicature account of *even-if*, where the completion defeats the implication that the speaker will miss the train, but it does not cast doubt on the truth of the assertion.

Given that the effects that we have observed result from implicature rather than the semantics of *even* as the quantifier account would have it, the question is that of what sort of account would be needed to explain our data. The points on which we would agree with current implicature accounts is that *even-if* calls up a scale, where this scale consists of conditionals with alternative antecedents. The scale can be characterized as being based upon unexpectedness, where unexpectedness can be characterized in terms of conditional probability. Hence, when a speaker asserts (11) they are stating that, contary to expectations, the probability of failing the exam given that they read the full reading list is high, but there are a range of alternative antecedents lower on the scale of unexpectedness where the probability of failing the exam is higher. Hence, the speaker is asserting their belief that P(q/p) is high, but P(q/not-p) is higher, leading to the interpretation, seen in both experiments, that the speaker is signalling their belief that the consequent will occur.

A similar probabilistic account explains the difference between MP and DA observed in Experiment 2. Here we find that participants are more likely to infer the consequent from DA than MP irrespective of content. In our view this finding falls directly out of an account based upon conditional probability, where P(q/p) is lower than the P(q/not-p) consequently leading to MP inferences that are less secure than DA inferences. Essentially participants recognize the potential for the outcome to be undone, given that the antecedent condition is fulfilled, and hence reason with less certainty that the consequent would necessarily occur in these circumstances. This is particularly the case for the exams content where studying to whatever degree is widely viewed as an action that has the potential to lead to exam success.

In summary the data from these studies provide good support for an implicature account of *even-if* and in our view are less amenable to an explanation based upon universal quantification. As we have seen, the data suggest that *even-if* brings to mind a representation of the assertion that is being denied, whilst at the same time calling up alternative antecedents where the P(q/alternatives) is higher than P(q/focal antecedent). The inferences that are drawn from *even-if* and the degree to which the consequent is inferred varies as a function of the point on the probability scale that the focal antecedent occupies. Within the context of our materials the higher up on the scale the more likely participants are to infer the consequent. Within the narrow range of materials we have considered, *even-if* also cues inferences about the antecedent. Currently these kinds of effects have not been discussed in the context of linguistic accounts, and they suggest that the pragmatic effects of *even-if* extend beyond those that have been identified to this point.

5.2 Psychological accounts

In this chapter we hope to have demonstrated the value of an experimental approach in illuminating the nature of the inferences that people draw from *even-if*. We intentionally employed methods that are commonly used to investigate and test theoretical models of human reasoning. Recently it has become increasingly apparent that the data that is collected using these sorts of approaches cannot be explained fully by accounts that assume that such reasoning is accomplished by a dedicated logical reasoning system. Instead, it is now widely recognized that pragmatics explains much, if not all, of the patterns of performance observed. This seems to be particularly the case for tasks that involve understanding and reasoning from conditional assertions (see, for example, Sperber et al., 1995). This has led a number of researchers to develop accounts of conditionals and conditional reasoning that incorporate pragmatic components. In this section we consider a psychological account of *even-if* and indicate the sort of modifications that would need to be made to such an account in order to explain our data.

Recently an account of *even-if* has been proposed that draws on the mental model theory of human reasoning (see Moreno-Rios, Garcia-Madruga and Byrne, 2003). According to this theory people understand conditionals by generating models corresponding to the possible situations described by an assertion and reasoning proceeds by the manipulation of these models. However, due to processing limitations, the initial models reasoners construct are often incomplete. Hence, when people are presented with an assertion of the following kind:

(15) Even if Jimmy studies hard then he will fail the exam.

they generate a single model:

(16) Jimmy studies   Jimmy fails the exam

which corresponds to the possibility in which the antecedent and consequent conditions are fulfilled. Recently the mental model theory has been modified to incorporate a pragmatic component whereby background knowledge may serve to add or remove possibilities from a model set (Johnson-Laird and Byrne, 2002). In the context of the assertion above we may access knowledge concerning the relationship that we hold to be true with regard to studying and passing exams; in this case a possibility in which

Jimmy doesn't study and fails the exam. This results in a representation consisting of the following two possibilities:

(17) Jimmy studies   Jimmy fails the exam
      Jimmy doesn't study   Jimmy fails the exam

Given that Jimmy fails the exam in both of the possibilities represented, such a model set will lead to the inference that Jimmy will fail the exam. Similarly the model set supports many of the conditional inferences observed in the present studies, specifically the tendency to infer the consequent from both the affirmation (MP) and denial (DA) of the antecedent, and the indeterminate responses observed when the consequent is affirmed (AC).

Whilst a mental models account of this kind provides some fit to our data, it is in our view over-simplistic and fails to account for many of the more subtle findings observed. The finding that people make inferences concerning expected actions is not captured by a simple model-based representation of the kind shown above. In our view this finding reflects the multi-dimensional nature of the representation (see, for example, Zwaan and Radvansky, 1998). It suggests that participants have represented, in some form, the speaker's intention to act. Given that there are two courses of action both of which lead to the same outcome, it appears that participants have inferred that the speaker is intending to choose the course of action that involves the least effort (i.e., not studying). This suggests, at the very least, that some form of information concerning the relative cost of action and the impact that this may have on the speaker's intentions is encoded into the representation. The proposal that representations of conditional assertions encode the relative costs and benefits of alternatives is a view that is echoed in discussions of our understanding of everyday conditional assertions. Recent work has suggested, for example, that perceived violations of rules governing permissions and obligations differ as a function of the perspective and goals of the individual (see Evans and Over, 1996, for a discussion of perspective effects). This has led some researchers to suggest that the representation of certain types of conditionals encode utilities that may vary as a function of an individual's goals (Manktelow and Over, 1991). Our findings here similarly suggest that certain uses of *even-if* invoke representations that incorporate information relating to costs and benefits of action.

Another aspect of our findings suggests that the representation of *even-if* also encodes information concerning the likelihood or probability of various outcomes. Hence, with MP, a significant proportion of our participants gave an indeterminate response rather than inferring the consequent. This finding suggests that the mental representation of the focal conditional is probabilistic. Although the speaker is stating that the probability of the consequent, given the antecedent condition is high, it is by no means certain, and our background knowledge provides us with a context which may also cast doubt on the certainty of the conditional. Recently, Oaksford, Chater and Larkin (2000) have argued that the certainty with which people are willing to endorse the conclusion in a conditional arguments task is predicted by conditional probabilities. In this model, the conclusion of MP equates with the conditional probability of q given p. Similarly, using a different paradigm, Evans, Handley and Over (2003) have shown that judgements concerning the truth of a conditional are predicted by conditional probability. In addition, as we have seen, Evans and Twyman-Musgrove (1998) have shown that the confidence with which an inference is drawn is influenced by the control that a speaker has over the antecedent event. Here, more secure inferences are drawn in cases where the speaker is perceived to have greater influence over the antecedent condition. These findings and the results of our studies suggest very strongly that the representation of conditional forms is probabilistic in nature.

In summary, whilst a model-based theory of *even-if* may be promising, the nature of the representation that is constructed and the cognitive processes involved in drawing inferences from *even-if* are more complex than such an account can currently manage. Any psychological model of these processes must incorporate an account of the sort of knowledge activated in the course of interpreting the utterance and the means with which this knowledge is integrated into a representation of the assertion. Whilst the representation constructed may well be model-based, we would contend that the models constructed are multi-dimensional in nature, encoding such things as the costs or benefits associated with alternative possibilities and the probabilistic characteristics of these alternatives.

5.3 Towards an Experimental Pragmatics

For the remainder of this discussion we will consider some of the general issues arising from the present research from both a methodological and theoretical perspective. In considering psychological accounts of the pragmatics of *even-if* we have discussed theoretical accounts of conditionals that have been developed in the context of understanding the processes underlying deductive reasoning. In addition the methodological approach that we adopted similarly drew upon the methods commonly employed in reasoning research. In standard reasoning research participants are requested to make necessary inferences, that is, inferences that must follow given that the premises are true. This is a very special case and arguably a very unusual case of inference; after all, most of our everyday reasoning entails inferring what is possible or probable given an assertion rather than what has to be the case (see, for example, Evans, Handley, Harper and Johnson-Laird, 1999). Whilst these methods and the theoretical accounts developed from them can be useful in informing our understanding of pragmatic inference, we do not believe that judging the quality of inferences against the normative criteria of logic will provide any significant insight into the processes involved in understanding everyday conditional assertions.

As we have argued with *even-if*, the inferences that are drawn reflect both an understanding of an individual's intentions together with a representation of the likelihood that various outcomes will occur. These inferences are plausible, but are defeasible, so given seemingly contradictory information (as is the case with MT), we still infer that particular outcomes are more likely than others. What is important here is not whether reasoners draw the appropriate logical inferences, but the pattern of inferences drawn and what this pattern can tell us about the way in which the connective is represented. Whilst abandoning logic as a criteria for judging performance on inference tasks requires a very different mind set, it is our contention that this step is unavoidable if we are to understand anything meaningful about the inferences that underlie our understanding of everyday assertions.

In considering the pragmatics of *even-if* it is clear that an understanding is arrived at through the combination of a representation of the speaker's assertion with background knowledge about more probable relationships. An important question with regards to *even-if*, and a more general question in terms of developing a research strategy in Experimental Pragmatics, concerns the appropriate research methods that will allow investigators to identify exactly what knowledge is activated in the course of interpreting an utterance. In the case of *even-if* proposals concerning the sort of information activated have intuitive appeal and some indirect support, although more direct evidence is required. A number of possible experimental strategies suggest themselves to us. These include the use of post-task sentence verification methodologies commonly used in the text comprehension literature (see, for example, Zwaan and Radvansky, 1998) or the use of priming methodology that has been employed in investigating on-line inference making (Lea, 1995). Whilst these approaches by no means exhaust the possible research strategies available, they may suggest ways in which researchers can approach experimentally the difficult task of identifying the knowledge that becomes activated in processing and making pragmatic inferences.

A second, more general issue that arises from the present research, and from research in pragmatics generally, is the extent to which pragmatic inferences are automatic or default, or entail explicit processing effort (see Noveck, 2001; and Chapters 12–14 in this volume). In the reasoning literature a number of authors have made the distinction between explicit processes that are conscious, effortful and limited by cognitive capacity, and implicit processes that are pre-conscious, automatic and not capacity limited (see, for example, Stanovich, 1999; Evans and Over, 1996). In most of these accounts the explicit system is linked to deductive reasoning, novel problem solving and intelligence, whilst the implicit system is associated with pragmatic, language-based processes. On this analysis pragmatic inferences are viewed as automatic default processes that may be overridden by the explicit system given sufficient processing resources. Hence, under this view the inferences that people make from *even-if* may be seen as automatic, rapid and unconscious.

Some of the most powerful evidence for the system-based distinction between pragmatic and logical processes has come from individual differences research carried out by Stanovich (1999) and his colleagues (see, for example, Stanovich and West, 1998a). This work shows that participants who give logical responses on reasoning tasks are of higher ability, or higher cognitive capacity, than those that give pragmatic responses. For example on the abstract selection task (Wason, 1966) participants who select the p and not-q cards are of significantly higher cognitive capacity than those that give the classic p and q pragmatic response. In contrast on deontic versions of the task, where the pragmatic and logical response coincide, there is no relationship between responses and ability (Stanovich and West, 1998b). In Stanovich's terms higher capacity translates into a greater ability to resist pre-potent or automatic pragmatic responses, and consequently to respond accurately on the task. These findings are not restricted to reasoning tasks, but also hold for judgement tasks such as Tversky and Kahneman's (1983) 'Linda' problem, where participants of higher capacity are less likely to make the pragmatically cued conjunction fallacy. What these findings appear to demonstrate is that logical reasoning requires explicit processing effort, whilst pragmatic reasoning does not.

These findings stand in marked contrast to recent developmental findings (see Noveck, this volume; Noveck, 2001), that demonstrate logical interpretation of quantifiers amongst younger children, but pragmatic interpretations amongst adults. These findings suggest that the logical interpretation is basic, and pragmatics requires an additional processing effort, that is successfully engaged in by older children and adults, but not younger children. Whilst reconciling these contrasting accounts may be difficult, each suggests a potentially fruitful experimental approach in developing an understanding of the nature of the pragmatic processes involved in understanding everyday conditionals. If it is the case that pragmatic inferences involve processing effort, then one might expect cognitive capacity to predict the extent to which reasoners draw appropriate pragmatic inferences in a particular context. On the same account the developmental data should suggest a developmental sequence in the sophistication of the pragmatic inferences drawn. Whilst this work remains to be done, the use of converging experimental, developmental and individual differences methodologies are likely to provide clearer insights into the pragmatics of everyday conditional assertions.

6 Conclusions

The major aim of this chapter was to illustrate how the methodological approaches employed in research on deductive reasoning may be used to distinguish between accounts and provide evidence concerning the pragmatics of everyday conditional assertions. The findings of our studies suggest that this approach can prove successful in illuminating the rich pattern of inferences that people draw from everyday conditionals. It is our contention that an approach that draws on the methods and theories of deductive reasoning will have much to offer in the development of the field of Experimental Pragmatics.

References

Adams, E. W. (1975). *The Logic of Conditionals: An Application of Probability to Deductive Logic*. Dordrecht: Reidel.

Barrouillet P., Grosset N., and Lecas J. F. (2000). Conditional reasoning by mental models: chronometric and developmental evidence. *Cognition* 75: 237–66.

Bennett, J. (1982). Even if. *Linguistics and Philosophy* 5: 403–18.

Byrne, R. M. J. (1989). Suppressing valid inferences with conditionals. *Cognition* 31: 61–83.

Carston, R. (1993). Conjunction, explanation and relevance. *Lingua* 90: 27–48.

Declerck, R., and Reed, S. (2001). Some truths and non-truths about even-if. *Linguistics* 39: 203–55.

Evans, J. St B. T., and Over, D. (1996). *Rationality and Reasoning*. Hove: Lawrence Erlbaum Associates.

Evans, J. St B. T., and Twyman-Musgrove, J. (1998). Conditional reasoning with inducements and advice. *Cognition* 69: B11–B16.

Evans, J. St B. T., Handley, S. J., and Over, D. E. (2003). Conditionals and conditional probability. *Journal of Experimental Psychology: Learning, Memory and Cognition* 29: 321–55.

Evans, J. St B. T., Handley, S. J., Harper, C., and Johnson-Laird, P. N. (1999). Reasoning about necessity and possibility: A test of the mental model theory of deduction. *Journal of Experimental Psychology: Learning, Memory and Cognition* 25: 1495–513

Fauconnier, G. (1975). Pragmatic scales and logical structure. *Linguistic Inquiry* 6: 353–75.

Francescotti, R. M. (1995). Even: The conventional implicature approach reconsidered. *Linguistics and Philosophy* 18: 153–73.

Geis, M., and Zwicky, A. M. (1971). On invited inferences. *Linguistic Enquiry* 2: 561–6.

Hilton, D. J. (1995). The social context of reasoning: Conversational inferences and rational judgement. *Psychological Bulletin* 118: 248–71.

Jackson, F. (1987). *Conditionals*. Oxford: Blackwell.

Johnson-Laird, P. N., and Byrne, R. M. J. (2002). Conditionals: a theory of their meaning, mental representation, and role in inference. *Psychological Review*.

Konig, E. (1986). Conditionals, concessive conditionals and concessives: Areas of contrast, overlap and neutralization. In E. C. Traugott, A. T. Meulen, J. S. Reilly and C. A. Ferguson (eds), *On Conditionals*. Cambridge: Cambridge University Press.

Lea, R. B. (1995). Online evidence for elaborative logical inference in text. *Journal of Experimental Psychology: Learning, Memory and Cognition* 21: 1469–82.

Lycan, W. G. (1991). Even and even if. *Linguistics and Philosophy* 14: 115–50.
Manktelow, K. I., and Over, D. E. (1991). Social roles and utilities in reasoning with deontic conditionals. *Cognition* 39: 85–105.
Moxey, L. M., and Sanford, A. J. (1993). *Communicating Quantities: A Psychological Perspective*. Hove:Lawrence Erlbaum Associates.
Newstead, S. E. (1995). Gricean implicatures and syllogistic reasoning. *Journal of Memory and Language* 34: 644–64.
Noveck, I. A. (2001). When children are more logical than adults: Experimental investigations of scalar implicature. *Cognition* 78: 165–88.
Oaksford, M., Chater, N., and Larkin, J. (2000). Probabilities and polarity biases in conditional inference. *Journal of Experimental Psychology: Learning, Memory and Cognition* 26: 883–9.
Moreno-Rios, S. M., Garcia-Madruga, J. A., and Byrne, R. M. J. (2003). The effect of linguistic mood on if: semifactual and counterfactual conditionals. Submitted for publication.
Rumain, B., Connell, J. and Braine, M. D. S. (1983). Conversational comprehension processes are responsible for reasoning fallacies in children as well as adults: IF is not the biconditional. *Developmental Psychology* 19: 471–81.
Sanford, D. H. (1989). *If p then q: Conditionals and the Foundations of Reasoning*. London: Routledge.
Sloman, S. A. (1996). The empirical case for two systems of reasoning. *Psychological Bulletin* 119: 3–22.
Sperber, D., Cara, F., and Girotto, V. (1995). Relevance theory explains the selection task. *Cognition* 52: 3–39.
Stanovich, K. E. (1999). *Who is Rational?* Mahwah, NJ: Erlbaum.
Stanovich, K. E., and West, R. F. (1998a). Individual differences in rational thought. *Journal of Experimental Psychology: General* 127: 161–88.
Stanovich, K. E., and West, R. F. (1998b). Cognitive ability and variation in selection task performance. *Thinking and Reasoning* 4: 193–230.
Tversky, A., and Kahneman, D. (1983). Extensional versus intuitive reasoning: The conjunction fallacy in probability judgement. *Psychological Review* 90: 293–315.
Wason, P. (1966). Reasoning. In B. M. Foss (ed.), *New Horizons in Psychology*. Harmondsworth: Penguin.
Zwaan, R. A., and Radvansky, G. A. (1998). Situation models in language comprehension and memory. *Psychological Bulletin* 123: 162–85.

Part III
The Case of Scalar Implicatures

12
Implicature, Relevance and Default Pragmatic Inference

Anne L. Bezuidenhout and Robin K. Morris

1 Introduction

Grice distinguished between generalized and particularized conversational implicatures. The latter he described as 'cases in which an implicature is carried by saying that *p* on a particular occasion in virtue of special features of the context'. The former he characterized as cases in which the 'use of a certain form of words... would normally (in the absence of special circumstances) carry such-and-such an implicature or type of implicature' (Grice, 1989, p. 37). Grice did not develop the notion of a generalized conversational implicature (GCI) to any great extent. When he introduces the terminology in his paper 'Logic and conversation' he gives a few examples of the following sort:[1]

(1) A man came to my office yesterday afternoon.
(2) Max found a turtle in a garden.
(3) Robert broke a finger last night.

In the case of (1) the hearer would be surprised to discover that the man was the speaker's husband, for the use of the indefinite noun phrase 'a man' implicates that the speaker is not intimately related to the man. Similarly, in (2) we assume that neither the turtle nor the garden was Max's own, for if they were, the speaker would surely have used the expressions 'his turtle' and 'his garden'. On the other hand, the use of an indefinite noun phrase does not always implicate the lack of an intimate relation between the

---

[1] At the end of 'Logic and conversation' Grice also suggests that the use of truth-functional connectives such as 'and' and 'or' give rise to GCIs. These suggestions of Grice's have been extensively and systematically explored by others, and will be discussed later under the heading of Q- and I-implicatures.

subject and the thing indicated by the noun phrase. In the case of (3) there is an implicature that it was Robert's *own* finger that Robert broke.[2]

Grice held that one mark of an implicature, whether generalized or particularized, is that it can be cancelled. A GCI will be *explicitly* cancelled when the speaker says something incompatible with the GCI. For example, the speaker of (1) could follow her utterance of (1) with the following assertion:

(1') I should confess that the man was my husband.

A GCI will be *implicitly* cancelled when contextual information that is mutually manifest to both speaker and hearer is incompatible with the GCI. For example, if Robert is a known Mafia enforcer, then we may very well not derive the GCI that Robert broke his own finger. Or if we know that Max has a lot of pet turtles that went missing, and that he has been searching everywhere for them, we may not assume that the turtle he found was not his own.[3]

The notion of a GCI has been extensively explored by neo-Griceans, such as Atlas and Levinson (1981), Gazdar (1979), Hirschberg (1991), Horn (1984), and especially Levinson (1983, 1987b, 1995, 2000). A lot of attention has been paid to so-called scalar and clausal implicatures, which are subclasses of what Levinson (2000) calls Q-implicatures. Another important class of implicatures is what Horn (1984) calls R-implicatures and Levinson (2000) calls I-implicatures. Levinson (2000) also identifies a third general class of GCIs that he calls M-implicatures. These three sorts of GCIs are derived from what Levinson calls the Q-, I-, and M-Principles respectively. Here we cannot give a detailed account of Levinson's treatment of these three classes of implicatures, or explain how Levinson thinks such implicatures are derivable from his three principles. We will rely on just a few examples to give some

---

[2] Grice's discussion of examples like (3) is somewhat indeterminate. He could be interpreted as claiming that the utterance of (3) leads to the GCI that the broken finger was Robert's own. On the other hand, he could be interpreted as claiming that in the case of (3) the normal GCI associated with the indefinite (namely the suggestion of a *lack* of intimate relation) is implicitly cancelled. Following Nicolle and Clark (1999) we will assume that there are at least two distinct types of GCI associated with the use of the indefinite, which can be labelled the alienable possession implicature and the inalienable possession implicature respectively. Fingers are inalienable possessions of people, so 'a finger' in (3) will lead to the GCI that the finger in question is Robert's own. Turtles, on the other hand, are alienable possessions of people, so 'a turtle' in (2) will lead to the GCI that the turtle in question was not Max's own.

[3] If one holds that the assumption that the man was not the speaker's husband is not an *implicature* of (1) but is part of *what is said* by the utterance of (1), then one may dispute that (1') is an explicit cancellation of a GCI. Rather, one will think of it as a *retraction* or *reformulation* of what was said.

hint of the view under discussion. These examples do *not* cover the whole range of cases that have been classified as GCIs:

(4) Some books had colour pictures.
    GCI: *Not all* books had colour pictures.
(5) We scored three goals.
    GCI: We scored *at most* three goals.
(6) There might be a parrot in the box.
    GCI: There *does not have to be* a parrot in the box.
(7) I believe that there is life elsewhere in the universe.
    GCI: For all I know there *may or may not be* life elsewhere in the universe.
(8) Laurent broke his ankle and went to the hospital.
    GCI: Laurent broke his ankle *and then* went to the hospital (This is a case of what Levinson (2000) calls *conjunction buttressing*).
(9) John caused the car to stop.
    GCI: John brought the car to a halt *in an unusual manner*.

(4)–(6) are examples of scalar implicatures, (7) is a clausal implicature, (8) is an I-implicature and (9) is an M-implicature.[4]

Levinson thinks of all these kinds of implicatures as depending on metalinguistic knowledge. In the case of a scalar implicature, the hearer will not derive the GCI unless she is aware that the speaker has used an expression that is the weaker member of a so-called Horn or entailment scale. For instance, the expressions 'all' and 'some' form the entailment scale <all, some> because there is a one-way entailment relation between sentences (of an appropriate sort) that contain the quantifier 'all' and those same sentences

---

[4] Some people include more under the heading of GCIs than Levinson would be willing to include. Gibbs and Moise (1997), in some experimental work that is intended to test between a neo-Gricean view of GCIs and a rival view according to which GCIs are in fact a part of what is said, include cases of quantifier domain restriction under the heading of GCIs. For example, the quantifier in 'Everyone is a vegetarian', might in context be understood to be restricted to *everyone whom we have invited to dinner tonight*. Gibbs and Moise also include what they call 'time-distance' sentences under the heading of sentences that give rise to GCIs. These are cases like 'It will take us some time to get to the mountains' or 'The park is some distance away'. For instance, in context the former can convey the proposition that *it will take longer than the hearer might have expected to get to the mountains*. Levinson (2000) does not discuss examples like these, so it is unclear how he would deal with them. They could be assimilated to cases like conjunction buttressing, on the grounds that expressions like 'everyone', 'some time' and 'some distance' are unmarked, minimal expressions that invite contextual enrichment. In other words, they would be cases falling under Levinson's I-Principle.

with 'some' substituted for 'all'. More generally, an entailment scale is an ordered n-tuple of expression alternates $<x_1, x_2, \ldots, x_n>$ such that if S is 'an arbitrary simplex sentence-frame' and $x_i > x_j$, $S(x_i)$ unilaterally entails $S(x_j)$. (Levinson, 2000, p. 79). When a speaker uses the weaker expression 'some', she implicates that she knows that the sentence with the stronger expression 'all' substituted for 'some' is false (or that she doesn't know whether the stronger statement is true). Clausal implicatures depend in a slightly different way on the existence of such entailment scales.

In the case of I-implicatures, the hearer must realize that the speaker has used an unmarked, minimal expression, which then licenses the hearer to use stereotypical information made available in the context to enrich the content of the speaker's utterance. For instance, the connective 'and' is a minimal expression, and its use licenses the hearer to infer the enriched 'and then' interpretation. M-implicatures are in some sense the opposite of I-implicatures. They are licensed when the speaker uses a marked or prolix form of expression. For instance, the phrase 'caused the car to stop' is a marked way of speaking, and the hearer is thus licensed to infer that the speaker is suggesting that there was something unusual or non-stereotypical about the way the car was stopped. Clearly, in both the case of I-implicatures and the case of M-implicatures, speakers and hearers need to be aware that there are marked and unmarked ways of saying (roughly) the same thing, which is why Levinson thinks there is a metalinguistic element to such GCIs.

Levinson proposes to treat GCIs as the result of 'default pragmatic inferences which may be cancelled by specific assumptions, but otherwise go through' (Levinson, 1987a, p. 723). He has been developing a set of heuristics or default inference rules that he says are used to generate GCIs (Levinson 1987b, 1995, 2000). These default inferences yield interpretations that represent a level of meaning that he calls utterance-type meaning, which is intermediate between sentence-type meaning and speaker meaning.[5]

In this chapter we will use Levinson's account of GCIs as the basis for our own speculations about how sentences that give rise to GCIs are processed. We also contrast the processing view derived from Levinson with an alternative processing account derived from the work of Sperber and Wilson (1986), Carston (1988, 1993, 1995, 1997, 1998a, 1998b), and Recanati (1991, 1993). This alternative account sees GCIs as interpretations that are arrived at by

---

[5] Levinson's account is similar to the one that has been developed by Kent Bach (1994a, 1994b, 1999), although the terminology Bach uses to describe his views is very different from Levinson's. To articulate the similarities and differences between these views would require a separate paper, but they seem to be in agreement about the default status of GCIs. What Levinson calls GCIs Bach calls implicitures, but just as Levinson thinks that GCIs are the result of default inferences, Bach thinks implicitures are derived on the basis of default inferences, which in turn are the result of a process of standardization.

the pragmatic development of semantically underspecified logical forms. This development occurs as the result of the operation of local pragmatic processes of enrichment and loosening. A hearer trying to understand a speaker's utterance will use the information that is semantically encoded in the words the speaker uses along with information that is mutually manifest to speaker and hearer in the conversational context. The lexical concepts the hearer accesses will be pragmatically enriched or loosened to yield *ad hoc* concepts. These *ad hoc* concepts become constituents of a representation the hearer is building of the proposition expressed by the speaker's utterance. In other words, these enriched meanings are attributed to the speaker as part of what the speaker said, not as something merely implicated. In this process the Gricean notion of what is strictly and literally said plays no role. Enriched interpretations are directly constructed via the local processes just described.

Clearly, what we have said so far cannot do justice to the rich and detailed accounts that have been offered by these rival theorists on the topic of GCIs. We have limited our description to those aspects of the theory that are relevant to two very simple pragmatic processing models, the Default Model (DM) and the Underspecification Model (UM), which are loosely inspired by the neo-Gricean and Relevance Theory views respectively.[6] Both these models are intrinsically interesting, and both have some *a priori* appeal. But we consider them to be incompatible with one another. They cannot both be true. In Section 3 we report data from an eye monitoring experiment that provides some initial data regarding on-line processing of sentences that give rise to GCIs and discuss how these data may inform further development of processing models derived from these two theoretical perspectives. Before proceeding to the experiments we flesh out the models in more detail.

2 Two competing pragmatic processing models

As we understand them, DM and UM are rival models of the pragmatic processes that are involved in understanding utterances of the sort that Grice claimed give rise to GCIs. One simple way to see how DM and UM differ is to look at a couple of examples and see the different accounts these models give of the process that leads to the recovery of GCIs. We will look first at the case of scalar expressions, which fall under Levinson's Q-Principle.

According to DM, expressions that give rise to scalar implicatures, like 'some', 'three', 'possibly' and so on, belong to entailment scales. An entailment scale is a set of lexicalized expressions that belong to the same semantic field and that are ordered according to 'strength'. For example, <all, many, some>

---

[6] There is some theoretical value to testing extreme versions of theoretical hypotheses. For one thing, these extreme versions make stark predictions, which may not follow from the more hedged theories that are their inspiration.

is an entailment scale, as are <..., four, three, two, one> and <necessarily, possibly>. According to DM, the use of a sentence containing a weak expression from such an entailment scale gives rise to a default GCI. It implicates the denial of that same sentence with any stronger expression from the same scale substituted for the weaker one. Thus 'Some dogs have spots' scalar implicates 'Not all dogs have spots' (as well as 'Not many dogs have spots').

Let us look in more detail at the case of cardinal sentences such as:

(10) Jane has three children.

    (a) Jane has *at least* three children.
    (b) GCI: Jane has *no more than* (i.e., *at most*) three children.
    (c) Jane has *exactly* three children.

According to DM 'three' means 'at least three'. Hence, what is said by (10) can be represented as (10a). The Q-Principle will be triggered by the cardinal expression 'three' in (10), which belongs to the entailment scale <... four, three, two, one>, thereby yielding the default GCI (10b).[7] What is said together with the GCI thus entails (10c). This process of retrieving the default GCI and combining it with what is said would be DM's explanation for why in many contexts the expression 'three children' in (10) is understood to mean *exactly three children*.

UM on the other hand claims that expressions such as 'some' and 'three' are semantically underspecified. They must be specified in context and how they are specified depends on the operation of a local pragmatic process of enrichment. In the case of (10), a pragmatic process of enrichment takes the semantically underspecified lexical concept *three*, and yields a contextually appropriate enrichment. Depending on the assumptions accessible in the context, the proposition expressed by (10) could be either (10a) or (10b) or (10c). In particular, to understand the speaker to have been communicating (10c), the hearer need not go through a process whereby (10a) and (10b) are retrieved as well.

The two models DM and UM also treat cases falling under Levinson's I-Principle differently. Consider the case of what Levinson calls *conjunction buttressing*:

(11) Laurent broke his ankle and went to the hospital.

    (a) (Laurent broke his ankle) ∧ (Laurent went to hospital).
    (b) GCI: Laurent broke his ankle *and then* he went to the hospital.

---

[7] Strictly speaking according to the Q-Principle, the utterance of (10) would generate a whole series of GCIs 'Jane does not have four children', 'Jane does not have five children' and so on, which can be subsumed under the summary GCI represented by (10b).

According to DM, 'and' has the same meaning as the conjunction symbol '∧' in truth-functional logic. A consequence is that as regards what is said by a conjunction, it can be interpreted simply as a list, the order of the conjuncts being irrelevant to its overall meaning. Since 'and' is an unmarked, minimal expression, DM claims that it triggers the I-Principle and gives rise to the default GCI in (11b). In other words, DM holds that a temporal interpretation of the conjunction 'and' is the default interpretation.

On the other hand, UM claims that 'and' is semantically underdetermined and it can be specified in many different ways, depending on the assumptions that are operative in the discourse context (see Carston, 1988, 1993). There will be cases in which it is appropriate to interpret 'P and Q' to mean *P and then Q*. But in other cases it may be appropriate to interpret it just as a list, or in some alternative enriched way as either:

P and as a result Q (e.g., 'She kicked him and he cried.')
P and for that reason Q (e.g., 'She cheated him and he sued her.')
P and simultaneously, Q (e.g., 'She took a shower and practised her singing.')
P and in the course of doing P, Q (e.g., 'She talked to him and found she liked him.')
etc.

DM of course can admit that all these alternatives are possible interpretations, in the sense that the default *and then* interpretation can be cancelled, either implicitly or explicitly in favour either of the 'logical' list meaning or one of these other enriched interpretations. According to DM, these alternative enriched understandings will not of course be GCIs, but particularized conversational implicatures (PCIs).[8]

One obvious difference between these models is that DM posits an additional step in the recovery of what Griceans think of as the GCI. First, the hearer must recover what is said by the utterance, and this will trigger a further inference to the GCI. For instance, on hearing (10) a hearer will first recover the meaning *Jane has at least three children* and only then will he recover the GCI *Jane has at most three children*. UM on the other hand suggests that if the interpretation *Jane has at most three children* is constructed, it will be directly constructed by adjusting an underspecified lexical concept, and the hearer will not first need to construct the interpretation *Jane has at least three children*. In such a situation, the only candidate for what is said is *Jane has at most three children*, given that DM's candidate for what is said is

---

[8] Another possible understanding of DM is that it is claiming that all these interpretations are default GCIs and hence all are accessed whenever the minimal expression 'and' is encountered. Context then selects the appropriate interpretation. We are assuming that this is not the version of DM that is most plausible, as we explain below.

not recovered. Similar remarks could be made with respect to sentences such as (11).

There has been some experimental work addressed to the question of whether DM's candidate for what is said must be constructed in the course of recovering the GCI. See Gibbs and Moise (1997), Nicolle and Clark (1999), Bezuidenhout and Cutting (2001). In the experimental literature, DM's candidate for what is said is, somewhat misleadingly, called the *minimal* meaning and the GCI is called the *enriched* meaning. The evidence overall seems to count against DM on this issue, since it has been shown that hearers are relatively insensitive to minimal meanings as compared with enriched meanings. This has been shown in off-line judgement tasks, where people, when asked to choose the meaning that they think best represents what is said, favour the enriched over the minimal meaning. It has also been shown in on-line self-paced reading tasks. These have shown that people are relatively slow to read sentences in contexts requiring only the minimal interpretation as compared to the time to read those same sentences when they are presented in contexts requiring that they be given enriched interpretations. If the minimal meaning must always be accessed first, reading times in contexts favouring enriched readings should not be faster than reading times in contexts favouring minimal readings.[9]

In this chapter we wish to focus on another aspect of DM; that is DM's claim that GCIs have the status of default inferences that are automatically triggered unless something in the context blocks them. We report on an eye monitoring experiment that is designed to explore the role of defaults in understanding sentences of the sort that Grice claims give rise to GCIs.

However, before turning to a description of our experiment, we believe that some clarification of terms is in order. In particular, we wish to explain how we understand the notion of default inference that is invoked by DM.

---

[9] Kent Bach (1995, 1998) has argued that by a process of standardization, the enriched meanings (i.e., the GCIs) have become entrenched. Thus, when a hearer processes a sentence such as (10) the minimal meaning (what is strictly and literally said) can be bypassed. In fact, Bach thinks this is exactly what it means to say GCIs are defaults. We attend to what is literally said only in unusual circumstances, when something in the context signals that the default should not be drawn. But in normal circumstances the complete inferential process that first derives what is said and then derives the GCI will be short-circuited. This standardization explanation works well in some cases, when there is a formulaic GCI associated with a certain lexical item. But where two or more enrichments are possible, which vary with the context, then the standardization explanation is more problematic. There will be no standard GCI associated with a sentence type. The standardization view could be supplemented with some sort of frequency-based account. The standard GCI could then be regarded as the one most frequently associated with a certain sentence-type. We have more to say below about the possible role of frequency information in processing sentences of the sort that we are focusing on.

Also, we try to clarify the notion of semantic underspecification that is invoked by UM.

2.1 Default inferences

What is generally meant by calling something a default is that it settles some issue *without* the need to make a choice among alternatives. For instance, consider what is meant by calling something a default *assumption*. Suppose we are in a restaurant and have been shown to a table and given some menus. We assume that a waiter or waitress will come to the table to take our order. This is a default assumption, and we don't have to think about alternative scenarios, unless something untoward happens. Something similar holds in the case of default *inferences*. For example, normally when something looks red to me I will infer that it is red. This is a default inference. It is one that will be drawn, all things being equal. If all things are not equal, say because I am alerted to the fact that there is something unusual about my perceptual circumstances, then I will consider alternative conclusions. For example, perhaps the object is actually white but I am seeing it through a red filter or perhaps the object is actually green and I am seeing it through colour inverting spectacles and so on.

In several places Levinson suggests that a given sentence-type will be associated with several different GCIs. Take a sentence such as 'John's book is on the table'. Levinson (2000, pp. 37, 207) grants that by his I-Principle each of the enrichments 'book John owns/bought/borrowed/read/wrote', as well as many more, are possible for 'John's book'. Similarly, Levinson (2000, pp. 38, 117) allows that temporal, causal and teleological interpretations are all possible under the I-Principle for a sentence such as 'Jane turned the key and the engine started'. And multiple possible GCIs seem possible also under his Q-Principle. For instance, a scalar sentence such as 'I ate some of the cookies' ought to generate the GCIs 'I did not eat all/most/many of the cookies' (see Levinson, 2000, p. 77).

However, if we allow multiple GCIs for a given sentence-type, and allow that they are all inferred whenever a sentence of that type is used, this will defeat the whole purpose of having a default in the first place. For the hearer will now be forced to decide between the multiple interpretations that have been inferred, since in many cases these interpretations will be incompatible with one another (e.g., John's book can't be both bought *and* borrowed). But, as already noted, the point of having a default is precisely to avoid the need for making choices unless forced to do so by unusual circumstances.

Levinson could retreat to the claim that although there are *potentially* many GCIs only one will *actually* be inferred and which one is actually inferred will be determined by context. However, if he adopts this position then it is no longer clear that we can talk about a system of default inferences. Such a view of defaults would be compatible also with UM, since all parties can agree that a sentence such as 'John's book is on the table' has many possible interpretations, including 'The book John owns/wrote/borrowed/bought/sold is on the table'. Which one of these interpretations will be selected depends on the context.

Thus it seems that to defend a view that can truly be said to be a default theory, we must select one from among all these alternatives as the single default. There are places where Levinson does seem to opt for a single default. For instance, in the case of conjunctions using 'and' Levinson (2000, p. 123) suggests that the default is the temporal understanding 'and then'. The problem with this is that it is unclear that this sort of lack of flexibility in pragmatic processing is ultimately a good thing. It may speed up processing in some cases, but it could substantially hinder processing in others. For instance, if the default for 'John's book' is the ownership interpretation, then if the speaker means to communicate that the book is one that John borrowed, this will cause some difficulties in interpretation, as the default will first have to be cancelled/ overridden/ suppressed. This will require processing resources, thus increasing the hearer/reader's processing load. Hence, the system of defaults could be said to be inefficient from a processing perspective.

Of course, Levinson will stress that a default inference is one that will be drawn only if things are otherwise normal. If there is something unusual about the situation, the default will not be drawn (just as when I am alerted to something unusual in my perceptual circumstances I do not infer from something's looking red that it is red). So Levinson might argue that if the context makes it clear that what we are talking about is the book John borrowed, then the default inference that the book is one that John owns will not be drawn. Hence, there is no processing cost associated with setting the default to the ownership interpretation. This is so, provided that the context that supports the borrowing interpretation *precedes* the use of the phrase 'John's book'. If it does not, then the default inference will be drawn and there will then be a need to cancel/suppress/override the default.[10]

It has been suggested to us that there is no reason to deny that there can be multiple default GCIs associated with a sentence-type. Just as an ambiguous expression is associated with many different meanings, all of which may be initially activated when the expression is used, so too a sentence-type can be associated with multiple GCIs, all of which are inferred when the sentence is used. However, we do not think that ambiguous expressions provide a good analogy for thinking about sentences that give rise to GCIs. GCIs, even on the default model, are not ready-made chunks of meaning that are accessed

---

[10] Besides, someone who believes in a system of defaults shouldn't stress too much those contexts that block the generation of defaults. For if situations in which default inferences are blocked occur very frequently, the rationale for the idea that we need defaults in the first place is weakened.

from the lexicon. They must be computed, even if the computation is fast and automatic. For instance, it is not part of the lexically encoded meaning of 'John ate some of the cookies' that John did not eat all of the cookies. This is something that must be inferred, even if, as DM claims, this inference is a default one. Thus, we continue to think that it is problematic to talk of multiple defaults. Moreover, the fact that ambiguity is not a good model for GCIs also means that lexical ambiguity resolution is not a good model for the process whereby a particular GCI is derived.

2.2 Semantic underspecification

The notion of semantic underspecification has been widely discussed in recent times, particularly by cognitive linguists such as Fauconnier (1985, 1997), Pustejovsky (1995, 1998), Van Deemter (1996), Coulson (2001), Tyler and Evans (2001) and others. It is also a crucial component of Sperber and Wilson's (1986) Relevance Theory. These theorists all stress that it is necessary to distinguish between the meaning that is encoded by a lexical expression and the interpretation that the item receives in an utterance context. Such interpretation frequently goes beyond the encoded meaning. It must be constructed on-line, on the basis of the encoded meaning of the expression together with non-linguistic contextually available information. This sort of meaning construction is a process of conceptual integration, which combines linguistic and non-linguistic concepts according to general cognitive principles. A semantically underspecified lexical expression is thus one whose encoded meaning does not fully specify its contextual meanings.

In order to fully flesh out a processing model that relies on the notion of underspecification, one would have to know a lot more about the cognitive principles that are involved in sense construction. Moreover, one needs to have a clearer idea of what a semantically underspecified meaning is. One suggestion is that an underspecified meaning is the meaning that is common to all its possible specifications. This idea doesn't seem very plausible. For example, consider a polysemous noun such as 'newspaper'. It can refer to either a publisher ('The newspaper fired its editors') or a publication type ('The newspaper today has an obituary for Nozick') or a publication token ('The newspaper is on the kitchen table'). One might want to argue that the term itself is semantically underspecified, and that the various meanings it has are constructed from the encoded meaning on the basis of contextual information. However, it is not plausible to say that the encoded meaning in this case corresponds to the meaning that all these uses have in common. The different uses refer to things that belong to different ontological categories, and thus there may be nothing they share in common (except something trivial, such as *being a thing*, or something unhelpful, such as *being able to be referred to by the term 'newspaper'*).

A more promising idea is to select one of the meanings as primary, and hence to regard its referent as the primary referent of the term. Other meanings will be secondary, and their referents will be things that stand in certain relations to the primary referent. Context will be needed to figure out what the relevant relation is in a particular case. For example, suppose one holds that 'newspaper' refers primarily to a publication token, then the underspecified meaning of 'newspaper' would be *thing(s) that stands in relation R to this token publication*. In context, the relation R could be specified as the identity relation, in which case the term refers to a publication token. Alternatively, it could be specified as the relation of publishing, in which case the term refers to the publisher of some publication token. This account is very similar to the one proposed by Recanati (1995) for possessive phrases such as 'John's book'. Recanati proposes to analyse this as making reference to a relation R holding between John and the book that has to be specified in context. If the relation is specified as one of ownership, then 'John's book' refers to the book John owns. If the relation is specified as one of authoring, then the phrase 'John's book' refers to the book John wrote, and so on.

In the course of their own work on the preposition 'over', Tyler and Evans (2001) lay out some helpful methodological principles to be followed in identifying the primary sense associated with such a polysemous term. They also suggest strategies to be followed in determining whether a sense is a distinct sense instantiated in semantic memory or is instead to be accounted for by a cognitive process of conceptual integration (i.e., is instead a meaning that must be constructed on-line). These methodological principles could be profitably applied in the current context, in order to give an account of the primary senses associated with terms such as 'some' and 'and', and to distinguish such primary meanings from senses that must be constructed on-line by a process of conceptual integration of lexical and non-lexical concepts.

3 An eye movement monitoring study

In this section we report on an eye monitoring experiment that we hope can clarify the role of defaults in pragmatic processing. The experiment is focused on the processing of sentences giving rise to scalar implicatures of one particular sort. (In other work of ours not reported here we have focused on the processing of sentences that give rise to I-implicatures.) Our experiment is focused on the processing of sentences of the form 'Some S are P', which, according to DM, give rise to the default GCI 'Not all S are P'. Such a GCI is an example of a scalar implicature, and according to DM is derived by means of the Q-principle.

Noveck (2001) has a very interesting study of scalar implicatures of this sort that some might read as lending support to DM. Noveck's study was aimed at uncovering a developmental trend in children's acquisition of certain modal and quantificational constructions. But the results he got for his adult control subjects are what are of interest to us here. Consider the following example:

(12) Some elephants have trunks.
   (a) Some *but not all* elephants have trunks.
   (b) Some *and possibly all* elephants have trunks.
   (c) Not all elephants have trunks. (GCI)

In Noveck's study 59 per cent of adults judged that sentences such as (12) are *false*. This means these adults must have interpreted (12) as (12a). DM has a ready explanation for this. Adults on hearing (12) automatically derived (12c), which is the scalar implicature triggered by the use of 'some' in (12). This GCI is combined with (12b), which is what is strictly and literally said by (12) according to DM, and thus people construe the speaker of (12) to have asserted (12a). This explanation requires DM to say that when people are asked to judge the truth or falsity of a statement, they actually end up judging the truth or falsity of something that is an amalgam of what is said and what is implicated.

Noveck himself does not construe the results of his experiments as lending support to DM. He says: 'Our evidence is compatible with either the neo-Gricean or Relevance account of scalar implicatures' (2001, p. 186). We think that Noveck is right to be cautious, since defenders of UM would presumably want to give their own explanation for Noveck's findings. According to UM when a speaker utters (12), the hearer will *directly* generate either the (12a) or (12b) reading from an underspecified form, and which one is generated will depend on the context. In particular, to generate the (12a) reading does not require first generating the (12b) and (12c) readings, as DM claims. So if adults interpreted (12) as (12a) in Noveck's experiment, there must have been something in the context that made that the most relevant reading.

The main point we wish to make here is that *post hoc* processing explanations can be given for Noveck's data both from the point of view of the DM model and of the UM model, even if this is perhaps more difficult from the UM perspective. This suggests that we need processing data to differentiate between these two models. Eye movement monitoring during reading is one way to collect data of the required sort. Eye movement data provide an on-line record of processing as it unfolds over time without having to rely on secondary task responses gathered after the fact.

There is a large literature validating the use of fixation time measures to assess higher order cognitive processing (see Rayner and Morris, 1990, for review). This literature has established a tight (albeit not perfect) link between where a reader is looking and what the reader is processing. For example, lexical factors such as word frequency and lexical ambiguity directly influence the initial processing time that readers spend on a word. Recent work by Poynor and Morris (2003) demonstrates that the gaze duration measures gleaned from the eye movement record are sensitive to the process of generating inferences. In situations in which readers have committed to a particular syntactic or semantic analysis and then find that they must abandon this analysis in order to successfully comprehend the text, readers may spend additional processing time on the new information and/or the reanalysis process may be reflected in second-pass reading time on critical words or phrases (see also Folk and Morris, 1995).

3.1 Scalars, cancellation and predictions of the two models

If we are to be able to detect the presence of a default inference, we need a situation in which a default inference will be generated, but in which the reader subsequently gets information that is inconsistent with the default, thereby suggesting the need for a retraction of the meaning assignment made on the basis of the default inference. If a default is generated and then retracted, there should be evidence of processing difficulties in the eye movement record. In our experiment, readers saw pairs of sentences. The first sentence of each pair was of the form 'Some N were/had P'. It was followed by a sentence that *explicitly* cancelled the GCI 'Not all N were/had P'.[11] The second sentence was of the form 'In fact all of them were/did...'. The word 'all' in the cancellation sentence is the first information that readers encounter that suggests that *some but not all* may not be the appropriate interpretation for the initial determiner phrase. The immediately following phrase 'them were/did' definitively rules out the *some but not all* interpretation. We then created two further versions of each item by replacing the word 'some' in the first sentence by either 'many' or 'the'. An example of the three versions of a typical item is given below:

(13) (a) Some books had colour pictures. In fact all of them did, which is why the teachers liked them.
(b) Many books had colour pictures. In fact all of them did, which is why the teachers liked them.
(c) The books had colour pictures. In fact all of them did, which is why the teachers liked them.

According to DM, the GCI 'Not all N' should be triggered automatically both when the reader sees 'Some N' and when s/he sees 'Many N' but not when s/he sees 'The N'. In general, a default is an alternative that we assume

---

[11] For the distinction between explicit and implicit cancellation of a GCI, See Section 1 above.

to be true unless we are told otherwise. Defaults are useful in that they allow us to proceed with processing in the face of incomplete information, and to avoid costly processing of multiple alternatives. However, this also implies a cost to the reader when those assumptions turn out to be wrong. In those conditions in which the GCI is derived, two things will occur when the reader gets to information that explicitly cancels the GCI. First, the reader must cancel the default GCI and, second, they must reanalyse their discourse representation in order to retrieve or construct the context appropriate interpretation. It does not make sense that readers would engage in this costly cancellation and reanalysis process until and unless they are presented with compelling evidence of the need to do so. In our materials the information that explicitly cancels the GCI occurs when the reader encounters the phrase 'them were/did'. It is only when the reader processes the anaphoric pronoun, which refers back to the things N referred to in the initial sentence, that s/he can know that there is an explicit contradiction between the GCI 'not all N' and the 'all N' claim of the cancellation sentence. When readers encounter the word 'all', it is possible that the 'all' here is a quantifier applying to something N*, different from the N referred to in the initial sentence. This applies even if readers get a preview of the word 'of' when reading 'all'. For example, the sentence pair could have been:

(13a*) Some books had colour pictures. In fact all of the pictures were highly coloured, which is why the children liked them.[12]

Hence DM predicts that there will be increased processing time in the 'them were/did' region in the 'Some N' and 'Many N' conditions compared to the 'The N' condition. This increased processing time should show up in first-pass time in this region. DM also predicts that there should be some indication that readers are engaging in re-analysis. This could come in a number of different possible forms. The increased processing time on 'them were/did' could spill over on to the following region of the sentence (which we call the end-of-sentence region, namely the region immediately following the

---

[12] We did construct 30 of our filler items to begin with some/many/the N sentences and then to continue in the divergent way that (13a*) does, instead of ending with a cancellation of a GCI. However, there were too few of these divergent filler items with the word 'all' in the continuation sentence to allow for a comparison with our 'Some N' condition. This is a comparison we intend to explicitly test in a follow-up study. We note also that the fact that we had 30 filler items that started in the same way as our cancellation sentences, but that did not end in a cancellation of the GCI 'not all' makes an Early Cancellation strategy very risky. If readers anticipated a cancellation every time they saw the words 'Actually', 'In fact', and so on, half the time they would be mistaken.

phrase 'them were/did' to the end of the sentence). Or the reader might return to the initial determiner phrase (which we call the 'Det N' region) in the 'Some N' and 'Many N' conditions. This could be observed in the number of regressions or in rereading time on that initial region.

In contrast to DM, UM claims that readers in the 'Some N' condition do not fully commit to the *some but not all* reading right away. Rather, they engage in an incremental process utilizing all available information at any given moment in time. Under this view readers rely on probabilistic information to develop their interpretation over time. Thus, according to UM readers may (or may not, depending on your point of view) be biased toward the *some but not all* interpretation when they encounter the word 'some' but the item remains underspecified (in either case) until more information accrues. The word 'all' provides information that is biased toward the *some and possibly all* interpretation of 'some'. This predicts increased processing time on the word 'all' to reflect the fact that the reader has registered information potentially relevant to the specification of an underspecified item.[13] When readers reach 'them were/did' that information is consistent with their current interpretation. Thus, there should be no increase in processing time in this region. There is no need for a re-analysis, since the incorrect interpretation never was assigned to 'some' and hence under this view there is no prediction of increased rereading time or regressions to the initial 'Det N' region. This account makes no firm predictions about differences in behaviour between the 'Some N' and 'Many N' conditions. What probabilistic information is deemed relevant to the specification process and at what points it is deemed relevant may differ for 'some' and 'many'.

These predictions are summarized in Table 12.1.

Table 12.1 Summary of the predictions made by the various versions of DM and UM

Behaviour in *Some N* Condition and Comparisons with and between Controls	DM's Predictions	UM's Predictions
Increased time on *all*?	No	Yes
Increased time on *them were/did*?	Yes	No
Regressions/rereading of *Some N*?	Yes	No
Some should behave like *Many*?	Yes	??
Some should behave like *The*?	No	No
Many should behave like *The*?	No	??

---

[13] We are presupposing here that the specification process uses resources to integrate the new information, and hence that it has some processing cost.

3.2 Method

Participants

24 participants from the University of South Carolina community were recruited for this experiment. They either received one experimental credit in a psychology course or were paid $5 an hour for their time. All participants had normal, uncorrected vision and were native speakers of English.

Materials and design

We created a series of 31 items, each consisting of a pair of sentences. The first sentence of each pair was of the form 'Some N were/had P'. It was followed by a sentence that explicitly cancelled the GCI 'Not all N are P'. Two additional versions of each item were then created by replacing the word 'some' in the first sentence with either 'many' or 'the'. An example of the three versions of a typical item is given in Section 3.1 above. The preface to the cancellation sentence was not always the same for all items. In constructing our materials we used a variety of phrases such as 'in fact', 'actually', 'as a matter of fact', 'in all truth', 'truth be told', 'of course' and 'in reality'. All participants saw ten items from each of the three conditions (Some/Many/ The). No one saw the same item in more than one condition. In addition, each person saw 63 filler items. Thirty-three of these fillers came from another, unrelated experiment (whose materials also consisted of sentence pairs). The other 30 fillers were created to begin in the same way as the experimental items. However, what followed the preface to the second sentence in these items was not a cancellation of the GCI associated with the first sentence, but was either an elaboration of the first sentence or the introduction of some new topic (e.g., 'Some/Many/The pictures were fuzzy. Actually their resolution was so bad that they were impossible to make out'). Materials were presented in random order, a different order for each participant.

Procedure

Items were displayed on a colour monitor 35 inches from participants' eyes. Text was centred on the screen, with three characters subtending 1° of visual angle. We monitored the movements of participants' right eyes during reading, using a Fourward Technologies Dual Purkinje Generation 5 eye-tracker. The software sampled the tracker's output every millisecond to determine fixation location within a single character position. A bite bar was prepared for each participant to minimize head movement. The experimenter calibrated the tracker for each participant and performed a check before each item. Participants were instructed to read for comprehension. A yes/no question (without feedback) followed 20 per cent of the items. All participants performed at 80 per cent or better on comprehension questions.

Data analysis

We measured the time spent reading the phrase 'some/many/the N' (the Det N region) in the initial sentence. We also measured the words 'all' and 'them were/did' in the cancellation sentence, as well as the end-of-sentence region in the cancellation sentence. For single words we measured gaze duration, and second-pass time. For multi-word regions we measured first-pass time and second-pass time. We also measured the number of regressions into (i.e., the number of times participants looked back at) the phrase 'some/many/the N', as well as regressions into the words 'all' and 'them were/did'. First-pass time or gaze duration is all the time spent in a region before exiting to either the left or the right of that region. Second-pass time includes all the time spent rereading in a region, excluding gaze duration or first-pass time.[14] The first two measures give some indication of early processing, whereas second-pass time and number of regressions into a region allow one to make inferences about the reanalysis and text integration processes (see Rayner, Sereno, Morris, Schmauder and Clifton, 1989). We report ANOVAs treating participants (F_1) and items (F_2) as random effects.

3.3 Results

Table 12.2 shows mean reading times (in milliseconds) for two of the regions of interest in the 'cancellation' sentences in each of the three conditions (Some/Many/The N).

The region 'all'. Readers spent more initial processing time (as measured by gaze duration) on the word 'all' following 'Some N' than following 'The N' (F_1 (1, 23) = 10.59, MSE = 423, $p < 0.01$; F_2 (1, 60) = 7.36, MSE = 1001, $p < 0.01$). This is consistent with the predictions of UM. DM predicts no increased processing time in the 'Some N' condition until the 'them were/did' region. There was no difference in gaze duration on 'all' between the 'Many N' and 'The N' conditions (Both F_1 and $F_2 < 1$).

Table 12.2 Mean initial reading times (in msecs) for 2 critical regions in 'cancellation' sentences

	Gaze Duration on 'all'	First-Pass Time on 'them were/did'
Some N	*275*	301
Many N	260	308
The N	256	*329*

---

[14] 'Gaze duration' is the term used to refer to initial time spent in a one-word region whereas 'first-pass time' is used to refer to initial time spent in a multi-word region.

The lack of reading time difference between the 'Many N' and 'The N' conditions is also inconsistent with DM as DM predicted that the 'Many N' and 'Some N' conditions would produce similar effects relative to the 'The N' condition. UM made no prediction regarding this comparison.

The region 'them were/did'. Initial processing time (as measured by first-pass time) on the words 'them were/did' in the 'Some N' condition is *faster* than in the 'The N' condition, although this difference is reliable only in the subjects analysis (F_1 (1, 23) = 11.44, MSE = 828, p < 0.01; F_2 (1, 60) = 3.16, MSE = 4135, p = 0.08). This effect was somewhat surprising. The effect is in the opposite direction of that predicted by DM, which predicted a slow down in the 'Some N' condition. The UM predicted no difference between these two conditions in this region. UM could provide a *post hoc* account for the speed up on 'them were/did'. That is, if the specification of 'some' is achieved in the 'Some N' condition when the reader encounters the word 'all', then the information in this subsequent 'them were/did' region simply confirms an interpretation that has already been made and the reader can pass swiftly through this region.

Relevance theorists provide another possible interpretation. They have claimed that the processing of redundant information is costly, and that when a speaker uses repetition or redundancy, the speaker intends the hearer to derive extra contextual effects to off-set the extra costs of processing the repetition or redundancy. When one uses a plural definite description such as 'the books', one implies/presupposes that one is talking about the totality of some contextually specified group of books. Thus, there is a felt redundancy in the sentence pair 'The books had colour pictures. In fact all of them did...'. Hence, there could be increased processing time in the cancellation sentence in the 'The N' condition. This would presumably be localized in the 'them were/did' region, as that is the point at which the redundancy manifests itself, since that is the point at which the reader knows that the two sentences are talking about the same things. Under this account it is not that processing is speeded up in the 'Some N' condition but rather that it is slowed down in the 'The N' condition. The data do not discriminate between these two interpretations. But it is important to note that neither explanation favours the DM.

There was no difference in first-pass time on 'them were/did' between the 'Some N' and 'Many N' conditions (Both F_1 and F_2 < 1). This appears to be consistent with DM. However, DM predicted that the overall reading patterns for the 'Some N' and 'Many N' would be similar to each other and different from the 'The N' condition. This prediction was not upheld. Readers spent more time on the word 'all' in the 'Some N' condition compared to the 'The N' condition. But there was no difference between 'Many N' and 'The N' in that region. The UM made no strong predictions regarding similarities or differences between these conditions.

Other regions of potential interest. DM predicted that there would be some evidence of increased processing effort reflecting the readers' retrieval or construction of a new interpretation following the cancellation of the default GCI. There was no evidence of this in our data. Readers did not differ in time spent in the end-of-sentence region, or rereading the initial 'Det N' region as a function of determiner condition. This is consistent with the UM assumption that the specification is made incrementally as relevant information accrues and thus there is no need to re-analyse in the circumstances portrayed in our materials.

3.4 Discussion

Overall, the model best supported by our data is the UM, since it predicted a slow-down on 'all' in the 'Some N' condition as compared to the 'The N' condition, and it predicted no overt reanalysis of the 'Det N' phrase. Moreover, it is able to give a reasonable explanation for the speed-up in processing on the 'them were/did' phrase in the 'Some N' condition. But given that the results presented above constitute the first demonstration of these processing time effects and given that one might raise questions regarding the appropriateness of 'The N' as the control condition, we ran a second version of our experiment with a new control condition.

In this second version, processing of the 'Some N' sentence pairs from Experiment 1 was compared to processing patterns on sentence pairs that began with the quantifier phrase 'At least some N'. This new version was intended as a case in which DM does not predict the triggering of the default GCI 'Not all N'. At the same time it does not involve a potential redundancy or repetition, as arguably the 'The N' condition does.[15] Thus, readers saw 33 sentences of the following sort in either its (a) or its (b) version:

(14) (a) Some books had colour pictures. In fact all of them did, which is why the teachers liked them.
(b) At least some books had colour pictures. In fact all of them did, which is why the teachers liked them.

Twenty-four participants from the University of South Carolina community (different from those who participated in Experiment 1) were recruited for this version. They either received one experimental credit in a psychology course or were paid $5 an hour for their time. All participants had normal vision and were native speakers of English. The procedure was the same as for Experiment 1. Again, a yes/no question (without feedback) followed 20 per cent of the items, and all participants performed at 80 per cent or

---

[15] Thanks to Kent Bach for the suggestion of this control.

Table 12.3 Mean reading times (in msec) for 2 critical regions in the 'cancellation' sentence in the 'Some N' and 'At least some N' conditions

	Gaze Duration on 'all'	First-Pass Time on 'them were/did'
Some N	264	303
At least some N	249	325

better on comprehension questions. The analysis and regions of interest were the same as for Experiment 1.

Table 12.3 presents our results. The results of most interest were again to be found in the 'all' and 'them were/did' regions in the cancellation sentences.

The regions 'all' and 'them were/did'. The most striking thing to note in Table 12.3 is the similarity in the pattern of data obtained here and in Experiment 1. Once again readers spent more initial processing time on 'all' following 'Some N' than in the control condition, in this case 'At least some N' ($F_1(1, 23) = 3.57$, MSE = 766, p < 0.07; $F_2(1, 62) = 3.83$, MSE = 1019, p = 0.05). In addition, readers spent less time in the region that forces the *some and possibly all* interpretation (namely on the words 'them were/did') in the 'Some N' condition. But this result was not statistically significant ($F_1(1, 23) = 1.07$, MSE = 5250, p < 0.31; $F_2 < 1$). Although the pattern and magnitude of these results replicates our previous results, the data were more variable and thus the statistical support for the results of this experiment alone is not as strong. As with the previous experiment we also looked at second-pass time and regressions into these regions and at second-pass times on the 'Det N' phrase. No differences between the two conditions were found.

If we assume that the predictions of the various models are similar to those displayed in Table 12.1, then again the UM is best supported. It predicts a slow down on the word 'all' in the 'Some N' condition as compared with the control 'At least some N' condition. The lack of evidence of any differential reanalysis effects between the two conditions is also consistent with the UM account. The numerical differences in the first-pass time in the 'them were/did' region suggest that the pattern observed in Experiment 1 was not because readers were slowing down in the 'The N' condition due to a felt redundancy. However, we would not want to make any strong claims in this regard since the difference observed in the second experiment was not statistically reliable.

4 General discussion and conclusions

We have presented two very basic models based on two contrasting theoretical explanations of GCIs and data from two eye movement experiments that test predictions generated from the models. As far as we are aware, there is no previous published research using eye monitoring methodology to study the way in which sentences that give rise to GCIs are processed.

Our results are most consistent with UM. However, while our results pose problems for the DM as we conceived it, this model could be modified to accommodate the results we obtained. For instance, a version of the Default Model in which readers abandon the default in the face of potentially conflicting information (the word 'all' in our materials) rather than waiting until forced to do so (the 'them were/did' region) would make predictions that are largely compatible with the data that we obtained. Unfortunately, this is accomplished at the expense of compromising much of the utility ascribed to the notion of defaults in the first place. If the default can be dislodged even before the reader has conclusive evidence against it, it is unclear what the utility is of deriving it in the first place. Moreover, this Early Cancellation version of DM still does not account for the pattern of data observed in the 'them were/did' region, and it faces the problem that we found no evidence that readers engage in the sort of reanalysis predicted by the DM. Perhaps the more fundamental point to be made here is that whichever theoretical view one favours, these experiments provide new evidence of how sentences that give rise to GCIs are processed over time. We believe that this level of description can be put to good use to advance theory development in this area.

There has of course been other experimental work directed to studying people's comprehension of scalar and other generalized implicatures. For instance, Noveck (2001) and Chierchia, Crain, Guasti, Gualmini and Meroni (2001) have investigated children's understanding of scalar implicatures. Also, several of the contributors to this volume, including Noveck, Guasti, and Politzer, report experimental investigations of scalar and clausal implicatures. These studies rely on judgement tasks in which people are asked to respond in some way to sentences that give rise to implicatures (e.g., people are asked to agree or disagree with these sentences, given some prior story context). The data produced by such experiments are very valuable. They can provide evidence that people are aware of implicatures, but cannot tell us at what point in the comprehension process implicatures are derived or whether implicatures are derived via default inferences. We have argued that what is needed in order to investigate the competing claims of DM and UM is data that is gathered during on-line processing of such sentences. Monitoring people's eye movements during reading can provide this sort of window on to on-line processing.

The eye monitoring methodology has been used successfully to study a variety of phenomena, including processes as diverse as lexical ambiguity resolution, sentence parsing, inferencing during reading, comprehension of metaphors and jokes, and comprehension of metonymies. There is every reason to think that it can also be fruitfully applied to investigate the processing of GCIs. It may even be that work applying this methodology in other areas can be mined for insights about how to deal with the processing of GCIs. For instance, Frisson and Pickering (1999) and Pickering and Frisson (2001) have used this methodology to study the processing of metonymies. The sorts of metonymy that they have been interested in are cases such as:

(15) The mayor was invited to talk to *the school*.
(16) My great-great-grandmother loved to read *Dickens*.

The first of these (15) is an example of a place-for-institution metonymy, whereas (16) is a producer-for-product metonymy. Frisson and Pickering have shown in a series of eye-tracking experiments that people have no difficulties with such metonymies. These are processed just as quickly as cases in which the noun phrases in question are used 'literally', such as in 'The demolition crew tore down *the school*' or 'My great-great-grandmother loved to see *Dickens*'. At least, this is the case when there is some familiar type of institution associated with the critical noun phrase, or when the person whose name is used is famously associated with the product referred to. On the other hand, they found that novel metonymies such as 'My great-great-grandmother loved to read *Needham*', when they are presented in neutral contexts, cause people processing difficulties compared with 'literal' uses and familiar metonymies. However, if in the preceding context it is mentioned that Needham is an author, readers experience no difficulties with a metonymical use of the name 'Needham', such as 'My great-great-grandmother loved to read *Needham*'.

Frisson and Pickering interpret their results as lending support to what they call the Underspecification Model. They argue that noun phrases of the sort they have studied express semantically underspecified meanings. But these underspecified meanings point to more specific meanings. When a word is first encountered, only the underspecified meaning is accessed initially. Which specific meaning (if any) is accessed will depend on the context. They compare their view to the one defended by Barton and Sanford (1993), according to which readers initially engage only in shallow semantic processing, in which only the 'core meaning' of an expression is accessed. Further meaning refinements are possible and which ones are made will depend on the context. Sometimes a reader may go no further than the core meaning, if it seems to fit with the discourse context (and Barton and Sanford use this fact to explain why the so-called Moses Illusion tricks people).

Although we are sceptical about the 'core meaning' interpretation of semantic underspecification (see Section 2.2 above), we think Frisson and Pickering's work is highly related to our own. They make a convincing case for the claim that only a methodology such as eye-tracking can help decide between the Underspecification Model and some of the rival models they identify. They argue that their eye-tracking results count against any theory that argues that one particular meaning is the default. This applies, whether one privileges the 'literal' meaning or one of the metonymical ones. Frisson and Pickering found that at the earliest moments of processing there is no difference in reading times between phrases that require a 'literal' interpretation and ones requiring a metonymical reading. They relied on measures such as first fixation and first-pass time, which give a picture of early processing. These are opposed to measures such as total time or number of regressions, which arguably give a picture of processing during the stage at which interpretations must be integrated into a discourse level representation. See Rayner, Sereno, Morris, Schmauder and Clifton (1989).[16]

Implicating something in a generalized way is clearly not the same thing as using a word or phrase metonymically, but metonymy is related to the sort of polysemy that has been of concern to cognitive linguists such as Pustejovsky (1995) and Tyler and Evans (2001). The notion of semantic underspecification is central to the account these cognitive linguists have given of polysemy, and they have done a lot to clarify the notion of semantic underspecification. The notion of semantic underspecification plays a crucial role in our own work on GCIs, given that we are concerned to explore the competing pictures painted by the UM and the DM. Thus, we think that work in cognitive linguistics and work in psycholinguistics can be fruitfully brought together, and the study of the processes involved in the derivation of GCIs is a good place at which to bring these fields together.

In this chapter we hope to have done three things. We hope to have convinced readers: (i) that it is worth investigating the way in which people process sentences of the sort that Grice thought give rise to GCIs; (ii) that there are at least two competing models that give an initially plausible picture of what this processing is like; and (iii) that processing data from eye monitoring can give us the sort of evidence that is needed to test and refine these models.[17]

---

[16] For a view opposed to Frisson and Pickering's, see Giora (2003). Giora argues that with polysemies there is no need to select between the alternative meanings, and so all possible meanings can be accessed simultaneously, without any impact on processing. Thus, she is sceptical that Frisson and Pickering have unequivocal evidence for their underspecification model.

[17] We would like to acknowledge the assistance of Carter Henderson, Sachiko Matsumoto, Rihana Williams, Beth Myers, Matt Traxler and Johnny Hancock. This paper was first presented to the Experimental Pragmatics Workshop held in Lyon, France in May 2001. We would like to thank those in the audience who gave us advice. Special thanks to Dan Sperber and Ira Noveck, both in their role as hosts of the Experimental Pragmatics Workshop, and as editors of this volume, and also to Rachel Giora, who read the penultimate version of this paper and gave us extensive and helpful feedback. This research was conducted with the help of a grant from the National Science Foundation (NSF BCS-0080929).

References

Atlas, J. D., and Levinson, S. C. (1981). It-clefts, informativeness, and logical form: Radical pragmatics (Revised standard version). In P. Cole (ed.), *Radical Pragmatics*: 1–61. New York: Academic Press.

Bach, K. (1994a). Conversational impliciture. *Mind and Language* 9: 124–62.

Bach, K. (1994b). Semantic slack: What is said and more. In S. L. Tsohatzidis (ed.), *Foundations of Speech Act Theory: Philosophical and Linguistic Perspectives*: 267–91. London: Routledge.

Bach, K. (1995). Standardization vs. conventionalization. *Linguistics and Philosophy* 18: 677–86.

Bach, K. (1998). Standardization revisited. In A. Kasher (ed.), *Pragmatics: Critical Assessment*. London: Routledge.

Bach, K. (1999). The semantics-pragmatics distinction: What it is and why it matters. In K. Turner (ed.), *The Semantics/Pragmatics Interface from Different Points of View*: 65–84. Oxford: Elsevier.

Barton, S., and Sanford, A. J. (1993). A case-study of pragmatic anomaly-detection: Relevance-driven cohesion patterns. *Memory and Cognition* 21: 477–87.

Bezuidenhout, A. L., and Cutting, J. C. (2002). Literal meaning, minimal propositions and pragmatic processing. *Journal of Pragmatics* 34: 433–56.

Carston, R. (1988). Implicature, explicature and truth-theoretic semantics. In R. Kempson (ed.), *Mental Representations: The Interface between Language and Reality*: 155–82. Cambridge: Cambridge University Press.

Carston, R. (1993). Conjunction, explanation and relevance. *Lingua* 90: 27–48.

Carston, R. (1995). Quantity maxims and generalized implicature. *Lingua* 96: 213–44.

Carston, R. (1997). Enrichment and loosening: Complementary processes in deriving the proposition expressed? *Linguistische Berichte*: 103–27.

Carston, R. (1998a). Informativeness, relevance and scalar implicature. In R. Carston and S. Uchida (eds), *Relevance Theory: Applications and Implications*: 179–236. Amsterdam: John Benjamins.

Carston, R. (1998b). *Pragmatics and the Explicit-Implicit Distinction*. Unpublished Ph.D. dissertation, University College London, London.

Chierchia, G., Crain, S., Guasti, T., Gualmini, A., and Meroni, L. (2001). *The Acquisition of Disjunction: Evidence for a Grammatical View of Scalar Implicatures*. Paper presented at the 25th Annual Boston University Conference on Language Development.

Coulson, S. (2001). *Semantic Leaps: Frame-Shifting and Conceptual Blending in Meaning Construction*. Cambridge: Cambridge University Press.

Fauconnier, G. (1985). *Mental Spaces: Aspects of Meaning Construction in Natural Language*. Cambridge, MA: MIT Press.

Fauconnier, G. (1997). *Mappings in Thought and Language*. Cambridge: Cambridge University Press.

Folk, J. R., and Morris, R. K. (1995). Multiple lexical codes in reading: Evidence from eye movements, naming time, and oral reading. *Journal of Experimental Psychology: Learning, Memory, and Cognition* 21: 1412–29.

Frisson, S., and Pickering, M. J. (1999). The processing of metonymy: Evidence from eye movements. *Journal of Experimental Psychology: Learning, Memory and Cognition* 25: 1366–83.

Gazdar, G. (1979). *Pragmatics: Implicature, Presupposition, and Logical Form*. New York: Academic Press.

Gibbs, R. W., and Moise, J. F. (1997). Pragmatics in understanding what is said. *Cognition* 62: 51–74.

Giora, R. (2003). *On Our Mind: Salience, Context, and Figurative Language.* New York: Oxford University Press.

Grice, P. (1989). *Studies in the Way of Words.* Cambridge, MA: Harvard University Press.

Hirschberg, J. (1991). *A Theory of Scalar Implicature.* New York: Garland Publishing.

Horn, L. R. (1984). Toward a new taxonomy for pragmatic inference: Q-based and R-based implicature. In D. Schiffrin (ed.), *Meaning, Form, and Use in Context: Linguistic Applications*: 11–42. Washington, DC: Georgetown University Press.

Levinson, S. C. (1983). *Pragmatics.* Cambridge: Cambridge University Press.

Levinson, S. C. (1987a). Implicature explicated? *Behavioral and Brain Sciences* 10: 722–3.

Levinson, S. C. (1987b). Minimization and conversational inference. In J. Verschueren and M. Bertuccelli-Papi (eds), *The Pragmatic Perspective*: 61–129. Amsterdam: John Benjamins.

Levinson, S. C. (1995). Three levels of meaning. In F. R. Palmer (ed.), *Grammar and Meaning*: 90–115. Cambridge: Cambridge University Press.

Levinson, S. C. (2000). *Presumptive Meanings: The Theory of Generalized Conversational Implicature.* Cambridge, MA: MIT Press.

Nicolle, S., and Clark, B. (1999). Experimental pragmatics and what is said: A response to Gibbs and Moise. *Cognition* 69: 337–54.

Noveck, I. A. (2001). When children are more logical than adults: Experimental investigations of scalar implicature. *Cognition* 78: 165–88.

Pickering, M. J., and Frisson, S. (2001). Why reading Dickens is easy (and reading Needham is hard): Contrasting familiarity and figurativeness in language comprehension. Ms.

Poynor, D. V., and Morris, R. K. (2003) Inferred goals in narratives: Evidence from self-paced reading, recall and eye movements. *Journal of Experimental Psychology: Learning, Memory and Cognition* 29: 3–9.

Pustejovsky, J. (1995). *The Generative Lexicon.* Cambridge, MA: MIT Press.

Pustejovsky, J. (1998). The semantics of lexical underspecification. *Folia Linguistica* 32(3/4): 323–47.

Rayner, K., and Morris, R. K. (1990). Do eye movements reflect higher order processes in reading? In R. Groner, G. d'Ydewalle and R. Parham (eds), *From Eye to Mind*: 179–91. North Holland Press.

Rayner, K., Sereno, S. C., Morris, R. K., Schmauder, A. R., and Clifton, C. E. (1989). Eye movements and on-line comprehension processes. *Language and Cognitive Processes* 4: 21–50.

Recanati, F. (1991). The pragmatics of what is said. In S. Davis (ed.), *Pragmatics: A Reader*: 97–120. New York: Oxford University Press.

Recanati, F. (1993). *Direct Reference: From Language to Thought.* Oxford: Blackwell.

Recanati, F. (1995). The alleged priority of literal interpretation. *Cognitive Science* 19: 207–32.

Sperber, D., and Wilson, D. (1986). *Relevance: Communication and Cognition.* Oxford: Blackwell.

Tyler, A., and Evans, V. (2001). Reconsidering prepositional polysemy networks: The case of *over*. *Language* 77(4): 724–65.

Van Deemter, K., and Peters, S. (eds). (1996). *Semantic Ambiguity and Underspecification.* Stanford: CSLI Publications.

13
Semantic and Pragmatic Competence in Children's and Adults' Comprehension of *Or*

Gennaro Chierchia, Maria Teresa Guasti,* Andrea Gualmini, Luisa Meroni, Stephen Crain and Francesca Foppolo

1 Introduction

The interpretation of language is a complex phenomenon. One of the best established models maintains that language interpretation arises from the interaction of two major components. On the one hand, sentences are assigned truth conditions, which provide a characterization of propositional content and constitute the domain of semantics. On the other hand, use of propositional content (i.e., truth conditions) in concrete communication is governed by pragmatic norms. In speaking, not only do we pay attention to truth conditional content, we also aim at being cooperative and at saying something relevant to the situation. One way to study this intricate interplay between semantics and pragmatics is by looking at the way adults and children interpret logical words, for example, connectives and quantifiers. In particular, we would like to concentrate on Scalar Implicatures, inferences that we draw when we interpret sentences including certain logical words and that allow one to go beyond what is literally said in the sentence. For example, following Grice and much literature inspired by him, it can be argued that if a speaker says 'Some students passed the exam' the hearer is likely to assume that the speaker intended to convey that 'Some students passed the exam, *but not all did*'. The addition of 'but not all did' is not, however, part of the truth conditions, but an implicature that arises from the way we use language. Literally speaking, or as far as semantics is concerned, a sentence like 'Some students passed the exam' can be true in a situation where, in fact, all students passed the exam.

In this article, our goal is twofold. First, we would like to present a model, the Semantic Core Model, that challenges a way of interpreting Grice's proposal that has come to be dominant in the field. According to the dominant view, one first retrieves the semantics of a whole root sentence and then processes

---

* Corresponding author.

the implicatures associated with it (in a strictly modular way). The Semantic Core Model proposes, instead, that semantic and pragmatic processing takes place in tandem. Implicatures are factored in recursively, in parallel with truth conditions. Our second goal is to present experimental evidence from adults and children that is consistent with this new model.

2 What's the problem?

At the semantic level, a sentence including 'or', like *A or B*, is true when either one or both disjuncts are true (inclusive meaning). However, when we interpret sentences including *or*, for example (1), we tend to take it as meaning that John learned French or English, but not both, that is, as in (2a) (exclusive meaning). Similarly, a sentence including *some*, like (1b), is certainly true when some boys learned French, but also if it turns out that all the relevant boys learned French. However, we tend to take it as meaning that some boys, but not all, learned French, as in (2b):

(1) (a) John learned French or English.
    (b) Some boys learned French.
(2) (a) John learned French or English, but not both.
    (b) Some boys, but not all, learned French.

Why do we tend to interpret words like *some* or *or* in a different way, depending on the context? We assume that these words are not ambiguous and that the different interpretations that we associate with them are to be explained by appealing to pragmatic norms on the use of language. We tend to interpret the sentences in (1) as in (2) because logical words, like *some* and *or*, are part of a scale and are usually associated with a scalar implicature (SI) (see Horn, 1972). A scale is an ordering among certain (logical) words based on informational strength and SIs are inferences that result in interpreting a speaker's utterance as meaning more than what is literally said. A statement like *A and B* is true in a subset of the situations in which a statement like *A or B* is true, and thus the logical words contained in these statements can be ordered in the subset/superset relationship (viewed as entailment generalized to non-propositional functors): *and* $\subseteq$ *or*, where a statement including *and* is more informative than one including *or*. Similarly, a statement including *all* (or *every*) is true in a subset of the situations in which a statement including *most* (or *many* or *some*) is true and thus these logical words can be ordered along a scale as follows: *every* $\subseteq$ *most* $\subseteq$ *many* $\subseteq$ *some*, where a statement including *all* is stronger (and more informative) than a statement including *most, many* or *some*. SIs exploit systematically Grice's Maxim of Quantity. The reasoning that is at the basis of our way of interpreting (1a) is the following. The statement in (3) is more informative than (1a) (Maxim of Quantity):

(3) John learned French and English.

If the speaker had evidence for (3), she should have uttered it, instead of (1a). Since the speaker did not utter (3) and under the assumption that she is being cooperative, the hearer is entitled to conclude that the speaker did not believe or had no evidence for (3). Thus, she can infer that (3) does not hold and that by uttering (1a) the speaker had (2a) in mind, where (3) is negated (Maxim of Quantity). Notice that by adding the negation of (3) to (1a), as in (2a), the original statement in (1a) gets strengthened (i.e., becomes more informative). This is so because semantics assigns an inclusive meaning to *or* and thus (1a) is true in three situations, as displayed in (4):

(4) S1 = John learned only French
 S2 = John learned only English
 S3 = John learned English and French

Once the implicature is added and *or* is construed as having an exclusive meaning, (1a) becomes true only in two situations: S1 and S2 in (4), and thus provides more information.

The interpretation of a sentence *S* including the logically weaker term (e.g., *or*) requires the comparison of *S* against a set of alternative propositions, *ALT*, which would be relevant given the same conversational background (e.g., Krifka, 1995). The choice of a sentence *S* from *ALT* implicates the negation of all stronger propositions in *ALT*. Given (1a), where *or* is present, the alternative is (3), where *or* is replaced with the stronger term of the scale. Choosing (1a) implicates the denial of (3), which amounts to (5) or equivalently to (2a):

(5) John learned French or English and it is not the case that (John learned French and English)

We call the interpretation in (5) the scalar meaning and we use the term plain meaning to refer to the semantic meaning (e.g., inclusive meaning of *or*).

3 The computation of scalar implicature: how do semantics and pragmatics interact?

According to the standard Gricean view, SIs are calculated by the pragmatic module once the semantic module has completed its computation and has thus assigned truth conditions recursively to sentences. This view was challenged by Chierchia (2001) on the grounds that SIs fail to arise in specific linguistic contexts and that there are embedded SIs that must be computed before the semantic module has completed its job. Let us start with embedded implicatures. Consider the example in (6) and more specifically let us concentrate on the implicature triggered by the second disjunct. Sentence (6) implicates sentence (7):

(6) Mary is either reading a paper or seeing some students.
(7) Mary is either reading a paper or seeing some students, but she is not seeing every student.

Now, according to the global view, the ALT relative to the second disjunct of (6) is (8), which is stronger and hence more informative than (6). By choosing (6), all stronger alternatives are denied, in particular (8) is denied. This would amount to (9) where the alternative in (8) is denied:

(8) Mary is either reading a paper or seeing *every* student.
(9) Mary is either reading a paper or seeing some students and it is *not the case* that (Mary is reading a paper or seeing every student).

But from (9) we get (10), which contradicts what is stated in (6) and this is unwanted:

(10) Mary is not reading a paper.

This example shows that if SIs are computed at the end of the semantic computation, or globally, we run into trouble. Clearly, (9) is not what one means by uttering (6). The intended meaning of (6) (i.e., (7)) seems to be calculated with respect to the embedded disjunct only. But this is not possible in the standard Gricean model, since implicatures must be computed after the recursive computation of truth conditions. Let us now move on to the next challenging problem for implicature computation. Consider the following pairs of sentences:

(11) (a) Paul invited John or Bill.
(b) Paul didn't invite John or Bill.
(12) (a) Every student will be invited to the party or to the city tour.
(b) Every student who plays or sings will be invited to the city tour.
(13) (a) No student with an incomplete or failing grade is in good standing.
(b) No student who missed class will take the exam or contact the advisor.

While (11a) tends to be interpreted as meaning that Paul invited either John or Bill, but not both, (11b) means that Paul invited neither John nor Bill. It cannot mean that he didn't invite either John or Bill or neither, that is, while in (11a) the SI arises and *or* tends to be interpreted exclusively, in (11b) it does not arise and *or* is interpreted inclusively. Similarly, the SI arises in (12a) but not in (12b). In the case of (12a) there is no expectation that the students did both. Instead, in the case of (12b), the SI does not arise, witness the fact that if every student did both (played and sang), we are nevertheless willing to accept that he could be invited to the party. Finally, no SI arises in (13a) and (13b). In fact, we do not interpret (13a) as meaning that a student who has both an incomplete grade and a failing grade is in good standing. Similarly, we do not take (13b) to mean that the student who missed class will do both (take the exam and contact the advisor). These examples show that there are certain structural contexts that favour the inclusive meaning of *or* and in which the SI does not arise, while in other contexts such implicature arises. Interestingly, the contexts in which SIs fail to arise are by and large the contexts that license the Negative Polarity Item *any*. This is shown below:

(14) (a) *Paul likes any other linguist.
     (b) Paul doesn't like any other linguist.
(15) (a) *Every student will be invited to any party.
     (b) Every student who knows any linguistics can come to the party.
(16) (a) No Italian eats any raw fish.
     (b) No Italian who eats any fish will eat it raw.

How are these contexts characterized? The contexts in which *any* is licensed and SIs fail to arise share a semantic property: they are downward entailing (DE) contexts, that is, contexts that license inferences from sets to their subsets. For instance, negation is DE such that *John did not buy a car* entails *John did not buy a red car*. The first argument of *every* and both the first and second argument of *no* are all DE contexts (*every student will come* entails *every blond student will come*, but *every student will come* does not entail *every student will come by train*).

In DE contexts, then, the SIs triggered by the presence of scalar terms tend to not arise and we interpret scalar terms according to the semantic principles (i.e., *or* is interpreted inclusively). Notice that the interpretation we assign to *or* and to any other scalar term in DE contexts is a *default interpretation*, that is, it is the interpretation that most people would give in circumstances in which the context is unbiased one way or the other. The default interpretation is highly favoured in DE contexts, witnessed by the fact that we tend to interpret disjunctions inclusively in (11b), (12b) and (13b) much more easily than we do in (11a), (12a) and (13a). However, there are cases in which the SI arises in spite of the fact that the structural context under investigation is DE. Consider the example in (17) (see also Levinson, 2000):

(17) It was a two course meal. But everyone who had skipped the first or the second course enjoyed it more.

We do not interpret (17) as meaning that a person who had skipped both the first and the second course enjoyed it more, that is, we do not take *or* inclusively in (17). Rather (17) means that a person who has eaten only one course enjoyed it more. Here the implicature arises even if the context is DE, the reason being that otherwise a contradiction would ensue (see Chierchia, 2001 for a treatment of such contexts). Thus, all things being equal, SIs do not arise in DE contexts unless something else forces them to arise.

The natural question that arises is: why do SIs not arise in DE contexts? The answer is that DE contexts reverse the scale of informativeness. While in non-DE contexts *and* is more informative than *or*, in DE contexts the reverse holds. In non-DE contexts, a statement including *and*, as (18a), is true in just one situation, S1, when both conjuncts are true. By contrast, a statement including *or* is true in three situations, S1, S2, S3:

(18) (a) Paul invited John and Bill.

S1 = Paul invited John and Bill

(b) Paul invited John or Bill.

S1 = Paul invited John and Bill
S2 = Paul invited John
S3 = Paul invited Bill

In DE contexts, a statement including *or*, like (11b) repeated in (19a), is true in just one situation, S1, when neither Paul nor Bill is invited. A statement like (11b), but with *and* instead of *or*, as in (19b), is true in three situations, S1, S2, S3. Thus, in non-DE contexts a statement including *and* is more informative than one including *or*; in DE contexts, it is the one including *or* that is more informative and thus the SI usually associated with the weaker term of the scale has no reason to arise. In a DE context, *or* becomes stronger and more informative than *and*, that is, the scale is reversed:

(19) (a) Paul didn't invite John or Bill

S1 = Paul invited neither John nor Bill

(b) Paul didn't invite John and Bill

S1 = Paul invited neither John nor Bill
S2 = Paul invited John
S3 = Paul invited Bill

The generalization that SIs fail to arise in DE contexts is not *per se* a problem for the neo-Gricean view (see Horn, 1989, and Levinson, 2000). It becomes one, however, when put together with the observation that there are embedded implicatures. This is so because to obtain embedded implicatures we need to compute them locally; but then we also need a filtering mechanism that eliminates them in DE contexts.

Let us now move on and ask how can one take account of embedded implicatures and of the fact that SIs are sensitive to particular contexts, DE contexts? To solve these problems, Chierchia (2001) has proposed a model, called the Semantic Core Model, according to which SIs are computed as part of the recursive interpretation of a sentence. This means that the computation of SIs occurs locally, as does the computation of truth conditions. There are two mechanisms operative in the interpretation of sentences: one is the mechanism that recursively computes the meaning (truth conditions plus alternative set, when necessary) and the other is the mechanism that takes care of the context incrementation (which statement is to be added to the context). Both mechanisms are guided by the search for maximal information and are essentially based on Grice's norms. Let us spell out the procedure employed to interpret sentences including scalar items:

(20) (a) For any sentence α including a scalar term, the *scalar* or *strengthened* interpretation is computed by adding to the plain meaning of the sentence an implicature which amounts to the negation of any stronger alternative statement.
(b) The plain value and the scalar value are compared.
(c) The scalar interpretation is adopted, only if it leads to a more informative interpretation (i.e., true in a narrower set of circumstances).

Consider (6) and let us see how (20) operates in a concrete case. To interpret a sentence S, we need to compute the plain and scalar meanings for every item that belong to scales. In the case of (6), we start with the second disjunct, let us call it $D2$. The plain value of the second disjunct in (6) is something like (21). To compute the scalar meaning we compare $D2$ against a set of alternative propositions, ALT, which are given in (21). Once we have truth conditions (plain meaning) and the set of alternatives, we keep track of them to compute the scalar value in (21c). This consists in adding to the asserted sentence the denial of the stronger alternatives, an operation that strengthens the asserted sentence in (6) repeated below. Finally, we compare the plain and scalar meaning and choose the more informative one in order to add it to the context. When we are done with the second disjunct, we go on with the recursive interpretation of the sentence and with the computation of the SI associated with the disjunction *or*, essentially by repeating the operations described in (21):

(6) ... seeing some students.
(21) (a) *Plain meaning*: some students$_j$ [seeing t$_j$]

(b) *Alternatives*: [some students$_j$ [seeing t$_j$]] ALT = {seeing many students, seeing most students, seeing every student}

(c) *Scalar meaning*: some students$_j$ [seeing t$_j$] and it is not the case that Mary is seeing every student

We established that in non-DE contexts SIs arise, while in DE contexts they tend not to and here sentences including scalar terms receive a default interpretation. This general claim that structural factors influence the default interpretation of implicatures needs experimental support (for, arguably, it constitutes a generalization that cannot be firmly established just by introspection).

We thus carried out two experiments aimed at establishing whether participants respond differently to sentences including scalar items depending on whether the context is DE or non-DE. The material used for these experiments were sentences introduced by the universal quantifier *every* or the negative quantifier *no*, as in (12) and (13) repeated below, and including the scalar term, *or*, in the second argument or in the first argument of the quantifiers. Below, we have italicized the argument of the quantifiers. In (12a) and (13a), *or* is in the second argument and in (12b) and (13b) it is in the first argument:

(12) (a) Every student *will be invited to the party or to the city tour*.

2nd argument of *every*

(b) Every *student who plays or sings* will be invited to the city tour.

1st argument of *every*

(13) (a) No student *who missed class* will take the exam or contact the advisor.

2nd argument of *no*

(b) No student *with an incomplete or failing grade* is in good standing.

1st argument of *no*

The first experiment was a sort of inference task. Participants were presented with a statement describing an expectation, a fact that happened, and were asked to establish how much the fact confirmed the expectation by giving a score between 1 and 5, where 1 stood for 'the expectation is highly confirmed' and 5 for 'the expectation is confirmed a little'. One exemplary item is presented in (22) with the quantifier *every*. The first statement, (22a), reports the expectation and this is always a sentence in which *or* is present, either in a DE context or in a non-DE context. The second statement describes what happened, (22b), and the third is a question, (22c). Under that, there is a list of scores among which participants had to choose. (23) is another example with the quantifier *no*:

(22) This is what I expect:
  (a) Every third year student will do a written summary or an oral presentation.
  (b) Paolo, a third year student, did both.
  (c) How much does the fact above support my expectation?

(23) This is what I expect:
  (a) No third year student will do a boring summary or a bad presentation.
  (b) Paolo, a third year student, did both.
  (c) How much does the fact above support my expectation?

A total of 125 Italian-speaking adults were asked to rate four sentences (two with an *every* of the kind in (12) and two with a *no* of the kind in (13)) interspersed with fillers. The results are given in Table 13.1, where we report the means and standard deviations.

It is evident from Table 13.1 that participants tended to accept the inclusive meaning of *or* much more when *or* was in the first argument of *every*, a DE context, than when it was in the second argument; that is, the implicature was raised much more in the latter case than in the former one. The difference between the two means is highly significant ($p < 0.00001$). In the case of the quantifier *no*, participants responded in the same way whether *or* was included in the first or second argument of *no*, since both contexts are DE, that is, in both cases they took *or* to have an inclusive meaning. The difference between the two means is not significant this time ($p = 0.4$). Thus, acceptance of the inclusive meaning of *or* is higher in DE than in non-DE contexts, as we anticipated.

In a second experiment, we used the Truth Value Judgement Task (TVJT) (see Crain and Thornton, 1998; see also below). Participants watched a video displaying a story, acted out with toys and props, at the end of which a blindfolded puppet uttered a sentence describing what happened in the

Table 13.1 Mean and standard deviation of adults' responses to *or-statement* in DE and non-DE contexts

	Mean	Standard Deviation
Every 1st arg. DE	1.89	1.11
Every 2nd arg non-DE	2.99	1.37
No 1st arg. DE	4.09	1.20
No 2nd arg. DE	4.13	1.30

story. Material again included sentences of the kind in (12) and (13), with *or* in the first or second argument of *every* and *no*. Two examples of the sentences used are in (24) and (25):

(24) Every boy that got a cake or a biscuit sat down.
(25) No monkey that got a banana or a biscuit took a shower.

Sixty-two Italian-speaking participants were asked to respond on a sheet of paper whether the heard sentence was good or bad. The results show that participants responded differently depending on whether *or* is in the first or second argument of *every*. When *or* was in the first argument of *every*, participants said that the sentence was good 83.8 per cent of the time (52/62) while when it was in the second they said that the sentence was good only 45 per cent of the time (28/62). This difference is significant ($\chi^2 = 20.29$, $p < 0.005$). In contrast, there was no difference in the responses participants gave for *or* in the first and second argument of the quantifier *no*; participants said that the sentences were good 17.7 per cent (11/62) and 11.2 per cent (7/62) of the time. This is expected since both arguments of these quantifiers are DE.

5 Children's understanding of the logical connective 'or'

The first question we would like to address is: how do children interpret sentences including logical words? One result from previous research is that second graders do not distinguish *and* from *or* (Paris, 1973). However, in an experiment we carried out, it was established that children access the full range of truth-conditions licensed by the semantic component and distinguish between *or* and *and*. In our experimental procedure, a set of characters and toys was introduced to the child and to a puppet and it was announced that these characters were going to do something with the toys. The puppet was then invited to make a bet about the outcome of the story. After that, the story was acted out in front of the child and the puppet. At the end of the story, the child was asked to establish whether the puppet had won the bet. A typical bet took the form in (26):

(26) *Puppet*: I bet that Batman will take a cake or an apple.

In a typical story for that sentence, it turned out that Batman took both a cake and an apple. In addition to sentences of the kind in (26), in which both disjuncts are true, there were sentences including *or* in which only one disjunct was true and sentences including *and* in which only one conjunct was true. Nine Italian 5-year-olds ranging in age from 5;1 to 6;0 (mean age: 5;5) and 22 adult controls participated in the experiment. Their percentages of acceptance (yes responses) are reported in Table 13.2 where underlining indicates that the disjunct/conjunct is true.

Table 13.2 Rates of acceptance (in percentage) of statements including a conjunction or disjunction

Type of Sentences	Children	Adults
A or B̲	95	60
A̲ or B̲	78	100
A̲ and B	16	0

Table 13.3 Rates of acceptance (in percentage) of statements including a conjunction or disjunction

Type of sentences	Children	Adults
A̲ or B	98	88
A̲ or B̲	86	90
A̲ and B̲	24	8

Acceptance of sentences in which both disjuncts are true (first row in Table 13.2) indicates that the inclusive meaning of *or* is accessible. To be sure that children respond as they do because they have access to inclusive meaning of *or*, we must be sure that they distinguish *and* and *or*. The second and third rows of Table 13.2 show that children distinguish the two logical words *and* and *or*. Children accept a statement as true 78 per cent of the time when just one disjunct is true, but they do so 16 per cent of the time when only one conjunct is true. This indicates that they treat statements with *and* and statements with *or* differently. Similar results were obtained by Gualmini, Crain and Meroni (2000) with 14 English-speaking children (age range 3;2 to 5;9; mean age: 4;8) and 26 adult controls. Their results are summarized in Table 13.3.

Thus, we can conclude that children access the inclusive meaning of *or* and distinguish between *or* and *and*. One fact that is evident in both Tables 13.2 and 13.3 is that adults accept the inclusive meaning of *or* much less than children. A similar finding is discussed in Noveck (2001) who found that English and French learners are less likely to detect a violation of the quantity maxim than adults in contexts including the scalar terms *some* and *might*. A similar conclusion is invited by the findings of other experiments using sentences with scalar items to describe events that had already taken place (Crain, Gualmini and Meroni, 2000). In such contexts, some children accepted the scalar terms *some* and *or* in contexts in which adults rejected them. These results may suggest that children are more logical than adults and are less likely than adults to compute scalar implicatures. To investigate this question, we carried out a new series of experiments.

6 Children's pragmatic knowledge associated with the logical connective 'or'

Thus far, we have established that children can access the full range of truth conditions associated with the logical connective *or*. We also noticed that children may be less likely to compute scalar implicatures. What we do not know is whether structural factors influence the computation of scalar implicatures in children as they do in adults. Therefore, our next concern is to establish how children behave in DE and non-DE contexts with respect to SIs. Do they obey the generalization that SIs arise in non-DE contexts, but fail to do so in DE contexts? Using the Truth Value Judgement methodology (Crain and McKee, 1985; Crain and Thornton, 1998), Chierchia, Crain, Guasti, Gualmini and Meroni (2001) tested English speaking participants with sentences including *or* in the first or in the second argument of *every*. A sample of the material used is presented in (27). Recall that the first argument of *every* is a DE context, while the second is a non-DE context. Therefore, we expect higher acceptance of the inclusive meaning of *or* in the first argument of *every* than in the second one:

(27) (a) Every dwarf who ate a strawberry or a banana got a jewel.
(b) Every space-guy took a strawberry or an onion ring.

Children and adult controls were divided in two groups, one group was tested with sentences of the kind in (27a) and another with sentences of the kind in (27b). The first group included 15 English-speaking children ranging in age from 3;7 to 6;3 (mean age: 4;11) and 11 adult controls who were presented with sentences such as (27a). The second group included 15 children (ages ranging from 3;5 to 6;2; mean age: 5;2) and 8 adult controls who heard sentences such as the one in (27b). Each subject was presented with four target sentences interspersed with four fillers. An experimenter acted out a story, at the end of which a puppet, manipulated by another experimenter, was invited to say what happened in the story, by using a sentence of the kind in (27). The child's task was to say whether the puppet had said 'the right thing' about the story. A story for the sentence in (27a) featured Snow White and four dwarves going to a picnic. When it was time to eat, Snow White invited the dwarfs to choose healthy food, promising them a jewel if they did so. Three of the dwarves wanted to receive a jewel, so they chose both a banana and a strawberry. One of the dwarves said he didn't care about jewels, and he chose non-healthy food, potato chips. Snow White only gave a jewel to the dwarves who had chosen a banana and a strawberry, as depicted in Figure 13.1. The story is then described by the puppet with the sentence in (27a).

Semantic and Pragmatic Competence: Or   295

A story for the sentence in (27b) featured four space guys who were choosing something to eat. There was a lot of food – some onion rings, some strawberries, and some bananas. After considering the possible choices, the four space guys took both an onion ring and a strawberry, as seen in Figure 13.2.

Figure 13.1   Scene for sentence (27a): *Every dwarf who ate a strawberry or a banana got a jewel*

Figure 13.2   Scene for sentence (27b): *Every space-guy took a strawberry or an onion ring*

Table 13.4 Rate of acceptance (in percentage) of the target sentences by adults and children

	Children	Adults
Every 1st arg. DE	92	95
Every 2nd arg. non-DE	50	0

The results for the two groups are summarized in Table 13.4. As a group, children accepted sentences like (27a) 92 per cent of the time and adult controls did so 95.5 per cent. As for sentences like (27b) acceptance was 50 per cent in the children group and 0 per cent in the adult group, that is, children rejected the target sentence in half of the trials, while adults always did so.

These results show that adults and children consistently access the inclusive reading of disjunction when *or* appears in the first argument of *every*, a DE context. Adults do not accept the inclusive reading when *or* appears in the second argument of *every*, a non-DE context. As for children, although the acceptance of the inclusive reading of the disjunction in a non-DE context is considerably lower than in a DE context, the inclusive reading of disjunction was still accessed on half of the trials. In keeping with the Semantic Core Model, adults do not compute the SI in DE contexts, while they do in non-DE context. Children behave as adults with respect to the first argument of *every*. They clearly differentiate DE and non-DE contexts, but they still display a high level of acceptance of the inclusive reading of disjunction in the latter contexts, that is, they seem to calculate the SI only half of the time. When we look at the individual subject data of children tested with sentences like those in (27), we see that the vast majority of children did not behave by chance. Seven children correctly rejected the target sentence 26 times out 28 trials (92.8 per cent), seven children rejected the target sentence only twice out of 28 trials (7.2 per cent) and just one behaved by chance. In short, there is one group of children that behaved like the adult controls, while the other consistently tolerated the violation of the SI when the disjunction operator occurred in a non-DE environment. In summary, adults conform to the predictions of the Semantic Core Model and compute SIs in non-DE contexts, while they don't in DE contexts. Some children are like adults and compute SIs in non-DE contexts, while others appear to not compute SIs even in non-DE contexts. These findings are consistent with those mentioned in Section 5 showing that children are less likely than adults to compute SIs. Why do children appear to be more logical than adults? Do these children lack pragmatic knowledge? What is the source of this different behaviour between adults and children? In the next section, we address these problems.

7 The felicity judgement task

Why do some children not compute SIs in non-DE contexts? One hypothesis is that children have access to semantic knowledge that ensures the inclusive interpretation of disjunction, but lack a piece of pragmatic knowledge, Grice's Maxim of Quantity, and are thus insensitive to SIs. We call this the Pragmatic Delay hypothesis. To test this hypothesis, Chierchia and colleagues developed a new experimental technique, called the Felicity Judgement task. The Felicity Judgement task involves the presentation of two sentences as alternative descriptions of a specific situation. Importantly, the two sentences have the same truth value in the context under consideration, but they differ in appropriateness. The aim of the Felicity Judgement task is to determine if children detect any difference between the two descriptions. In the particular case under consideration, this is relevant to determine whether children know Grice's Maxim of Quantity. If they do, this would be evidence against the Pragmatic Delay hypothesis: children can compare sentences and know how to increment the context, that is, they know how to apply the operations in (20b) and (20c) above involved in the computation of SIs. If they don't, this would be evidence that children lack a basic piece of pragmatic knowledge involved in the computation of SIs, that is, they would be unable to compare and establish which sentence is most informative in the context.

In the case at hand, after a story in which all farmers decided to clean a horse and a rabbit, among their animals, two puppets each provided a different description of the story, as in (28); one sentence contains the disjunction operator *or* and the other the conjunction operator *and*:

(28) (a) Every farmer cleaned a horse or a rabbit.
    (b) Every farmer cleaned a horse and a rabbit.

Fifteen children (ranging in age from 3;2 to 6;0; mean age: 4;8) participated in the experiment, and were asked to reward the puppet who 'said it better'. Children correctly rewarded the puppet who had used the conjunction *and* on 56 cases out of the 60 trials (93.3 per cent). This result shows that children can compare two sentences and know which is most appropriate in the context. This evidence suggests that the Pragmatic Delay hypothesis cannot be correct: when the relevant alternative representations are readily available, children consistently indicate the puppet who had provided the most felicitous description of the situation, that is, the most informative statement. Of course, the findings do not prove that children have access to every piece of pragmatic knowledge, but merely that they know one piece of pragmatic knowledge, Grice's Maxim of Quantity. We can conclude, then, that children know how to increment the context, that is, they can apply the procedure described in (19)[1]. By inference, the results of the experiment reported in Section 6 showing that children over-accept the inclusive meaning of *or* in the second argument of *every*, a non-DE context, cannot be attributed to lack of knowledge of Grice's Maxim of Quantity.

How, then, can we explain the behaviour of children who over-accept the inclusive meaning of *or* in the second argument of *every*? We have seen that the computation of SIs involves two components: the recursive interpretation of a sentence (computation of truth conditions and of the set of alternatives and their maintenance in working memory) and the context incrementation (comparison of two representations, one corresponding to the plain meaning and the other corresponding to the scalar meaning and choice of the most informative statement to add to the context). We just established that the second component is not problematic. Therefore, the source of children's non-adult behaviour cannot be located at the level of the context incrementation. One possible hypothesis is that some aspect of the first component, the recursive interpretation of sentences, is responsible for children's failure to compute SIs. The operations involved in the recursive interpretation of a sentence may impose considerable demands on the language processing system, because they require children to build and maintain in working memory different representations of an assertion while another task is occurring, the recursive interpretation of the sentence. Thus, one can conjecture that children may fail to compute SIs because of limitations of the memory system. We call this hypothesis the Processing Limitation hypothesis. According to this hypothesis, children have to maintain some material in memory (two representations) and at the same time perform another task

---

[1] According to Grodzinsky and Reinhart (1993), a comparison of alternative representations is involved in the interpretation of sentences including non-reflexive pronouns, as in (i):

(i) John scratches him.

It is a well known fact, that children misinterpret sentences such as (i) and take the pronoun to be coreferent or anaphoric with the noun, that is, children take (i) to mean John scratches himself. Grodzinsky and Reinhart propose that, in order to reject the anaphoric reading of the pronoun in (i), participants have to establish if there is a more direct way of obtaining such a reading. In the case of (i), such a way consists in using a sentence that includes *himself* instead of *him*, as in (ii):

(ii) John scratches himself.

Given (i), children have to compute its representation and compare it with the representation of (ii). Such a comparison would result in rejecting the anaphoric reading of the pronoun in (i) because the grammar offers (ii) as a more appropriate way to convey such a reading. According to Grodzinsky and Reinhart, children know how to interpret pronouns, but are unable to compare two representations because this task exceeds their working memory. Therefore, they fail and make errors in interpreting sentences like (i). Our experiment shows that children do not have trouble comparing representations and thus suggests that children's working memory can hold two representations.

(the recursive interpretation of the sentence). It is the combination of the two tasks that may cause children to fail in computing SIs. Thus, we assume that children have the relevant semantic and pragmatic knowledge needed to interpret sentences including scalar items, that is, they have Grice's Maxim of Quantity, but they are indeed unable to calculate an implicature because of processing limitations. If this hypothesis is correct, we expect that children who fail in calculating SIs display lower performance in a task that evaluates their memory system and conversely that children who calculate SIs display higher performance in such a task. A task suited for this is one that requires children to keep some material in working memory and at the same time requires them to perform some other task. An alternative hypothesis is that children fail in the task described in Section 6 and over-accept the inclusive meaning of *or* in non-DE contexts because they focus on truthfulness or falsity of sentences and not on the felicity (or lack thereof) of a sentence in a given situation. Thus, their answers are usually driven by consideration of truthfulness or falsity. By contrast, adults are more skilled in focusing directly and preferably on the felicity of what is said. According to this hypothesis, children can compute SIs, but usually focus on truthfulness or falsity of sentences and answer accordingly. If this hypothesis is correct, we expect that, by manipulating the task, it is possible to have children focus on the felicity and to make them respond in a more adult-like way. Current experiments are focusing on these two alternative views of children's failure.

8 Conclusions

This chapter presented the results of various experiments investigating logical words. The relevance of the findings is twofold. First, they are consistent with the Semantic Core Model. Adults as well as a group of children compute the scalar implicatures in non-DE contexts, but fail to do so in DE contexts. Second, they show that children, at least some, appear to be more logical than adults and do fail to compute SIs, even in non-DE contexts. This failure may be due to the processing cost involved in computing scalar implicatures, in which different tasks compete for memory resources. Alternatively, it may arise because children do not focus directly on the felicity of sentences. Children do not lack any kind of pragmatic knowledge and in particular know which sentence is more appropriate to the context.

References

Chierchia, G. (2001). '*Scalar Implicatures and Polarity Phenomena.*' Ms. University of Milano-Bicocca.

Chierchia, G., Crain, S., Guasti, M. T., Gualmini A., and Meroni L., (2001). 'The acquisition of disjunction: Evidence for a grammatical view of scalar implicatures.' In *Proceedings of the 25th Boston University Conference on Language Development*: 157–68. Sommerville, MA: Cascadilla Press.

Crain, S., Gualmini A., and Meroni, L. (2000). 'The acquisition of logical Words.' *LOGOS and Language* 1: 49–59.

Crain, S. and McKee, C. (1985). 'The acquisition of structural restrictions on anaphora.' *Proceedings of NELS* 15: 94–110.

Crain, S., and Thornton, R. (1998). *Investigations in Universal Grammar: A Guide to Experiments on the Acquisition of Syntax and Semantics*. Cambridge, MA: The MIT Press.

Gazdar, G. (1979). *Pragmatics: Implicature, Presuppositions and Logical Form*. New York: Academic Press.

Grice, P. (1975). 'Logic and conversation.' In P. Cole and J. Morgan (eds), *Syntax and Semantics. Volume 3: Speech Acts*. New York: Academic Press. Also in Paul Grice (1989), *Studies in the Way of Words*. Cambridge, MA: Harvard University Press.

Grodzinsky, Y., and Reinhart, T. (1993). The innateness of binding and coreference. *Linguistic Inquiry* 24: 69–101.

Gualmini, A., Crain, S., and Meroni, L. (2000). 'Acquisition of disjunction in conditional sentences.' *Proceedings of the 24th Boston University Conference on Language Development*: 367–78.

Gualmini, A., Meroni, L., and Crain, S. (2001). 'The inclusion of disjunction in child grammar: Evidence from modal verbs.' *Proceedings of NELS* 30: 247–57. Amherst, MA: GLSA, University of Massachusetts.

Horn, L. (1972). *On the Semantic Properties of Logical Operators in English*. Ph.D. Dissertation, UCLA.

Horn, L. (1989). *A Natural History of Negation*. Chicago: University of Chicago Press.

Kadmon, N., and Landman F. (1993). 'Any.' *Linguistics and Philosophy* 16: 353–422.

Krifka, M. (1995). 'The Semantics and Pragmatics of Polarity Items.' *Linguistic Analysis* 25: 209–57.

Ladusaw, W. (1979). *Negative Polarity Items as Inherent Scope Relations*. Ph.D. Dissertation, University of Texas at Austin.

Levinson, S. C. (2000). *Presumptive Meaning*. Cambridge, MA: MIT Press.

Noveck, I. (2001). When children are more logical than adults: Experimental investigations of scalar implicature. *Cognition* 78: 165–88.

Paris, S. (1973). Comprehension of language connectives and propositional logical relationships. *Journal of Experimental Child Psychology* 16: 278–91.

14
Pragmatic Inferences Related to Logical Terms
Ira A. Noveck

1 Introduction

Paul Grice was concerned with the way logical terms such as *some, or* and *and* take on extralogical meanings in conversational contexts. To take one example, Grice (1989) described *or* as having a weak word meaning identical to formal logic's inclusive disjunction (which is false only in the case where both disjuncts are) but as conveying in conversation a speaker's stronger meaning corresponding to the exclusive disjunction (which is false in the case where both disjuncts are false and where both are true). Grice used the term *implicature* to describe the pragmatic inference linking word meanings to speaker's meanings and laid the foundations for nearly all of the linguistic-pragmatic studies found in this volume.[1]

The idea of submitting Grice's hypotheses to experimental investigation is extremely attractive. But what would testable predictions (and especially processing predictions) look like precisely? Grice assumed that a hearer expects a speaker, in producing an utterance, to obey a set of maxims following from a general cooperative principle. When the hearer's initial interpretation of an utterance fails to confirm that the speaker has obeyed the maxims, or at least the cooperative principle, the hearer derives implicatures so as to reconcile the overall interpretation of the utterance with his expectation. How does this work exactly? In general, one would expect that the derivation of an implicature should involve extra processing, but Grice does not provide enough detail. Moreover, Grice suggests that some implicatures – his so-called 'generalized conversational implicatures' linked

---

[1] The pragmatic literature has fine-tuned the notion of implicature (Bach, 1994; Sperber and Wilson, 1986/1995) making it a confusing term for describing the inferences to be discussed here. I will refer to the derived extra-logical meanings generically as pragmatic inferences or specifically as either a scalar inference (e.g., when *but not both* is derived from *or*) or as pragmatic enrichment (when *and* is treated as *and then*).

in particular to words such as *some*, *and* or *or* – are derived by default, and may be contextually cancelled. For these generalized implicatures, it is their cancellation that should involve extra processing. Efforts to clarify Grice's position has led to more recent theoretical work in pragmatics which has both sharpened Grice's ideas and provided alternative accounts, bringing us closer to formulating experimentally testable hypotheses.

In what follows, I briefly review the proposal from neo-Griceans, focusing on Levinson (2000), before turning to Relevance Theory (Sperber and Wilson, 1986/1995). I will then show how the developmental psychological literature had been investigating implicatures over 20 years ago, albeit unwittingly. The rest of the chapter describes several experiments that I and colleagues of mine have been carrying out in my lab, with children and adults, in order to better understand how pragmatic inferences linked to logical terms are generated.

2 Two post-Gricean approaches

According to neo-Griceans, such as Horn (1973) and Levinson (Levinson, 1983, 2000), the scalar inference illustrated by Noemi's response in (1) is a case that works on *terms* that are relatively weak:

(1) [*Knock at the door*]
    Isaac: Is that Mama and Papa?
    Noemi: It's Mama or Papa.

Noemi's choice of a weak term *or* implies the rejection of the stronger term *and*. More specifically, the connectives *or* and *and* may be viewed as elements of a scale (<or, and>), where *and* constitutes the more informative element (since *p and q* entails *p or q*). In the event that a speaker chooses to utter a disjunctive sentence, *p or q*, the hearer will take it as suggesting that the speaker either has no evidence that a stronger element in the scale, that is, *p and q*, holds or that she perhaps has evidence that it does not hold. Presuming that the speaker is cooperative and well informed the hearer will tend to infer that it is not the case that *p and q* hold, thereby interpreting the disjunction as exclusive. This neo-Gricean analysis can be extended to other logical terms. For example, if a speaker uses the weak quantifier *some* (as in *some triangles are equilateral*), it implies that the stronger quantifier *all* is not appropriate. If one uses the modal *might* (as in *Bill might be in the office*), it implies that the speaker had reason *not* to say the stronger-sounding *must* (as in *Bill must be in the office*). However, the neo-Gricean account is not limited to logical terms and it has been applied to a host of scales initially described by Horn (for a review, see Levinson, 1983, 2000). Other possible scales are frequency (where the use of *sometimes* excludes *always*) and epistemic status (where the weaker *think* implies that it is not the case that *know*).

In each case, scales range from less to more informative and the speaker's use of a less-informative term implies the exclusion of a more-informative one. Levinson (2000) motivates his most recent account by pointing out how 'the remarkably slow transmission rate of human speech', leads to a 'bottleneck in the efficiency of human communication' (p. 28). His proposal for surmounting the bottleneck is to profit from one's relatively high speed of comprehension in order to treat pragmatically enriched meanings of certain terms as a 'default' or as a 'preferred' meaning. These preferred meanings are put in place as a result of heuristics. For example, scalars are considered by Levinson to result from a Q-heuristic, dictating that 'What isn't said isn't (the case)'. It is named Q because it is directly related to Grice's (1989) first Maxim of Quantity: *Make your utterance as informative as is required*. In other words, Levinson assumes that scalar inferences are general and automatic. When one hears a weak scalar term like *or*, *some*, *might* and so on, the default assumption is that the speaker knows that a stronger term from the same scale is not warranted or that she does not have enough information to know whether the stronger term is called for. This means that relatively weak terms prompt the inference by default – *or* becomes *or but not both*, *some* becomes *some but not all* and so on. Scalar inferences by default can be cancelled, but the very idea of cancellation (as opposed to, for instance, inhibition) implies that it must occur *subsequent* to the production of the inference.

The other account comes from Relevance Theory (Sperber and Wilson, 1986/1995), which assumes that the interpretation of an utterance can be inferentially enriched in order to better capture the speaker's intention, but such pragmatic enrichment is not achieved through context-insensitive default inferences triggered by the mere presence of a weak scalar term. According to Relevance Theory, the so-called scalar inferences are ordinary pragmatic inferences drawn by a hearer in order to arrive at an interpretation of an utterance that meets his expectations of relevance. How far the hearer goes in constructing an utterance's interpretation is governed by considerations of effect and effort; hearers expect the intended interpretation to provide satisfactory effect for minimal effort.

A non-enriched interpretation of a scalar term (the one that more closely coincides with the word's meaning) can often lead to a relevant-enough interpretation of the utterance in which it occurs. Consider *Some monkeys like bananas*. This utterance with an interpretation of *Some* that remains in its weaker form (this can be glossed as *Some and possibly all monkeys like bananas*) can suffice for the hearer and not require further pragmatic enrichment. In contrast, the potential to derive a scalar inference comes into play when an addressee has higher expectations of relevance. A scalar inference could well be drawn by a hearer in an effort to make an utterance more informative and thereby more relevant. Common inferences like scalars are those that optionally play a role in such enrichment; they are not steadfastly linked to the words that could prompt them. When a scalar inference takes place and renders an underinformative utterance more informative, it ought (all things being equal) to involve extra effort.

One can better appreciate the two accounts by taking an arbitrary utterance (2) and comparing its linguistically encoded meaning (3a) to the meaning inferred by way of scalar inference (3b):

(2) Some X are Y.
(3) (a) Some and possibly all X are Y (logical interpretation).
    (b) Some but not all X are Y (pragmatic interpretation).

Note that (3a) is less informative than (3b) because the former is compatible with four possibilities: (i) X is a subset of Y; (ii) Y is a subset of X; (iii) X and Y overlap; and (iv) X and Y coincide, whereas interpretation (3b) is compatible only with possibilities (ii) and (iii); the pragmatic interpretation reduces the range of possible states of affairs. According to Levinson, the interpretation in (3b) is prepotently adopted through the Q-heuristic. This becomes the default meaning unless something specific in the context leads one to cancel the inference giving rise to (3b) and to *then* adopt the reading in (3a).

According to Relevance Theory, a hearer starts with the most accessible interpretation, which, in the absence of contextual cues, is provided by the plain linguistic meaning of a word such as 'some', as in (3a); if that reading is satisfactory to the hearer, he will adopt it. However, if interpretation (3a) fails to meet the hearer's expectation of relevance, he may enrich it and adopt interpretation (3b) instead. Given that (3b) arrives by way of a supplementary step (an inference), there is a cost involved (i.e., cognitive effort). This amounts to deeper processing, but at a cost.

3 Classic experimental findings

As this volume exemplifies, only recently has there been a concerted effort to tackle linguistic-pragmatic issues experimentally. However, it is important to point out that there are some classic reasoning studies (Braine and Rumain, 1981; Evans and Newstead, 1980; Paris, 1973; Smith, 1980; Sternberg, 1979) that serve as a prelude for work addressing the issues here. These prior studies yield two kinds of results. One is that when adult participants are presented weak scalar utterances in a context in which a stronger scalar expression is justified, they are often equivocal between logical and pragmatic interpretations. For example, consider two studies on disjunction: one conducted only with adults and a developmental one that included adults. In the adult study Evans and Newstead (1980, Experiment 2) presented participants with a disjunctive rule on a screen along with a letter–number pair. Consider 'Either there is a P or a 4' along with letter–number pairings such as a P with a 4, a P with a 9, or a Q with a 4 and so on. When the letter–number pair is P *and* 4 (presented as 'P 4'), the authors report that 57 per cent of participants reply that such a rule is true of this combination (and given that the task required a forced choice, 43 per cent said that it was false). In the developmental study, Paris (1973) showed that 67.5 per cent of adults respond True when presented with two images, for example a boy next to a bicycle and a monkey in a tree, and told to evaluate 'Either the boy is next to the bicycle or the monkey is in the tree' (and 75 per cent of adults respond true if the formulation excludes *Either* as in 'The boy is next to the bicycle or the monkey is in the tree').

The other, more remarkable result is that, provided identical situations, children are more likely than adults to provide logically correct responses. In the Paris (1973) study, upwards of 90 per cent of 8-year-old children accept (as true) cases where both disjuncts are true; moreover, these children were significantly more likely than adults to respond True to these cases and the developmental trend (ranging across five ages) is monotonic. This developmental effect appears robust since it is found elsewhere in the literature (Braine and Rumain, 1981; Sternberg, 1979; see also Smith, 1980). Here is what Sternberg (1979, p. 492) plainly said after coming across the same kind of developmental result that Paris reported:

> The data show an interesting interaction between age and interpretation of *or*... children at the lowest grade level use the inclusive interpretation of *or* in preference to the exclusive interpretation... At the higher grade levels, children show a strong tendency to use the exclusive interpretation in preference to the inclusive interpretation.

What is absent in these papers is an explanation for this effect. Although researchers recognized its curious nature, they were generally mystified by it. After all, it was rare to find children behaving more logically than adults.

Viewed through the prism of pragmatics, however, the effect becomes obvious: the linguistically encoded meanings of weak terms (like *or, some* and *might*) are compatible with minimal interpretations of underinformative items while pragmatic inferences increase with age. That is, the minimal interpretation for each of these terms is compatible with a logical one (e.g., when someone says *or*, it means that at least one disjunct is true, when someone says *some*, it means that at least one of several quantified objects is the case etc.). Pragmatic enrichments provide for the adult responses. This pragmatic insight could readily solve a small mystery in the developmental literature. The next section shows just how general this effect is.

4 Establishing the developmental-pragmatic effect

Recent efforts to better understand the developmental-pragmatic effect have led to the creation of a small cottage industry (Noveck, 2001; Doitchinov,

2004; Papafragou and Musolino, 2003; Papafragou and Tantalou, in press; Feeney et al., in press; Meroni, Gualmini and Crain, 2001; Chierchia et al., this volume; Guasti et al., 2003). My own efforts began by investigating children's responses to weak scalar utterances that were expressed as modals or quantifiers (Noveck, 2001) and was inspired by investigations into children's modal reasoning performance (Noveck, Ho and Sera, 1996). For modals, the critical test item was *There might be a parrot* when something like *There must be a parrot* would be more appropriate. For quantifiers, the test items were statements like *Some elephants have trunks* (and we know that *All elephants have trunks*). The scenarios from Noveck (2001) along with its main results are presented in the next two sections.

4.1 The modal *might*

Consider a reasoning task involving three boxes. One is open and has a toy parrot and a toy bear in it (the Parrot + Bear Box), the second is open and has only a parrot (the Parrot-only Box), and the third stays covered (Box C). Participants are told that Box C has the same content as either the Parrot + Bear Box or the Parrot-only Box. A puppet presents eight statements and it is the participant's task to say whether the puppet's claim is right or not. The critical statement that allows us to study the pragmatic inference is *There might be a parrot in the box* when the evidence shows that there *must* be a parrot. On the one hand, if the participant adopts a logical interpretation of *Might* (where *Might* is compatible with *Must*), one would expect an affirmative reply ('the puppet is right'). On the other hand, if the participant adopts a pragmatic, restrictive interpretation for *Might* (where *Might* is not compatible with *Must*) one would expect a negative reply ('the puppet is wrong') or at least some equivocation. Seven-year-olds' rate of logical interpretations with respect to *There might be a parrot in the box* (80 per cent) is intriguing not only because they respond at rates that are significantly above chance levels, but because they do so at a rate that is significantly higher than that of the adults' (35 per cent), which does resemble chance levels.

4.2 Existential quantifiers

In another experiment investigating quantified statements, 8-year-old and 10-year-old children and adults were confronted with various statements including a set that can be exemplified with *Some elephants have trunks*. These kinds of utterances have the potential to generate scalar inferences of the sort *Not all elephants have trunks*, which leads to a contradiction with one's stereotypical knowledge about elephants. Two features of these studies are worth pointing out. One is that there were six kinds of statements overall based on there being Existential and Universal Quantifiers (*Some* and *All*) and three kinds of relations: (a) Absurd (e.g., Some chairs tell time/All crows have radios); (b) Appropriate (e.g., Some houses have bricks/All elephants have trunks); and (c) Inappropriate (e.g., Some giraffes have long necks/All dogs have spots). It is the *Inappropriate* condition for *Some* that presents us with the infelicitous, Underinformative statements. The other important feature is that the test was conducted under double-blind conditions. The experimenter was simply told to present the utterances in an even tone of voice before soliciting an Agree/Disagree response (and the experimenter later told me that she thought the absurd statements were the critical ones). Aside from near perfect responses to the five indisputable statements, the results showed that all the children were very likely to agree (at rates at or above 85 per cent) with the Underinformative statements and significantly more so than adults, who were split in their responses (41 per cent of participants consistently agreed with the Underinformative statements).

5 Better characterizing the developmental-pragmatic effect

I have presented just two examples from the more recent literature on the developmental-pragmatic effect. Others have reported similar findings, making this a robust effect (references here include: Chierchia et al., this volume, Chapter 13; Guasti et al., 2003; Papafragou and Musolino, 2003; Papafragou and Tantalou, in press; Meroni, Gualmini and Crain, 2001). When a relatively weak term is used in scenarios where a stronger term is justified, younger children are typically more likely than adults to find the utterance acceptable.

Several issues remain. For one thing, some colleagues have reacted to my data as if I mean to say that children are pragmatically delayed at young ages (e.g., Meroni, Gualmini and Crain, 2001). This is very far from what I claimed because I had, in fact, anticipated that 'one would find the same effect among even younger children if a task were made easy enough' (Noveck, 2001, p. 184). Another issue concerns the relevance of this effect to resolving the debate between the two opposing accounts of scalar inferences compared earlier. These developmental findings do not favour one account over another because both could explain it. From the Default perspective, it could be claimed that scalar inferences become automatic with age and that our results are simply revealing how such inference-making matures. In contrast, Relevance Theory would suggest that children and adults use the same comprehension mechanisms but that greater cognitive resources are available for adults, which in turn encourages them to draw out more pragmatic inferences.

One way to address both of these issues is by finding a link between scalar-inference production and task complexity. If one assumes that cognitive effort is indeed a critical factor in such inference-making, a simpler task ought to make its production more likely (for similar arguments, see Noveck, Chierchia, Chevaux, Guelminger and Sylvestre, 2002). Such a finding would have to be considered favourable to Relevance Theory because it would indicate that the simplification of contexts facilitates the production of a specific interpretation of an utterance. If scalar implicatures were automatic and linked to particular words, task complexity ought not to matter. This is what Nausicaa Pouscoulous, Guy Politzer, Ann Bastide and I aimed to investigate in a series of studies conducted with much younger children.

5.1 Existential quantifiers again

In a set of studies, we employed a reasoning scenario that tested quantifier comprehension but that (unlike in the above study on *Some*) did not require long-term knowledge. In our first experiment, we presented a standard paradigm that placed four cardboard boxes in front of participants with different plastic animals placed in and around the boxes. The 9-year-old and adult participants were asked whether they agreed with a puppet that made statements about the scenario. Among numerous control items was the critical test utterance statement – 'Some turtles are in the boxes' (*Certaines tortues sont dans les boîtes*) when in fact there was a turtle in each of the boxes. Responses to this statement are then used to assess whether participants make the scalar inference. If participants were to make the scalar inference, they would disagree with the puppet (because *all* the turtles are in the boxes), whereas if they treated the word *some* in a logical way they would agree with the statement. The results confirmed the previous findings (91 per cent of children responded logically whereas 53 per cent of adults did). This is the scenario that was modified so that younger children could be tested.

We made three changes which we believed would reduce the effort necessary to perform the task, thus encouraging the pragmatic responses in children. First, we used the French word *quelques* instead of *certains*. Although both words mean *some*, teachers indicated that children were more comfortable with the former. Second, the presentation concerned only tokens that were in boxes; there were no utterances concerning tokens left outside the boxes (as was the case for the animals in the standard paradigm). Third, we asked participants to perform an action on the basis of the puppet's instructions rather than make a judgement on the validity of the puppet's statements. That is, participants were asked to fulfil a wish made by a puppet concerning the items in the boxes. For example, the puppet would say 'I would like all of the boxes to have a token', and the participants would have to determine whether or not they should alter the scenario in order to comply with the wish. The utterance of interest arose when the puppet said 'I would like some boxes to have a token' (*Je voudrais que quelques boîtes contiennent un jeton*) when all of the boxes already contained a token. If participants believed that *some* was compatible with *all* they ought to leave the boxes unchanged (which is actually difficult to do given that a request implies that a change is called for); otherwise they could remove some of the tokens.

When very young children (4-year-olds, 5-year-olds, and 7-year-olds) are presented with a scenario having (this time) five boxes and each with a token, one finds many more responses indicating scalar inference generation than in the prior experiments. Only 32 per cent of the 4-year-olds and

27 per cent of the 5-year-olds gave a response that indicated that they chose the logical response (i.e., they left the boxes untouched); even fewer seven-year-olds did so (17 per cent). Scalar inference making is more apparent in the children even as the developmental ordering remains marked. Pragmatic inference making is affected by task ease (and appears to continue to be affected by the level of sophistication on the part of the hearer). This is evidence that the ability to draw scalar inferences is not uniquely linked to maturity. By making a task easy enough, we have encouraged children to apply their available resources to drawing the scalar inference.

5.2 Pragmatically enriching 'and'

Does this developmental effect extend to non-scalar cases? Here, we focus on a pragmatic inference linked to *and* which prompts the same developmental tendency as those described for scalars. To appreciate the pragmatic enrichment in question, consider the two conjunctive utterances in (4):

(4) (a) Mary got married and got pregnant.
    (b) Mary got pregnant and got married.

The two are equivalent from a logical point of view because they both contain the same two components, that is, $P \& Q = Q \& P$. However, the sequence of the conjuncts in each of the two utterances conveys two very different sets of implications. Whereas it would be considered a normal occurrence to hear about someone getting married before getting pregnant (4a); in some parts of the world, it would be considered scandalous to get pregnant before getting married (4b). Without the implicit sequential interpretation, the statement in (4b) would not seem scandalous.

The pragmatic inference linked to *and* prompts the same debate as the one for scalars.[2] On the one hand, Levinson (1983, 2000) argues that interlocutors 'buttress' conjunctions by interpreting *and* to mean *and then* and that they do this by default. On the other, Carston (1996, 2002) points out that there are a host of ways in which *and* can be enriched and argues that none dominate, that context determines which pragmatic enrichment to make and that there is nothing automatic about it. Below, are just five of the kinds of implications that a conjunctive utterance can convey:

(2) (a) *Contrast*: It's autumn in the US and it's spring in Chile.
    (b) *Sequential*: She took the scalpel and made the incision.
    (c) *Containment*: We spent the day in town and went to Macy's.

---

[2] The linguistic intuition that the pragmatic enrichment of *and* is comparable to scalar inference making is not universally shared. Recanati (2003) assumes that the pragmatic enrichment of *and* is sublocutionary, making it (in contrast to scalars) not as readily available to consciousness.

(d) *Causal*: She shot him in the head and he died instantly.
(e) *Indirect Causal*: He left her and she took to the bottle.

Carston's account is largely corroborated in two separate studies. Noveck and Chevaux (2001) presented 7- and 10-year-old children as well as adults a set of 12 stories about everyday events. Among these were four stories that ultimately presented a conjunctive sentence as a comprehension question. For half of the conjunctive comprehension questions, the two events in the question were presented in a sequence that respected the order in the stories and in the other half the sequence was inverted. To illustrate, consider the short story in (5) and its two kinds of follow-up questions in (6a) and (6b):

(5) While sitting on her couch, Julie was reading a comic book.
Suddenly, the phone rang.
She went out of the living room and ran to answer.
It was Isabelle who was inviting Julie to celebrate her birthday Saturday.
Since they were very good friends, Julie accepted the invitation.
(6a) Julie answered the phone and accepted an invitation?
(6b) Julie accepted an invitation and answered the phone?

Whereas the rates of agreement to (6a) are high and accurate for all participants, we found that linguistically competent children are less fussy than adults about sequence in conjunctive sentences. Roughly 85 per cent of 7-year-olds, 63 per cent of 10-year-olds, and about 29 per cent of the adults say 'Yes' to (6b). This developmental curve resembles the one for scalars. Younger children are more likely to agree with a statement's minimal interpretation because they are less likely to pragmatically enrich the meaning of *and*.[3]

Another piece of evidence supporting Carston's account comes from processing studies that followed up on the paradigm above. Noveck, Chevaux and Bott (2004) enlarged the list of stories and presented them line-by-line on a screen to 10-year-olds and adults.[4] The dependent measures were the Yes/No response to the question as well as the response time it

---

[3] It is important to point out that only the 10-year-old children appeared adult-like in all other respects. The 7-year-olds tended to respond 'Yes' to control, inverted-order questions that employed *and then* explicitly as its conjunction.
[4] The stories were highly similar to those in Noveck and Chevaux (2001), but were slightly modified for computer-presentation purposes and were not joined by filler items. There were nine stories altogether and each could be followed with any one of three kinds of comprehension questions: Order-preserved and Order-inverted as before, plus a new control Order-preserved with a false second conjunct. This new type of category can be created by changing a detail of the story. For example, the new control for the story in (5) had the story conclude with: *Since they were not close friends, Julie declined the invitation*. With such a conclusion, a question like the one in (6a) merits a 'No' response.

required. The developmental curve for the categorical responses was as before, although there was more pragmatic inferencing in general: The 10-year-olds were significantly more likely than adults to accept the inverted order as true (46 per cent vs 18 per cent, respectively) and this was the only adult–child comparison to yield significant effects.

Of further interest were 10-year-olds' response times to the two sorts of responses because one could readily draw out processing predictions from the two opposing accounts. On the one hand, if the sequential (buttress) interpretation occurs by default (or is automatic in some form), then 'No' responses ought to be quickest and 'Yes' responses ought to take at least as long because the latter response requires the cancellation of the buttress. On the other hand, a Relevance Theory account predicts that those who respond 'Yes' to the inverted-order questions ought to answer more quickly than those who answer 'No' because the pragmatic enrichment occurs subsequent to minimal semantic treatments of *and*. The data fall in favour of a Relevance Theory account. Children who answer 'No' to the inverted-order questions require (on average) 2 seconds more than those who answer 'Yes'. And it cannot be argued that the children's slow response here is due to hitting the 'No' key because the control question prompts a (correct) 'No' response and it led to the fastest response times overall (see Note 4). Unfortunately, there were too few adult responses that would justify a similar analysis among them here. However, in the next section I describe response times in sentence verification tasks that investigate scalar inferences among adults and their response times tell a similar story.

6 Adult studies

While the developmental studies are illuminating, it could be argued that they impose certain limitations. Perhaps children are categorically different from adults or perhaps their data are not as reliable as the adults'. Although I do not share these criticisms, support for Relevance Theory claims would be stronger if the same sorts of effects could be found among adults. How could we investigate that? One way is to uncover the time course of pragmatic enrichments a bit more carefully. If one could provide evidence showing that pragmatic interpretations of scalars are the first to arise and that interpretations that require their cancellation occur subsequently, then the default inference view would be supported. However, if one could show that minimal interpretations are at the root of initial interpretations and that pragmatic interpretations arise only later, that would be further support for Relevance Theory.

6.1 Existential quantifiers

Lewis Bott and I (Bott and Noveck, in press) set up four carefully controlled experiments which investigated the time course of statements like *Some monkeys are mammals*. The design was drawn from Smith (1980) and Noveck (2001), but it benefited from several rigorous controls. I describe three of these below.

First of all, our experiments included six kinds of sentences, all of the form *[Quantifier] A are B*, with two possible quantifiers (*Some* and *All*) and three kinds of set relationships between the As and the Bs – one in which the As are a proper subset of the Bs (*Some/All monkeys are mammals*), one in which the Bs are a proper subset of the As (*Some/All mammals are monkeys*) and a third where the As and the Bs form two disjoint sets (*Some/All monkeys are fish*). As can be seen in Table 14.1, these materials provide us with Underinformative items as well as a range of controls that include both True and False items. This way we could compare both kinds of anticipated responses for the Underinformative items with many different controls.

Second, the experiment was designed so that any one of nine subordinate categories (e.g., monkeys) could be randomly joined in a quantified statement with any one of six superordinate categories (mammals, reptiles, fish, insects, shellfish and birds). For example, one could insert 'monkeys' to be part of *Some monkeys are mammals, Some mammals are monkeys, Some monkeys are fish, All monkeys are mammals, All mammals are monkeys* or *All monkeys are fish*. Each subordinate category was used just once and randomly per experiment which means every participant viewed a unique set of materials.

Finally, an on-line investigation allowed us to be creative with our presentation. One could present the sentences one word at a time or an entire sentence at a time. We could also require participants to hurry by putting time limits or encourage them to take their time. All told, this form of

Table 14.1 Six sentence types in the time course studies of correct (justifiable) responses produced, and their concomitant reaction times (from Bott and Noveck, Experiment 3)

Sentence	Example	Justifiable Responses	Reaction Time
T1	Some elephants are mammals	True (41 %) False (59 %)	2617 (Logical) 3360 (Pragmatic)
T2	Some mammals are elephants	True (89 %)	2644
T3	Some elephants are insects	False (93 %)	2610
T4	All elephants are mammals	True (87 %)	2875
T5	All mammals are elephants	False (97 %)	2558
T6	All elephants are insects	False (92 %)	2340

experimentation allows us to create various experimental contexts as we ask participants to respond to Underinformative items and their controls.

Before continuing the description of the experiments, it is important to point out that classic categorization studies (much like the classic developmental studies mentioned earlier) did pay some attention to the infelicitous Underinformative items. However, most dealt with the two interpretations of *Some* largely by sidestepping it. Response-time experiments have generally instructed their participants to interpret *Some* in a strictly logical way (i.e., without the scalar inference). For example, Meyer (1970) told participants to treat *some* as meaning *some and possibly all* in a sentence verification task with sentences such as *Some pennies are coins*. To the best of our knowledge, there is only one psychological study to take an interest in the potentially conflicting interpretations of Underinformative sentences (Rips, 1975). Rips investigated how participants make category judgements by using sentence verification tasks with materials like *Some congressmen are politicians*. He examined the effect of the quantifier interpretation by running two experiments, one in which participants were asked to treat *some* as *some and possibly all* and another where they were asked to treat *some* as *some but not all*. This comparison demonstrated that the participants given the *some but not all* instructions in one experiment responded more slowly than those given the *some and possibly all* instructions in another. Despite these indications, Rips modestly hedged when he concluded that 'of the two meanings of Some, the informal meaning *may* be the more difficult to compute' (italics added). His reaction, of course, is not uncommon. Many colleagues share the intuition that the pragmatic interpretation seems more natural. In any case, this is an initial finding that goes in favour of the relevance account.

Our studies picked up where Rips left off. In Experiment 1, we replicated Rips (1975, Experiments 2 and 3) in one overarching procedure. Participants took part in two sessions. In one, participants were instructed to interpret the quantifier *Some* to mean *some and possibly all*, which we refer to as the Logical condition, and in the other they were told to interpret *Some* to mean *some but not all*, which we will refer to as the Pragmatic condition (and, of course, session order was counterbalanced). Central to our interests was the accuracy and the speed of response to the Underinformative sentences (e.g., *Some monkeys are mammals*) under the two conditions. According to the Default Inference account, a False response should be faster than a True response because the latter ought to occur as a result of the default inference's cancellation. In contrast, Relevance Theory would predict that a False response occurs more slowly than a true response because the False response would arise when Relevance Theory conditions are applied more stringently, resulting in the production of the scalar inference.

The results showed that when participants were under instruction to, in effect, draw the scalar inference, they required significantly more time to evaluate the Underinformative sentences than when they were under instructions to provide a Logical response. In fact, a response that reflects the presence of a scalar inference (i.e., to say False to *Some monkeys are mammals*) is extraordinarily slow (i.e., slower than any other True or False response in the task). This will become a staple finding in this series of experiments.

The data also show that participants have greater difficulty providing the 'correct' response when they are given Pragmatic instructions. Participants are accurate on approximately 85 per cent of the Underinformative items under Logical instructions and accurate on about 60 per cent of the Underinformative items under Pragmatic instructions. Moreover, the rate of correct responses for the Underinformative item under Logical instructions is comparable to the rates of correct responses to all of the control items, whereas the rate of 'correct' responses for the Underinformative items under Pragmatic instructions appears exceptionally low. Thus, at least some of the extra time required under Pragmatic instructions is due to the processing requirements of making and maintaining the inference. This much confirms Rips's initial findings. There are no indications that turning *some* into *some but not all* is an effortless step.

Experiment 3 used the same paradigm but we provided neither explicit instructions nor feedback about the way to respond to the Underinformative sentences. Instead, we expected participants to answer equivocally to these types of sentences – some saying False and some True. This means that we should have two groups of responses: one in which the inference is drawn (what we call Pragmatic responses) and another where there is no evidence of inference (Logical responses). We can therefore make a comparison between the two as we did in the previous experiment. Once again, if Logical responses are made more quickly than Pragmatic responses, we have evidence against a default system of inference. We can also use the control sentences to verify that under these more neutral conditions, responses which involve a pragmatic inference require more time than responses that do not (as we found above).

As can be seen in Table 14.1, the main finding here is that mean reaction times were longer when participants responded pragmatically to the Underinformative sentences than when they responded logically. Furthermore, Pragmatic responses to the Underinformative sentences appear to be slower than responses to all of the control sentences, indicating that the scalar inference prompts an evaluation that is characteristically different from all the other items. Collectively, these two experiments provide further evidence against the default inference view because there is no indication that Underinformative items prompt participants to take more time to arrive at a true response than they do to a false response. All indications point to the opposite being true: a Logical response is an initial reaction to Underinformative sentences and it is indistinguishable from responses to control sentences while a Pragmatic response to Underinformative items is significantly slower than a Logical response as well as to the other items in the task.

Some colleagues wonder whether the Pragmatic responses we record are slow because in general judging a statement to be false takes more time than judging it to be true. We argue against that by pointing out that three of the control sentences also require a 'false' response and their reaction times are significantly faster than the Pragmatic response. Consider the sentences we classify as T5, which is exemplified by *All mammals are elephants* (and which includes many of the same elements as *Some elephants are mammals*). Such items prompt 97 per cent of participants to respond False correctly and at a speed that is significantly faster than when they respond False to Underinformative sentences.[5] It also cannot be argued that the Pragmatic responses are simply due to error (where participants intend to say 'True' but hit the wrong button) because the percentage of participants making Pragmatic responses is of a characteristically different order when compared to those making errors in the control conditions (roughly 60 per cent choose False to the Underinformative item as opposed to between 3 and 13 per cent who make errors across all the control conditions). We argue that these results indicate that the scalar inference is at the root of the extraordinary slowdown in this paradigm. It is drawn specifically in reaction to the Underinformative items and prompts an unusually large number of participants to ultimately choose False. Furthermore, it arrives as a secondary process relative to a justifiable Logical interpretation; it does not appear to arrive by default.

Although our experiments provide evidence against the idea that scalar inferences become available as part of a default interpretation, they do not necessarily provide evidence that directly supports the Relevance account. Our final experiment, which combined the general procedure of both Experiment 1 (sentences were presented one word at a time) and Experiment 3 (participants were free to treat *Some* as they wished) was designed to test predictions from Relevance Theory concerning the processing of scalar inference. The crucial manipulation was the time available for the response; in one condition participants had a relatively long time to respond (3000 msecs, this is referred to as the Long condition), while in the other they were given a short time to respond (900 msecs, this is referred to as the Short condition). By requiring participants to respond quickly in one condition, we limited the cognitive resources they had at their disposal.

Expectations based on the Relevance Theory account are that participants would be more likely to respond with a quick 'True' response when they

---

[5] Bott and Noveck (in press, Experiment 2) addresses the key-press issue directly. While using Rips's paradigm, the response options become 'agree' or 'disagree' and participants first read a statement such as 'Mary says that the following is false' or 'Mary says that the following is true'. Those who *agree* with Mary when she says 'false' to a T1 sentence while keeping a Pragmatic interpretation still take significantly longer to respond than those who *agree* when she says 'true' while keeping a Logical interpretation.

have less time than when they have more. If one wanted to make predictions based on default interpretations, *Some* should be interpreted to mean *some but not all* more often in the short condition than in the long condition (or at least there should be no difference between the two conditions). The results show that while the rate of correct performance among the control sentences either improves or remains constant with added response time, responses were such that there were significantly more Logical responses to the Underinformative items in the Short condition than in the Long condition: 72 per cent True in the Short condition and 56 per cent True in the Long condition. This trend is in line with predictions made by Relevance Theory.

5.2 Disjunctions

This time-enriched pragmatic effect has been extended to disjunctions. Imagine that an experiment presents a five-letter word and participants are required to respond with a *Yes* or *No* to statements such as *There is a T or a B*. As one could imagine, all possible truth-conditions can be introduced, including the one synonymous with the Underinformative items above. That is, if the word that preceded the statement above were TABLE, then the statement could provide a Logical interpretation (to say 'True' because indeed there is a T; there is a B) or a Pragmatic interpretation (to say 'False' because there is both a T and a B). In work carried out in collaboration with Valentina Lanzetti, Lewis Bott, Coralie Chevallier, Tatjana Nazir and Dan Sperber, one sample reveals that when participants are encouraged to respond within a second (they are told that their responses are not recorded in cases where they take longer than a second), participants' rates of Logical responses are roughly 84 per cent. When they are given an unlimited amount of time to decide, their rates of Logical responses drop to around 55 per cent. As before, it is important to point out that the unambiguous control problems (in this case, half of all included statements use the connective *and*) yield rates of correct responses that are very high across all conditions. The evidence indicates that indeed minimal interpretations serve as the basis for quick judgements and that Pragmatic responses arrive subsequently.

6 Neuropsychological measures and pragmatic inference making

6.1 Evoked potentials

In order to get at immediate reactions that might run deeper than reaction times, colleagues and I have also investigated Evoked Response Potentials (ERP; Noveck and Posada, 2003). ERP studies typically present specific anomalies in a sentence in order to capture a characteristic pattern that follows (see Coulson, Chapter 9). Kutas and Hillyard (1980) pointed out how semantic anomalies give rise to a central parietal negative-going component that peaks about 400 msec after the appearance of an inappropriate word, like *socks* in (7); this is known as an N400:

(7) John buttered his bread with *socks*.

Kutas and Hillyard discovered that N400s are steeper when a target word (which would be the final word here) is: (a) not associated with the prior context; or (b) is just unanticipated. Noveck and Posada (2003) investigated N400s as participants were presented Smith's (1980) task (the one that presents *Some elephants have trunks* as an Underinformative item). It included adults only (19 of them) and the quantifier *Some* was the only one used. The task was expanded to include 25 Underinformative sentences along with 25 sentences that are Patently True (e.g., *Some houses have bricks*) and 25 that are Patently False (e.g., *Some crows have radios*). Before getting to the ERP data, which focused on the final word of each sentence, it is important to highlight the behavioural data.

The reaction-time data are compatible with those found in Bott and Noveck in that False responses to the Underinformative statements take much longer than (nearly twice as long as) the True responses and in that Patently False items yielded the fastest response times overall. Thus, the relative slowness of the False responders to the Underinformative items occurs despite evidence of preparedness for the Patently False items. This finding makes it difficult to argue that Underinformative items, by representing one-third of the stimuli, allowed for a rote response among the Pragmatic responders. What is clear in this study (and less so in Bott and Noveck) is that participants manifest two sorts of strategies: There are seven who respond True to the Underinformative items by responding literally and quickly overall and there are 12 who respond False by responding pragmatically and slowly overall. It is also apparent that the two strategies prompt spillover because those who respond True to the Underinformative items are also significantly faster in responding correctly to the two other conditions. The behavioural data thus indicate that those who give a False response to the Underinformative items undertake deeper processing that is, in turn, evident in the responses to the other items in the task. Again, the deeper processing linked to the False responses in the Underinformative condition is consistent with expectations based on Relevance Theory because it assumes that pragmatic inference making arrives as a result of an effortful process.

The ERP data were instructive because they indicated that the Underinformative items generally led to flat N400s (indicating little semantic integration) and were flatter than *both* the Patently True and Patently False items. This is seen regardless of adopted strategies. For those (seven participants) who generally responded True to the Underinformative items and for those (12 participants) who were generally *pragmatic* in their responses, the N400s were comparable. The fact that the ERP profiles for the pragmatic group of participants in the Underinformative condition remains unremarkable, even as their responses and response times indicate much deliberation, is further evidence that participants' immediate reaction to underinformativeness (and signalled by the final word in each sentence) is benign. This indicates that the scalar inference, which requires more effort and prompts participants to respond False, is part of a late-arriving, effort-demanding decision process (see Heinze, Muente and Kutas, 1998, for a similar argument with respect to a categorization paradigm).

6.2 fMRI

More comprehensive work using imaging techniques is underway with Vinod Goel. The interest here is to see which brain regions are involved in the True versus False responses (the Logical vs Pragmatic responses) to Underinformative items. One study that is now being carried out includes both *All* and *Some* as quantifiers and a large set of stimuli (similar to those in Bott and Noveck, such as *Some monkeys are mammals*).

These are some of the questions that this fMRI work is ultimately meant to address: Right frontal areas are known to be the centres of reconciliation of conflicting information (see Goel and Dolan, 2003; Noveck, Goel and Smith, in press) as well as for 'contextual' and figurative meanings (Bookheimer, 2002); will a negative reply to Underinformative items prompt activity in an area such as in the right anterior cingulate, which is engaged in tasks that require planning and satisfying goals and subgoals (Koechlin, Basso, Pietrini, Panzer, and Grafman, 1999) or will pragmatic inferences prompt activity uniquely in language regions like Broca's area? Will a False response to *Some monkeys are mammals* prompt activity that is different from what is found for a patently false item like *Some monkeys are fish*? If so, one might be able to localize brain regions that concern themselves with pragmatic inference making. Is neural activity prompted by pragmatic inferencing with scalars similar to other kinds of pragmatics-induced neural activity (e.g., to metaphor and other pragmatic enrichments)? This is an area of research that is in infancy, but will surely be tackled with the expansion of experimental pragmatics.

7 Conclusions

The developmental data show that the production of scalar inference is possible for children (consider 7-year-olds who are largely 'pragmatic' with simple tasks in Pouscoulous, Noveck, Politzer and Bastide) but effortful (now consider 7–10-year-olds on similar but slightly more difficult tasks in Pouscoulous et al., 2004; and in Noveck, 2001, who are 'logical'). If a scalar were indeed a default inference linked to a lexical item, adding a little complexity to a task ought not to block it so readily. It is hard to argue that children are more likely to cancel scalar inferences than adults because that implies that children are applying more effort than adults. The evidence shows that young children whose responses do not reveal the drawing of the scalar inference have simply not considered it.

The adult time-course data tell a similar story. Adults are generally equivocal about statements such as *Some elephants are mammals, Some elephants have trunks* or that *There is a T or a B* in the word TABLE. It becomes relevant to know which interpretation is the first, competent and measurable one. A default treatment of scalars would predict that it is the upper-bounded meaning for *Some* (*some but not all*) and an exclusive interpretation for *or*, but this is not what we find. We consistently find that – in terms of time course – the earliest treatment of such expressions is the minimal meaning. The same appears to extend to cases concerning the pragmatic enrichment of *and*.[6] This fits with Relevance Theory's account.[7]

7.1 A final word about the role for experimentation in linguistic-pragmatics

As these data reveal, experimentation provides a start for a, potentially long, process that can eventually adjudicate between two theoretical approaches. This is not the only use for experimentation in pragmatics. A subtext of this work is that experimentation can have two other functions in linguistic pragmatics. One is that linguistic intuitions can be usefully complemented with other kinds of evidence. This much might seem obvious enough. The other is that experimentation can also help identify linguistic-pragmatic categories. For instance, the fact that the temporal aspects of the conjunction are attained over the course of development much like scalar implicatures might indicate that these two belong to the same linguistic-pragmatic category. Whether one wants to strengthen or contradict linguistic intuitions,

---

[6] One could argue that it is the production of negative information (*not both, not all* etc.) that is the very cause of the slowdown in a scalar inference. However, note how the pragmatic enrichment of *and* also prompts a slowdown (when compared to a 'logical' interpretation) yet it cannot be attributed to negative information. The enrichment in these studies concerns adding a sequential marker such as *then*. This is an indication that it is the pragmatic step itself that slows participants down and not necessarily its output.

[7] The grammatical phenomena that Chierchia highlights in Chapter 13 and that led to the investigation in Noveck, Chierchia, Chevaux, Guelminger and Sylvestre (2002) is compatible with an account that (a) assumes that there is one minimal meaning of scalar terms that is selectively enriched and that (b) denies a role for defaults concerning scalar inferences. Relevance Theory can account for Chierchia's grammatical cases when one views the production of scalar inference as a relatively effortful step that makes an utterance more informative. Informativeness also explains why the lack of scalar inference is more common in the antecedent of conditionals, as in *If P or Q then R*; a truth table analysis shows that there are fewer true cases when the disjunction is inclusive.

to establish typologies, or devise tests to compare two theories, experimentation has much to offer linguistic pragmaticists and, as the present work shows, linguistic pragmatics has much to offer us experimentalists.

References

Bach, K. (1994) Conversational impliciture. *Mind and Language* 9: 124–62.
Bookheimer, S. (2002). Functional MRI of language: New approaches to understanding the cortical organization of semantic processing. *Annual Review of Neuroscience* 25: 151–88.
Bott, L., and Noveck, I. A. (in press). Some utterances are underinformative: The onset and time course of scalar inferences. *Journal of Memory and Language*.
Braine, M. D., and Rumain, B. (1981). Development of comprehension of 'or': Evidence for a sequence of competencies. *Journal of Experimental Child Psychology* 31(1): 46–70.
Carston, R. (1996). Enrichment and loosening: Complementary processes in deriving the proposition expressed. In *UCL Working Papers in Linguistics* 8: 61–88.
Carston, R. (2002). *Thoughts and Utterances*. Oxford: Blackwell.
Doitchinov, S. (2004). Naturalistic and experimental data in language acquisition: The case of epistemic terms. Proceedings from the *International Conference on Linguistic Evidence: Empirical, Theoretical and Computational Perspectives* (in Tuebingen): 36–40.
Evans, J. S. B., and Newstead, S. E. (1980). A study of disjunctive reasoning. *Psychological Research* 41(4): 373–88.
Feeney, A., Scrafton, S., Duckworth, A., and Handley, S. J. (in press). The story of some: Everyday pragmatic inference by children and adults. *Canadian Journal of Experimental Psychology*.
Goel, V., and Dolan, R. J. (2003). Explaining modulation of reasoning by belief. *Cognition* 87(1): B11–B22.
Grice, H. P. (1989). *Studies in the Way of Words*. Cambridge, MA: Harvard University Press.
Guasti, M., Chierchia, G., Crain, S., Foppolo, F., Gualmini, A., and Meroni, L. (2003). Why children and adults sometimes (but not always) compute implicatures. Ms.
Heinze, H. J., Muente, T. F., and Kutas, M. (1998). Context effects in a category verification task as assessed by event-related brain potential (ERP) measures. *Biological Psychology* 47(2): 121–35.
Horn, L. R. (1973). *Greek Grice: A Brief Survey of Proto-Conversational Rules in the History of Logic*. Paper presented at the Proceedings of the Ninth Regional Meeting of the Chicago Linguistic Society, Chicago.
Koechlin, E., Basso, G., Pietrini, P., Panzer, S., and Grafman, J. (1999). The role of the anterior prefrontal cortex in human cognition. *Nature* 399: 148–51.
Kutas, M., and Hillyard, S. A. (1980). Reading senseless sentences: Brain potentials reflect semantic incongruity. *Science* 207: 203–5.
Levinson, S. (1983). *Pragmatics*. Cambridge: Cambridge University Press.
Levinson, S. (2000). *Presumptive Meanings: The Theory of Generalized Conversational Implicature*. Cambridge, MA: MIT Press.
Meroni, L., Gualmini, A., and Crain, S. (2001). *Conversational Implicatures and Computational Complexity in Child Language*. Paper presented at the LSA, Washington, DC.
Meyer, D. E. (1970). On the representation and retrieval of stored semantic information. *Cognitive Psychology* 1: 242–99.

Noveck, I. A. (2001). When children are more logical than adults: Experimental investigations of scalar implicature. *Cognition* 78(2): 165–88.

Noveck, I. A., and Chevaux, F. (2001). The Pragmatic Development of *and*. Paper presented at the BUCLD, Boston.

Noveck, I. A., Chevaux, F., and Bott, L. (2004). Semantic and then pragmatic influences in enriching the meaning of *and*. Ms.

Noveck, I. A., Chierchia, G., Chevaux, F., Guelminger, R., and Sylvestre, E. (2002). Linguistic-pragmatic factors in interpreting disjunctions. *Thinking and Reasoning* 8(4): 297–326.

Noveck, I. A., Goel, V., and Smith, K. W. (in press). The neural basis of conditional reasoning with arbitrary content. *Cortex*.

Noveck, I. A., Ho, S., and Sera, M. (1996). Children's understanding of epistemic modals. *Journal of Child Language* 23(3): 621–43.

Noveck, I. A., and Posada, A. (2003). Characterizing the time course of an implicature: An evoked potentials study. *Brain and Language* 85: 203–10.

Papafragou, A., and Musolino, J. (2003). Scalar implicatures: Experiments at the semantics-pragmatics interface. *Cognition* 86(3): 253–82.

Papafragou, A., and Tantalou, N. (in press). Children's computation of implicatures. *Language Acquisition*.

Paris, S. G. (1973). Comprehension of language connectives and propositional logical relationships. *Journal of Experimental Child Psychology* 16(2): 278–91.

Pouscoulous, N., Noveck, I. A., Politzer, G., and Bastide, A. (2004). Evidence of the production of scalar implicates in young children. *Ms*.

Rips, L. J. (1975). Quantification and semantic memory. *Cognitive Psychology* 7(3): 307–40.

Smith, C. L. (1980). Quantifiers and question answering in young children. *Journal of Experimental Child Psychology* 30(2): 191–205.

Sperber, D., and Wilson, D. (1986/1995). *Relevance: Communication and Cognition*. Oxford: Basil Blackwell.

Sternberg, R. J. (1979). Developmental patterns in the encoding and combination of logical connectives. *Journal of Experimental Child Psychology* 28(3): 469–98.

15
Conversational Implicatures: Nonce or Generalized?[1]

Anne Reboul

1 Introduction

Ever since its beginning, pragmatics has been plagued with a dissension as to its status: is it or is it not a part of linguistics on a par with phonology, syntax and semantics? The debate, as debates tend to do, has been going back and forth, with this or that side taking a momentary advantage until the pendulum swings back to the other side again. It is dubious whether the question can be answered in general, if only because some pragmatic phenomena may be more dependent than others on linguistic conventions. Thus, it seems reasonable to look at the problem from the vantage point of a specific pragmatic phenomenon. This is what I intend to do in this chapter by concentrating on *conversational implicatures*.

Conversational implicatures, first described by Grice (1989), can be best described from examples:

(1) Anne has four children.
(2) Anne has exactly/at most four children.
(3) Anne has at least four children.
(4) Anne has four children and even five.

It is generally considered that such utterances as (1) license (2) rather than (3). This is intriguing in as much as (1) is logically compatible with (3): if it is true that Anne has more than four children, it is *a fortiori* true that she has four children. Thus, the inference from (1) to (2) is not logical, which is why it was called by Grice a *conversational implicature*. Grice noted that

---

[1] I want to thank Ira Noveck for invaluable help in devising the experiments and his rereading of the paper. I also want to thank Dan Sperber for rereading the paper. All remaining mistakes are of course my own.

conversational implicatures are *defeasible* (they can be cancelled), as shown by the fact that utterances such as (4) are not contradictory. These, then, are the facts on which everyone seems to agree: conversational implicatures are a pragmatic phenomenon; they are not logically licensed; they are defeasible. This is also the point beyond which agreement stops and controversy starts.

The present debate over conversational implicatures concerns whether they are due to *nonce* (one-off) inferences or are *default* inferences, triggered by lexical items or sequences of them. If the first, they can only be accessed at the end of the utterance, that is, at a *global* (sentential) level. If the second, they are accessed as soon as their trigger is met in the course of the utterance, that is, at a *local* (subsentential) level. In what follows, I will call advocates of conversational implicatures as nonce-inferences *globalists* (and their theories *global theories*) and advocates of conversational implicatures as default-inferences *localists* (and their theories *local theories*). Though I will refrain from going into theoretical details, let me just say that Sperber and Wilson's Relevance Theory (see Sperber and Wilson 1986/1995) is a good example of global theories, while Levinson's theory of Generalized Conversational Implicatures (see Levinson 2000) is a good example of local theories.

Regarding the relevance of the localist/globalist views of conversational implicatures to the status of pragmatics (integrated to or independent from linguistics), it should be clear that localists are partisans of integration while globalists are partisans of independence.

Both localists and globalists have tried to defend their respective theories on the general grounds of economy. Localists claim that default inferences are less costly than nonce inference in as much as they do not have to be made anew on every new instance. However, it might be argued that, be that as it may, default inferences can, in some cases, lead to costly interpretive dead-ends (see Bezuidenhout 2001 for such an argument). Thus, it seems that no simple economical argument is going to decide between localists and globalists. However, the remarks on interpretive dead-ends suggest that such cases would make a good testing ground for a choice between local and global theories because these theories would make very different predictions in such cases.

2 Comparatives as a test case

Some sentences semantically impose strong constraints on their components or on their components interpretation. This is the case of comparative sentences, which only make sense if the things being compared are different. No one would say: *George W. Bush is as/more/less intelligent than George W. Bush*. By contrast, it makes perfect sense to say: *George W. Bush is as/more/less intelligent than Bill Clinton*. This constraint seems to apply in the two following sentences: *Better red wine than no white wine; Better no red wine than no white wine.* I will take it in what follows that the speaker of each of these sentences is expressing his general preference for either red or white wine. I will also take it that there are three relevant situations to be considered: one in which there is **only** red wine; one in which there is **only** white wine; one in which there is **both** red and white wine. Let us begin with the semantic analysis: the expressions *red wine, no white wine* and *no red wine* are interpreted relative to the three situations just described. Interpretations of the two negative expressions are pretty straightforward: *no white wine* can only designate the situation in which there is only red wine; and *no red wine* can only designate the situation in which there is only white wine. The only problematic interpretation is that of *red wine* which may designate either the situation in which only red wine is available or the situation in which both red and white wines are available. This is where the general constraint on comparative sentences comes into play, eliminating the interpretation according to which *red wine* designates the situation in which there is only red wine. If it were licensed, the interpretation of the sentence *Better red wine than no white wine* would come out as *A situation in which only red wine is available is better than a situation in which only red wine is available*, which is nonsensical.

So let me sum up on the interpretation of these two sentences. On the present analysis, the sentence *Better red wine than no white wine* is to be interpreted as *A situation in which there is both red wine and white wine is better than a situation in which there is only red wine*, which seems to indicate a general preference for white wine, given that a situation in which there is only red wine is worse than a situation in which white wine is also available. The sentence *Better no red wine than no white wine* is to be interpreted as *A situation in which there is only white wine is better than a situation in which there is only red wine*, again indicating a general preference for white wine. These, then, are the interpretations of the two comparative sentences. It should be noted that, given the straightforward interpretation of the expressions in the sentence *Better no red wine than no white wine*, it can be predicted that it should, on the whole, be less costly to interpret than *Better red wine than no white wine*, given the 'ambiguity' of the expression *red wine*.

So far so good, but let us come back to localist and globalist views. Globalists should not demur in any relevant way at the purely semantic account given above. In particular, global theories would not consider that conversational implicatures are accessed from the sentence *Better red wine than no white wine*. By contrast, localists would predict a conversational implicature, to the effect that *red wine* should be interpreted as *only red wine*, that is, as designating the situation in which only red wine is available. This would lead to the nonsensical interpretation, *A situation in which only red wine is available is better than a situation in which only red wine is available*. What this means is that localists should predict that such a sentence would be interpreted as leading on a first interpretation (involving the dubious conversational implicature) to an interpretive dead-end. Even if, by restoring the regular semantic process described above, the correct interpretation were generated, this means that the localist accounts and the globalist accounts differ on how costly it would be to interpret such a sentence: though both predict it to be more costly than the other comparative sentence, localists see it as a lot more costly. Whether this difference in prediction would be enough to allow a test of both theories is not, however, entirely clear and I think that a closer look at how the conversational implicature is triggered on local theories might help there.

On local theories, the problematic conversational implicature is triggered through the *Q-Principle*, derived by Levinson (2000) from Gricean maxims (see Grice, 1989). The Q-Principle, in an abbreviated form, can be stated as follows (Levinson, 2000, p. 76):

Q-principle

Speaker's maxim. Do not provide a statement that is informationally weaker than your knowledge of the world allows (...). Specifically, select the informationally strongest paradigmatic alternate that is consistent with the facts.

Recipient's corollary. Take it that the speaker made the strongest statement consistent with what he knows.

There are two ways to trigger a Q-implicature, both of them lexical: by choosing an expression from a *Horn scale* or by choosing an expression from a *contrast set*. The triggering of the conversational implicature in the sentence *Better red wine than no white wine* falls under the second possibility.

Let us look at the following examples:

(5) The flag is white.
(6) The flag is not white and red.

(5) gives rise to the conversational implicature in (6) through the Q-Principle because if the flag had been white and red, the speaker would have said it was. Given the Q-Principle and the fact that he did not, the hearer is entitled to conclude that the flag is not white and red. *White* and *red* belong to the contrast set of colours.

Regarding the sentence *Better red wine than no white wine*, it seems legitimate to consider that *white wine* and *red wine* are part of the same contrast set. If this is the case, on a localist account, the sentence *Better red wine than no white wine* should indeed trigger the Q-implicature, given that the speaker could have said – and did not – *Better **red wine and white wine** than no white wine*. Thus the conversational implicature leading to an interpretive dead-end is triggered lexically through a contrast set.

This suggests that a test between localist and globalist theories should compare the interpretation of sentences such as *Better red wine than white wine* with the interpretation of sentences where the lexical expressions in the contrast set concerned are replaced by non-words. Such sentences with non-words should not lead to the triggering of conversational implicatures, which would avoid the interpretive dead-end. Hence, they should be much easier to interpret than sentences with words from contrast sets.

3 Experiment 1

The first experiment was intended to test whether the sentences *Better red wine than no white wine* and *Better no red wine than no white wine* are indeed understandable and that their interpretations are those given above. On the global approach, though the sentence *Better red wine than no white wine* may be harder to interpret than the sentence *Better no red wine than no white wine* (because *red wine* is 'ambiguous' while *no red wine* is not), it should be successfully interpreted as indicating a preference for white wine. Hence, although the globalists predict that the percentage of correct responses for the sentence *Better red wine than no white wine* should be lower than it is for the sentence *Better no red wine than no white wine*, it should nevertheless be higher than chance levels. Regarding the local approach, predictions are more tentative: again it should predict that the sentence *Better red wine than no white wine* will be more difficult to interpret than *Better no red wine than no white wine*, though for partly different reasons than do globalists. Localists predict that the implicature from *red wine* to *only red wine* (or *red wine and no white wine*), being automatic, will be made and then found to yield an incorrect interpretation at which point it should be abandoned and the semantic interpretation should come into play yielding the correct interpretation. This is a more costly process than is the process predicted by globalists and hence, one might expect the percentage of correct answers for the interpretation of the sentence *Better red wine than no white wine* to be not only lower than that for *Better no red wine than no white wine*, but, indeed, to be barely above chance.

3.1 Method

Subjects. A total of 328 first and second year students in psychology at the University Lumière-Lyon 2 participated.[2] There was a large majority of girls (90 per cent) and, with a few exceptions, subjects were in their late teens and early twenties (oldest: 63; youngest: 19). Their mean age was 23.6.

Materials. There were two conditions, the first one corresponding to the putative contrast set <*white wine, red wine*> and the second to the putative

---

[2] I want to thank O. Koenig for allowing me to test his students.

contrast set <coffee, tea>. Each condition included four sentences inserted into an appropriate simple scenario:

The wine condition
A man arrives very late at a party. There isn't much left to drink. Someone brings him a glass of wine. The man says:

Better red wine than no white wine.
Better no red wine than white wine.
Better no red wine than no white wine.
Better red wine than white wine.

The hot drink condition
A man arrives very late at a condominium meeting. Everyone is having a hot drink but there isn't much left. Someone brings him a mug. The man says:

Better coffee than no tea.
Better no coffee than tea.
Better no coffee than no tea.
Better coffee than tea.

The last sentence in each condition is a control sentence, a straightforward comparative. The second sentence in each condition is included for the sake of symmetry. It should raise the same difficulty as the first sentence and for the same reason: they both include an expression (respectively, *white wine* and *tea*) which can be interpreted as referring to either of two situations. Their interpretation, by parity of reasoning with that of the first sentences, are, respectively, *A situation in which there is only white wine is better than a situation in which there is both red and white wine* and *A situation in which there is only tea is better than a situation in which there is both tea and coffee*. It is however rather harder to pinpoint the general preference expressed by the speaker.

To avoid repetition effects, each subject was tested on a single sentence in a single condition with no preliminary training. Three questions were asked:

Q_1: What was he given to drink?
Q_2: What does he prefer?
Q_3: Justify your answers.

Answers to questions Q_1 and Q_2 were given through forced choice, subjects being offered three answers: *red wine* (*coffee*), *white wine* (*tea*), *don't know*. Question Q_1 was mainly a distracting question to avoid people guessing that Q_2 was the central one. Q_3 was intended to get some idea of the kind of heuristics that people use in answering Q_2.

3.2 Results and Discussion of Experiment 1

Predictions regarding answers to the question *What was he given to drink?* were made on an intuitive basis. Regarding answers to the question *What does he prefer?* predictions were based on the preceding analyses. These predictions are indicated in Table 15.1 and the results are indicated in Table 15.2:

The most important results are those on the right of Table 15.2. As can be seen, answers to the question *What does he prefer to drink?* agree with the above hypotheses concerning the interpretation of the utterances. It also highlights the fact that the utterances *Better no red white than no white wine* and *Better no coffee than no tea* are easier to interpret than *Better red wine than no white wine* and *Better coffee than no tea*, respectively, and that there is no

Table 15.1 Predicted answers for Experiment 1

Utterances	What was he given to drink?	What does he prefer to drink?
Better red wine than no white wine	Red wine	White wine
Better no red wine than white wine	White wine	?
Better no red wine than no white wine	White wine	White wine
Better red wine than white wine	Red wine	Red wine
Better coffee than no tea	Coffee	Tea
Better no coffee than no tea	Tea	Tea
Better no coffee than tea	Tea	?
Better coffee than tea	Coffee	Tea

Table 15.2 Results from Experiment 1 (figures marked in bold are the correct answers; DK = 'Don't Know')

Utterances	What was he given to drink?			What does he prefer to drink?		
	Red	White	DK	Red	White	DK
Better red wine than no white wine	**75**	7	18	12	**60**	28
Better no red wine than white wine	13	**59**	28	46	33	21
Better no red wine than no white wine	19	**62**	19	5	**83**	12
Better red wine than white wine	**79**	5	16	**74**	8	18
	Coffee	Tea	DK	Coffee	Tea	DK
Better coffee than no tea	**88**	2	10	7	**77**	16
Better no coffee than tea	7	**73**	20	78	15	7
Better no coffee than no tea	14	**60**	26	8	**90**	2
Better coffee than tea	**70**	14	16	81	5	14

clear answer to that question for sentences *Better no red wine than white wine* and *Better no coffee than tea*. Experiment 1 showed that sentences, however difficult they may seem to interpret, are nevertheless quite understandable and can thus be used as a testing ground for the respective claims of localists and globalists.

4 Experiment 2

The second experiment was designed to further test between local and global theories by using non-words instead of real words in the sentences. The idea is that localists suppose implicatures are triggered by lexical items, replacing lexical items by non-words should prevent implicatures from being drawn and make a participant's ultimate interpretation less complicated than those from corresponding sentences having lexical items that do trigger the implicature. In other words, on a localist account, the percentage of correct answers for the sentence *Better pekuva than no luveka* should lead to a higher rate of correct answers than for the sentence *Better coffee than no tea*. Globalists would not predict any significant difference as a function of the lexicality of the drinks.

4.1 Method

Subjects. A total of 128 students in history at the University Lumière-Lyon 2 were tested on two occasions.[3] Again there was a majority of women (66 per cent) and the subjects were, with a few exceptions, in their late teens and early twenties (oldest: 46; youngest: 17). Their mean age was 19.2.

Materials. Given that results were better for the condition based on the putative contrast set <*tea, coffee*>, the same condition was reproduced in this experiment along with an additional condition, where non-words substituted the words *tea* and *coffee*. Again, it contained four sentences, inserted in a scenario:

Non-word condition
An anthropologist arrives very late to a feast in Papouasy-New Guinea. There is not much left to drink. Someone brings him a gourd. The anthroplogist says:

Better pekuva than no luveka.
Better no pekuva than luveka.
Better no pekuva than no luveka.
Better pekuva than luveka.

---

[3] I would like to thank M. Martinat and E. Lynch for providing me with access to their classes.

I presented the same questions as those in Experiment 1. As before, answers to questions *What was he given to drink?* and *What does he prefer to drink?* were provided in a forced choice format. The participants were presented three options: *coffee (pekuva); tea (luveka)* and *don't know*. Subjects were tested on a single sentence in a single condition and precautions were taken to ensure that no communication would be possible between participants.

4.2 Results

The predictions regarding the correct answers to the questions for the non-words condition are indicated in Table 15.3:

In this experiment, localist and globalist theories would make different predictions not so much regarding the correct interpretations (they would agree on that point), but regarding the percentage of correct answers. Localists would predict a greater percentage of correct answers to the question *What would he prefer to drink?* with respect to the first sentence in the non-word condition as compared with the tea/coffee condition, given that no conversational implicature should be triggered by non-words and hence no interpretive dead-end should be reached. By contrast, globalists would not predict a significant difference across the two conditions. The results are presented in Table 15.4.

Let us concentrate on the answers to the question *What does he prefer to drink?* and, more specifically, on the first sentence in each condition. The rates of correct answers are quite similar; 63 per cent chose *tea* for the sentence *Better coffee than no tea* and 67 per cent chose *luveka* for the sentence *Better pekuva than no luveka*. Indeed, with the exception of the answers to the second sentence for which no correct answer to the preference question can be given, the rates of correct answers are remarkably similar across the coffee/tea and the non-words conditions. This is especially the case for the third sentence (respectively, *Better no tea than no coffee* and *Better no pekuva than no luveka*), where the difference in correct answers is 1 per cent. Indeed, it is across the control sentences in each condition that the difference is greatest; the rate of correct answers for the coffee/tea condition being 75 per cent while it is only 69 per cent in the non-words condition.

Table 15.3 Predictions for the non-word condition in Experiment 2

Utterances	What was he given to drink?	What does he prefer to drink?
Better pekuva than no luveka	Pekuva	Luveka
Better no pekuva than luveka	Luveka	?
Better no pekuva than no luveka	Luveka	Luveka
Better pekuva than luveka	Pekuva	Pekuva

Table 15.4 Results from Experiment 2 (figures marked in bold are the correct answers, DK = 'Don't Know')

Utterances	What is he given to drink?			What does he prefer to drink?		
	Coffee	Tea	DK	Coffee	Tea	DK
Better coffee than no tea	**69**	13	19	31	**63**	6
Better no coffee than tea	12	**65**	23	**65**	6	29
Better no coffee than no tea	13	40	**47**	0	**93**	7
Better coffee than tea	**63**	12	25	**75**	12.5	12.5
	Pekuva	Luveka	DK	Pekuva	Luveka	DK
Better pekuva than no luveka	**80**	7	13	7	**67**	26
Better no pekuva than luveka	6	**75**	19	**44**	37	19
Better no pekuva than no luveka	18	71	**11**	6	**94**	0
Better pekuva than luveka	**56**	25	19	**69**	19	12

4.3 Discussion of Experiment 2

The results largely support global theories of conversational implicatures. However, it should be clear that they are limited. For one thing, they only concern the triggering of Q-implicatures through contrast sets. The experiment has nothing to say about Q-implicatures triggered by Horn scales. To say anything about Q-implicatures in general, the experiment would have to be adapted (if possible) to Horn scales. For another, it is not, of course, sufficient to give a definite and general answer to the question of the status of pragmatics relative to linguistics. However, given these limitations, such experiments might be a step in the right direction.

5 Conclusions

The experiments presented above were designed to test the relative validity of two hypotheses regarding access to some conversational implicatures. Global theories claim that all conversational implicatures are nonce-implicature triggered by pragmatic inferential processes at a global or sentential level. Local theories claim that some conversational implicatures (those with which we are concerned above) are triggered at a local level (subsententially) through lexical items belonging to a contrast set. In some linguistic contexts, and in this specific case comparative sentences involving narrow negation, these two types of theories make different predictions on how difficult the interpretation of such sentences would be, local theories claiming that the occurrence of lexical items belonging to contrast sets would lead to interpretive dead-ends, while global theories claim that they would not make any difference. So far, the experimental results presented above seem to support global over local theories about this specific pragmatic phenomenon.

References

Bezuidenhout, A. (2001). 'Implicature, relevance and default pragmatic inferences'. In *Acts of the Workshop Towards an Experimental Pragmatics*, 17–19 May 2001, Lyons, France.

Grice, P. (1989). *Studies in the Way of Words*. Cambridge, MA: Harvard University Press.

Levinson, S. (2000). *Presumptive Meaning: The Theory of Generalized Conversational Implicature*. Cambridge, MA: MIT Press.

Sperber, D., and Wilson, D. (1986/1995) *Relevance: Communication and Cognition* (2nd edn). Oxford, Basil Blackwell.

Appendix to Chapter 15
Example of original experimental materials used in the wine condition of Experiment 1

Scénario:

Un homme arrive très en retard à une fête. Il n'y a plus grand chose à boire. Quelqu'un lui amène un verre de vin. Il dit: (*A man arrives rather late to a party. There is not much left to drink. He says*:)

1.1 Mieux vaut du vin rouge que pas de vin blanc
1.2 Mieux vaut pas de vin rouge que du vin blanc
1.3 Mieux vaut pas de vin rouge que pas de vin blanc
1.4 Mieux vaut du vin rouge que du vin blanc

Questions

Question 1: Que lui a-t-on donné à boire ? (*What was he given to drink?*)

Vin rouge ❏   Vin blanc ❏   ne sait pas ❏

Question 2: Qu'est-ce qu'il préfère ? (*What does he prefer?*)

Vin rouge ❏   Vin blanc ❏   ne sait pas ❏

Question 3: Justifiez vos réponses. (*Justify your responses*).

Name Index

Adams, E. W.   231n
Akmajian, A.   27n
Alksnis, O.   100
Alter, K.   203
Anscombre, J.-C.   7
Antos, S.   75
Aristotle 96
Astington, J. W.   208, 210, 215, 225
Atiyah, P.   52
Atlas, J. D.   4, 121, 258
Austin, J. L.   51, 207

Bach, K.   7, 62, 187, 260n, 264n, 301n
Bangerter, A.   14, 25–46
Baratgin, J.   109
Bard, E.   120
Barrouillet, P.   228
Barsalou, L.   79n, 89
Barss, A.   194
Barton, S.   279
Barwise, J. P.   116, 120, 121, 187
Bashore, T. R.   191
Basso, G.   318
Bastide, A.   308, 318
Bates, E.   174, 207, 224
Bavelas, J. B.   35n
Begg, I.   96
Bennett, J.   230, 233
Bentin, S.   192
Bernicot, J.   17, 207–25
Bersick, M.   193
Besson, M.   192, 200
Beun, R.-J.   42
Bezuidenhout, A. L.   5, 17–18, 257–80, 323
Bianco, M.   141
Binder, K. S.   178n
Blakemore, D.   5, 8
Blasko, D. G.   75
Blass, R.   5, 8
Bloom, F. E.   188
Blutner, R.   7
Bonnotte, I.   224
Bookheimer, S.   318
Bookin, H. A.   75, 83

Bott, L.   310, 311, 312, 315n, 316, 317, 318
Bourmaud, G.   103, 104
Braine, M. D. S.   99, 146, 228, 304, 305
Brennan, S. A.   38
Brennan, S. E.   40, 42
Brewer, W. F.   29
Bronckart, J. P.   216
Brown, C.   194, 196
Brown, P.   59
Bruner, J. S.   207, 215, 223, 224
Buchler, J.   41–2
Buttrick, S.   31–2, 42
Byrne, R. M. J.   98, 100, 103, 147, 148n, 228, 247

Camac, M.   78
Cara, F.   106, 141, 158, 163, 228
Carles, L.   141, 165–6
Carlson, T. B.   33n
Carroll, P.   85
Carston, R.   5, 63, 73, 88–9, 90 172, 187, 230, 260, 263, 309, 310
  automaticity 73, 88–9, 90
Castry, A.   141
Chafe, W. L.   27n
Chater, N.   249
Cheng, P. N.   146, 158, 160, 161
Chevallier, C.   316
Chevaux, F.   307, 310, 319n
Chierchia, G.   18, 278, 283–99, 306, 307, 319n
Chomsky, N.   27n
Clark, B.   63, 141, 258n, 264
Clark, H.   10, 11, 13, 14, 25–46, 125
  psycholinguistic experiments 56, 59, 60, 62
Clifton, C.   274, 280
Coates, L.   35n
Cole, P.   55
Coles, M.   189, 191, 196
Coley, J. D.   143
Connell, J.   99, 228
Connine, C. M.   75
Cooper, R.   116, 120, 121

Corblin, F. 119
Cormier, P. 110
Cosmides, L. 146, 160, 161
Coulson, S. 16–17, 187–204, 267, 316–17
Cox, J. R. 157, 158
Crain, S. 18, 278, 283–99, 306, 307
Cremers, A. H. M. 42
Cummins, D. D. 100
Custer, W. L. 110
Cutting, J. C. 264

Dagenais, Y. 110
Dascal, M. 7
Davis, S. 7
Dawydiak, E. J. 118, 119, 121, 122, 125, 126, 127
Declerck, R. 231, 241, 243
Delaney, S. 52, 208
Dennis, I. 100
Doitchinov, S. 305
Dolan, R. J. 318
Donchin, E. 191, 194
Donnellan, K. 55, 57
Ducrot, O. 7, 98
Duffy, S. A. 173, 174n
Dulany, D. E. 107
Duranti, A. 208n

Ellis, M. C. 100
Enfield, N. J. 38
Ervin-Tripp, S. 207, 224
Estes, Z. 79, 85, 90
Evans, J. St B. T. 147, 158, 228, 248, 249, 250, 304
  reasoning and judgement 98, 100, 103
Evans, V. 267, 268, 280

Fabiani, M. 196
Fairley, N. 100
Fauconnier, G. 7, 55, 187, 200, 230, 267
Faulconer, B. A. 85
Fay, N. 119, 121, 124, 132
Fayol, M. 224
Federmeier, K. D. 190, 191
Feeney, A. 17, 228–52, 306
Fein, O. 16, 172–84
Fiddick, L. 160
Fillenbaum, S. 101
Fleischman, S. 209, 216
Fodor, J. 73, 74, 90, 174, 175n, 184

Folk, J. R. 270
Foppolo, F. 18, 283–99
Forster, K. I. 194
Foss, D. J. 85
Fox Tree, J. E. 35n
Francescotti, R. M. 230, 231
Francik, E. 60
Frazier, L. 174n
Friederici, A. D. 194, 203
Frisson, S. 279, 280

Gaeth, G. J. 132
Gagné, C. L. 86
Garcia-Madruga, J. A. 247
Garnham, A. 117
Garrett, M. F. 194
Garrod, S. C. 12, 29, 117
Gazdar, G. 4, 258
Geis, M. 98, 232
Gentner, D. 74
George, C. 101, 103
Gernsbacher, M. A. 78, 176n
Gibbs, R. 10, 15, 50–70, 141, 174, 208, 259n
  electrophysiology 199, 200, 202
Gibson, T. 119
Gildea, P. 75, 83
Giora, R. 16, 172–84, 200, 280n
Girotto, V. 106, 141, 146, 158, 160, 163, 228
Glucksberg, S. 14–15, 68, 72–91, 200
Goel, V. 318
Goldvarg, Y. 78, 81, 82, 83
Gombert, E. 110
Gombert, J. 224
Goodman, N. 74
Goodwin, G. 41
Graesser, A. C. 11
Grafman, J. 318
Gratton, G. 196
Green, D. W. 160n
Grice, H. P. 2–4, 5–7, 8, 11, 14, 18, 98, 229
  automaticity 73, 74, 87
  context effects 174, 175, 184
  conversational implicature 322–3, 325
  electrophysiology 199, 200
  implicature, relevance and default inference 257–8, 261, 263, 280
  inference and logical terms 301–2, 303

Grice, H. P. (*Continued*)
  psycholinguistic experiments 63, 64, 68
  reference 27, 30n, 31, 32, 33, 40, 41, 45
  semantic and pragmatic competence of 'or' 283, 284, 285, 286, 289, 297–8, 299

Griggs, R. A. 96, 157, 158
Grodzinsky, Y. 298n
Groothusen, J. 194
Grosset, N. 228
Grosz, B. 42
Gualmini, A. 18, 278, 283–99, 306, 307
Guasti, M. T. 18, 278, 283–99, 306, 307
Guelminger, R. 307, 319n
Guerts, B. 119
Gutt, E.-A. 5

Hagoort, P. 194, 196
Hahne, A. 194
Halgren, E. 192
Halliday, M. A. K. 27, 29, 207, 224
Hamblin, J. 65, 68
Hampton, J. A. 83, 85
Handley, S. J. 17, 228–52
Happé, F. 141
Hardman, D. 141, 160n
Hare, R. 52
Harman, G. 30n, 147
Harner, L. 209, 216
Harnish, R. 7, 62
Harper, C. 249
Harris, G. 96
Hasson, U. 79
Haviland, S. 27, 28, 29, 56
Hawkins, J. A. 31
Hayes, B. K. 143
Heath, C. 41
Heinze, H. J. 318
Henle, M. 13
Herman, R. 173, 177
Hess, D. J. 85
Hickmann, M. 224
Hillert, D. 175n
Hillyard, S. 189, 190, 192, 316–17
Hilton, D. J. 95, 107, 228
Hindmarsh, J. 41
Hirschberg, J. 258
Ho, S. 306
Hoffman, J. E. 195

Holcomb, P. 192, 203
Holyoak, K. J. 146, 158, 160, 161
Horn, L. R. 4, 99, 111n, 258, 284, 288, 302
  quantifiers 119, 123, 124, 125, 128, 135

Ifantidou, E. 5
Inhelder, B. 109

Jackendoff, R. 27n, 84
Jackson, F. 231
Jefferson, G. 37
Johnson, A. T. 76
Johnson, T. 35n
Johnson-Laird, P. N. 17, 73, 85, 89, 117, 147n, 148n, 247, 249
Jorgensen, J. 141
Joshi, A. K. 117
Jucker, A. 5

Kahneman, D. 15, 106, 108, 132, 251
Kamp, H. 117
Kasher, A. 7
Kattunen, L. 29n
Katz, J. J. 7
Kawamoto, A. H. 173
Keane, M. 84
Keating, E. 208n
Keenan, E. L. 116, 120
Kellas, G. 173, 177
Kemmelmeir, M. 106, 141, 160
Keysar, B. 75, 76, 77, 78
Kibble, R. 121, 124, 127
King, J. 190, 191, 192, 194, 197, 203
Kintsch, W. 88
Klar, Y. 160n
Klima, E. S. 125
Kluender, R. 192, 193
Koechlin, E. 318
Koehler, J. J. 108
Konig, E. 231, 233
Krauss, R. M. 34, 35, 37
Krifka, M. 285
Kripke, S. 55
Krych, M. A. 35n, 43–4
Kutas, M. 316–17, 318
  electrophysiology 189, 190, 191, 192, 193, 194, 196, 197, 198–9, 203

Lakoff, R. 59
Landis, S. E. 188
Lanzetti, V. 316
Larkin, J. 249
Larking, R. 160n
Laval, V. 17, 207–25
Lawson, D. 193
Lea, R. B. 250
Lecas, J. F. 228
Leinonen, E. 141
Lerner, G. H. 38
Levin, I. P. 132
Levinson, S. 4, 17, 59, 287, 288, 323, 325
  implicature, relevance and default inference 258, 259, 260, 261–2, 265–6
  inferences and logical terms 302, 303, 304, 309
Levy, J. N. 82n
Lewis, D. 7, 30n, 31, 32, 33, 45, 54
Liberman, N. 160n
Lilje, G. W. 98
Lindamood, T. E. 192
Lubart, T. 100
Lucy, P. 10
Lycan, W. G. 230, 231, 244

McCallum, C. 191
McCarthy, G. 191
Macchi, L. 108
McClelland, J. L. 173
McGlone, M. S. 77, 82
McIsaac, H. 192
McKee, C. 294
Mackie, J. L. 101
McKinnon, R. 193
McKoon, G. 11, 85
McRae, K. 174
Magliero, A. 191
Majid, R. 123
Manfredi, D. 77, 82
Manktelow, K. I. 100, 156, 248
Mannes, S. 195
Marcos, H. 207, 215, 223
Marcus, S. L. 99
Markovits, H. 100
Marr, D. 188
Marshall, C. R. 30, 31, 44
Marslen-Wilson, W. D. 173
Martin, C. 173

Matlock, T. 200
Matsui, T. 5, 8, 11–12, 28, 29, 141
Mecklinger, A. 194
Medin, D. L. 143
Meroni, L. 18, 278, 283–99, 306, 307
Metcalf, K. 173, 177
Mey, J. 208n, 225
Meyer, D. E. 313
Miller, G. 73, 74, 141
Miller, S. A. 110
Mills, D. 193
Mitchell-Kernan, C. 207, 224
Mitchiner, M. 192
Moeschler, J. 5, 7
Moise, J. 63, 64–5, 67, 141, 202, 259n
Moreno-Rios, S. M. 247
Morgan, J. 62
Morris, R. 17–18, 174n, 257–80
Mosconi, G. 13
Moxey, L. M. 15, 116–35, 229
Mueller, H. 203
Mueller-Lust, R. 55, 57, 59
Muente, T. F. 190, 191, 196, 318
Munro, A. 10
Murphy, G. L. 83, 84, 86
Musolino, J. 306, 307

Nassau, G. 110
Nazir, T. 316
Neale, S. 7
Neville, H. 193, 194
Newsome, M. R. 78
Newstead, S. E. 96, 98, 100, 147, 229, 304
Nicol, J. 194
Nicolle, S. 63, 141, 258n, 264
Ninio, A. 207, 224
Noh, E.-J. 5
Noordman, L. G. 11
Noveck, I. A. 1–19, 87–8, 107, 109, 141, 268–9, 278, 301–20
  reasoning: 'even if' 229, 250, 251

Oaksford, M. 249
O'Brien, D. P. 146
Olson, D. R. 26–7, 32, 34, 35, 36, 38, 39, 40, 42
Orne, M. T. 95
Ortony, A. 74, 75, 77
Osherson, D. N. 84

Name Index

Osterhout, L. H. P. 193, 194
Over, D. 55, 228, 248, 249, 250

Pacht, J. M. 173, 174n
Panzer, S. 318
Papafragou, A. 5, 306, 307
Paris, S. 292, 304–5
Parks, M. 203
Paterson, K. B. 117, 118, 119, 126
Paul, S. T. 173, 177
Peetz, V. 52
Peirce, C. S. 41–2
Peleg, O. 16, 172–84
Percus, O. 119
Peters, S. 29n
Piaget, J. 15, 109, 228
Pickering, M. J. 279, 280
Pietrini, P. 318
Pilkington, A. 5
Plante, E. 203
Poli, J. 200
Politzer, G. 13, 15, 94–113, 141, 153, 278, 308, 318
Posada, A. 141, 316–17
Potter, M. C. 85
Pouscoulous, N. 308, 318
Poynor, D. V. 270
Prince, E. F. 27n, 117
Pustejovsky, J. 267, 280
Pynte, J. 200

Quine, W. 55
Quirk, R. 38

Radvansky, G. A. 248, 250
Ratcliff, R. 11, 85
Rawls, J. 52
Rayner, K. 173, 174n, 178n, 269, 274, 280
Reboul, A. 5, 7, 18–19, 322–33
Recanati, F. 7, 63, 65, 187, 260, 268, 309n
Reddy, M. 187
Reed, S. 231, 241, 243
Reinhart, T. 182, 298n
Reyle, U. 117
Reynolds, R. 75
Rips, L. J. 84, 146, 313–14, 315n
Rips, R. J. 99
Rist, R. 100

Ritter, W. 191
Roberts, J. L. 188
Robertson, D. 120
Robertson, R. R. 78
Robichon, F. 200
Rossi, S. 106
Rouchota, V. 5, 8
Rubin, S. 203
Rugg, M. 189, 192
Rumain, B. 99, 228, 304, 305
Rumelhart, D. E. 173
Russo, R. 80
Ryder, N. 141

Sacks, H. 36, 37
St George, M. 195
Sanders, T. J. 11
Sanford, A. J. 12, 29, 116–35, 229, 279
Sanford, D. H. 230
Sanford, T. 15
Schaefer, F. F. 33n, 34, 38
Schallert, D. 75
Schegloff, E. 36, 37, 42–3
Schelling, T. C. 31
Schiffer, S. R. 30n
Schiltz, K. 196
Schmauder, A. R. 274, 280
Schneider, S. L. 132
Schober, M. F. 39
Schreuder, R. 31–2, 42
Schriefers, H. 194
Schroyens, W. 106
Schunk, D. 59, 62
Schütze, C. T. 25
Searle, J. R. 7, 9, 14, 17, 73, 199
  promises 207, 210, 213, 216, 223
  psycholinguistic experiments 51–2, 53–4, 55, 56, 62, 65, 69
  reference 25, 27, 33, 36, 38, 39
Sera, M. 306
Sereno, S. C. 174n, 274, 280
Schwarz, N. 95
Sidner, C. 42
Simpson, R. 191
Singer, M. 11
Sloman, S. A. 228
Smith, C. L. 304, 305, 313, 317, 318
Smith, E. E. 84
Smith, M. E. 192
Snow, C. E. 207, 224

Name Index

Sorace, A. 120
Sperber, D. 1–19, 27, 63, 106, 141–69, 172, 323
  automaticity 73, 76, 84, 87, 89, 90
  electrophysiology 187–8, 201
  implicature and logical terms 301n, 302, 303, 316
  implicature, relevance and default inference 260, 267
  reasoning: 'even if' 228, 247
Spivey-Knowlton, M. J. 174
Spooren, W. P. 11
Springer, K. 83, 84, 85, 86
Squire, L. R. 188
Stalnaker, R. 7, 29n
Stanfield, R. A. 89
Stanovich, K. E. 228, 250, 251
Staudenmayer, H. 99
Stavi, J. 116, 120
Steinhauer, K. 194, 203
Sternberg, R. J. 304, 305
Stewart, A. J. 132
Storms, G. 143
Strawson, P. 55
Stroop, J. R. 75–6
Svartvik, J. 38
Sweetser, E. 7
Swinney, D. 174, 175n
Sylvestre, E. 307, 319n
Szeminska, A. 109

Tabossi, P. 85
Tanenhaus, M. K. 174
Tangram task 37, 39
Tantalou, N. 306, 307
Taplin, J. E. 99
Thompson, V. A. 100
Thornton, R. 291, 294
Tomasello, M. 207, 224
Torreano, L. 77
Trabasso, T. 11
Travis, C. 7
Treyens, J. C. 29
Tunstall, S. 119
Turner, M. 200
Tversky, A. 15, 78, 79, 106, 108, 132, 251
Twyman-Musgrove, J. 100, 228, 249
Tyler, A. 267, 268, 280
Tyler, L. K. 173

Uchida, S. 5
Unger, C. 8

Van Berkum, J. 196
Van Deemter, K. 267
Van der Auwera, J. 7, 99
Van der Henst, J.-B. 13, 16, 106, 141–69
Van Der Wege, M. M. 40
Van Naerssen, M. M. 216
Van Petten, C. K. M. 192, 193, 195, 200, 203
Vanderveken, D. 7, 17, 207, 210, 213
Vaughan, H. G. 191
Vu, H. 173, 177, 179, 181, 182, 184

Walker, C. H. 12
Walker, M. A. 117
Wason, P. C. 105, 106, 124, 156–64, 228, 251
Weber, H. 84
Weinheimer, S. 34, 35, 37
Wellman, H. M. 224
Werner, N. K. 78
West, R. F. 251
Westerståhl, D. 116
Wilkes-Gibbs, D. 36–7, 38–9, 40
Wilkins, D. P. 44
Williams, C. 119, 121, 124
Wilson, D. 5, 8, 12, 27, 63, 172, 323
  automaticity 73, 76, 84, 87, 89, 90
  electrophysiology 187–8, 201
  implicature, relevance and default 260, 267
  inference and logical terms 301n, 302, 303
  relevance 141, 146n
Wisniewski, E. J. 81
Wolff, P. 74

Yaxley, R. H. 89
Yekovich, F. R. 12
Yus, F. 5

Zeevat, H. 7
Zigmond, M. 188
Zwaan, R. A. 89, 248, 250
Zwarts, F. 116, 121
Zwicky, A. M. 98, 232

Subject Index

addressee-blind referring 27
adjective 83
adjective-noun 81, 83, 86
adverbs 209, 216
ahistorical referring 27
alternatives 26, 27
always 302
anchor 41
and 301, 302, 309–11
answer 94
antecedent 100, 101, 102–3
  explicit 56
  negation 99
  reasoning: 'even if' 233, 237, 238, 239, 240, 247, 249
  see also conditional reasoning, Denial of the Antecedent
approaches 1–7
armchair method 63, 124, 132, 158
  reference 25–6, 27, 29–30, 31, 33, 41, 46
Artificial Intelligence 105
assertion 208
attention 42, 44
automaticity of pragmatic processes 72–91
  automatic engagement 87–8
  conceptual combination 81–7
  conceptual representation: inference 88–90
  literal, primacy of in process models of comprehension 72–4
  metaphor comprehension, theories of 74–81
autonomous act 26, 33

base-rate expectation 128
base-rate fallacy 108–9
bridging 11–12
  inferences 28–9

cancellation 270–2
children
  class inclusion 109–11

see also 'or': competence in children and adults; speech acts in children class inclusion and categorization 109–12
  problem 15
  statements 78
classic experimental findings 304–5
clausal implicature 258, 259, 260
clause ending negativity 194
Closure Positive Shift 203
cloze probability 192, 198
clutter avoidance principle 147
cognition 5
cognitive effects 6
Cognitive Principle of Relevance 6, 16, 143–4, 152, 156, 169
  with relational reasoning tasks 146–56
collaboration 35–6, 37, 40, 43–4
colour-word interference effect 75, 76
commissive speech acts 207, 208, 212
common ground 29–33, 35, 41
  communal 30
  personal 30–1
communication 130–4
Communicative Principle of Relevance 6, 16, 144–5, 148, 162, 169
  speech production task 165–9
  with Wason selection task 156–64
comparatives 323–6
comparison 76–8
  theory of metaphor 74
complement set focusing 131
complement set reference 118
complementary necessary conditions 101–2, 103, 104
compounds understood compositionally/pragmatically 83–7
comprehension, process models of 72–4
compset index 127
conceptual combination 81–7
  compounds understood compositionally/pragmatically 83–7
  metaphorical interpretation generation 81–3

340

Subject Index 341

conceptual pacts 39–41
conceptual representation:
    inference 88–90
conceptualization 41
conclusion 148, 149–50, 156
    double-subject 151–2
    logical 154
    single-subject 151–2
conditional field 101
conditional inference task 239–40
conditional perfection principle 98
conditional reasoning 97–105
    invalid arguments 98–100
        Affirmation of the Consequent 98, 99
        reasoning: 'even if' 232, 234, 235, 236, 237, 242, 243, 248
        Denial of the Antecedent 98, 232, 234–7, 240–6, 248
    valid arguments and credibility of premises 100–3
    valid arguments and nonmonotonic effects 103–5
conditional rule 158
conditional statement 159
conditionals 15, 101, 102, 103
conjunction:
    conjunction buttressing 259, 262
    conjunction fallacy 106–8
    see also: and
consequent 100, 102–3, 159
    reasoning: 'even if' 233, 237, 238, 239, 240, 241–2, 245, 246–7
constitutive rules 52
Constructionists 11
constructive dilemma 232
context 16
    explicit 155–6
    implicit 155–6
    see also neutral context
context effects 172–84
    findings 176–83
        experiment 1 177–8
        experiment 2 178–80
        experiment 3 180–1
        experiment 4 181–3
    on initial processing 173–6
Context of Plausible Denial 124
contextual cue 214
contextual implication 154

contradictories 96
contraries 96
contrast model of similarity 78, 79
contrast set 325, 326–7, 329, 331
conversational implicatures 65, 66, 67, 68
    automaticity 73
    defeasible 323
    nonce or generalized 322–33
        comparatives as a test case 323–6
        experiment 1 326–9
        experiment 2 329–31
    cooperation 3, 5, 26–33, 40, 45, 301
    coordination 31–3, 34, 36, 38, 39, 45
    current issues 16–17

data 134–5
De Morgan conditions 121
declarative sentences 124
decoding 73, 88, 89, 90
deduction studies 96–105
    conditional reasoning 97–105
    quantifiers 96–7
default assumption 265
default inference 270, 313
Default Model 18, 261–8, 269, 270, 271–2, 274–5, 276, 278, 280
defective utterance 73
definite descriptions 54–9, 69
deictic expressions 45
demand characteristics 95
denial: judgements of tags 124–8
Denial Index 126, 127, 128, 134
describing 41, 44–5, 46
developmental-pragmatic effect 305–7
didactic contract 95
direct access view 173–4, 175, 177
direct speech acts 201
directing-to 42, 43
disabling conditions 100
disambiguation 16
    double 110
discourse models 59
Discourse Representation Theory 117, 119
disjunctions 316
double disambiguation procedure 110
downward-entailment 120–2, 127, 287–92, 294, 296, 299
dual-reference hypothesis 79

effect   16, 142–3, 146, 150–1, 153, 154, 162–4
effort   16, 18, 142–3, 145–6, 150–1, 158, 162–4, 166, 168
electroencephalogram   188–91, 194, 202–3, 204
electrophysiology   187–204
  electroencephalogram and Evoked Response Potentials   188–91
  Evoked Response Potentials   195–201
  Evoked Response Potentials, language-sensitive   191–5
  future directions   201–4
embedded implicature   285–6, 289
Engineer-Lawyer problem   15, 108–9
enriched meaning   264
enrichment process   89
entailment   201–2
  scale   259–60, 261–2
epistemic status   302
essential condition   208
'even if' see reasoning: 'even if'
Evoked Response Potentials   18, 188–91, 193, 195–201, 202, 203, 204, 316–18
  considerations in language research   196–7
  endogenous   191
  inferences and logical terms   317–18
  joke comprehension   197–9
  language-sensitive   191–5
    left anterior negativity   193
    lexical processing negativity   193
    N400 component   192–3
    P600   194
    slow cortical potentials   194–5
  metaphor comprehension   199–201
exhaustivity, law of   98
exogenous components   191
expectations   128–30
  base-rate   128
experiment see laboratory method
explicatures   201–2
explicit processes   250–1
external stimuli   141
eye movement monitoring study   18, 268–77
  method   273–4
  results   274–6
  scalars, cancellation and predictions   270–2

fat-free formulation   132–4
feedback   35
felicity condition   51–4
  Nonevident   52–3, 54
  Obligation   51, 52, 53, 54
felicity judgement task   297–9
field method   25–6, 27, 29, 33, 41, 42, 43, 46
fMRI   318
focus   27n, 117–19
formal reasoning   99
formalism   134–5
four-card problem   106
frame-shifting   197–9
frequency   175, 302
future
  act   53
  -action statements   212, 214, 215
  markers   215–16
  tense   209, 212, 216, 217, 224, 225
  see also immediate future; simple future

gaze duration measures   270
generalized conversational implicature   63–4, 257–73*passim*, 276–80, 323
gestures   41–5
Given–New distinction   27, 28
globalists/global theories   323, 324–5, 326, 329, 330, 331, 332
graded salience hypothesis   16, 174–5, 177, 179
Gricean   7
Gricean implicature   28
Gricean maxims   325
grounding   37–9

Hearer Preference felicity condition   52, 53
homonyms   177
Horn scale   259, 325, 331
hyperonym   109, 110, 111
hyponym   110, 111
hypothesis testing   105–6

I-implicature   258, 259, 260, 268
I-principle   4, 259n, 262, 263, 265
identification   38
immediate future tense   209–10, 218–24

Subject Index 343

implicature 19, 28, 63, 64, 69, 201–2
  cancelling 103
  clausal 258, 259, 260
  context effects 175
  embedded 285–6, 289
  Gricean 28
  I- 258, 259, 260, 268
  inferences and logical terms 301, 302
  M- 258, 259, 260
  necessity 243
  nonce 331
  particularized conversational 263
  Q- 258, 331
  quantifiers 135
  R- 258
  reasoning: 'even if' 230–1, 232, 237, 238, 244, 246
  reasoning and judgement 99, 101, 103, 107
  suspenders 123
  see also conversational implicature; generalized conversational implicature; implicature, relevance and default pragmatic inference; scalar implicature
implicature, relevance and default pragmatic inference 257–80
  Default Model and Underspecification Model 261–8
  default inferences 265–7
  semantic underspecification 267–8
  eye movement monitoring study 268–77
implicit processes 250–1
implied meaning 63
inclusive construal 107
index 41–2, 44
indicating 41, 46
  and describing in acts of referring 44–5
  methods 41–2
  within acts of referring 42–4
indicative expression 68
indicative utterance 67
indicatum 44
indirect inference model 56, 57, 58
indirect requests 14, 68, 69, 73
indirect speech 10, 59–62, 201
individual act 33

Inference Theory of Complement Focus 126
inferences 69, 73, 88–90
  bridging 28–9
  competence in comprehension of 'or' 283
  default 270, 313, 323
  immediate 96
  nonce (one-off) 323
  relevance 142, 143
  rules 146
  task, conditional 239–40
  see also implicature, relevance and default pragmatic inference; inferences and logical terms; scalar inference; truth of inferences
inferences and logical terms 301–20
  adult studies 311–16
  disjunctions 316
  quantifiers, existential 311–16
  characterizing developmental-pragmatic effect 307–11
  classic experimental findings 304–5
  developmental-pragmatic effect 305–7
  neuropsychological measures 316–18
  post-Gricean approaches 302–4
infinitive verb 209, 211
information 65
  focus 27
  primary 66, 68
initial processing 173–6
input 5–6, 142–3, 144
integers 105, 111
interactive process 33–41
  language and joint action 33–4
  reference and conceptual pacts 39–41
  referring and collaboration 35–6
  referring and grounding 37–9
  referring and interaction 34–5
interference effect 75, 76
interpretation task 239
interrogative sentence 109
intrinsic connections 42
intuition *see* armchair method
invalid arguments 98–100
issues 1–7

joint action 33–4
joint activity 34

joke comprehension 16, 197–9
judgement *see* reasoning and judgement
know 302
knowledge
  primary 65
  secondary 65, 66
  tacit 52
Koenig's puzzle 18, 19

laboratory method 25–6, 27, 29, 33, 41, 46
language 41–5
  and joint action 33–4
latency 190
left anterior negativity 193
lexical
  access 174
  ambiguity 269
  processes 173
  processing negativity 193
Linda problem 15, 106–8, 251
linguistic
  accounts 243–7
  act 41
  cue 214–15
  pragmatics *see* psycholinguistic experiments and lingustic pragmatics
literal
  mapping condition 200–1
  meaning and what speakers say 62–8
  primacy in process models of comprehension 72–4
literary models 36
local theories/localists 323, 324–5, 326, 329, 330, 331, 332
locative descriptions 45
logical
  instructions 314
  interpretation 315
  response 314, 316, 318
  terms *see* inferences and logical terms
Long condition 315–16
losing face 59

macroanalysis 113
magnitude estimation 120

mathematical hierarchies 111–12
maxims 4, 5, 8, 27, 325
  Gricean 325
  of manner 3, 27
  of quality 3
  of quantity 3, 27, 40, 98, 303
    competence in comprehension of 'or' 284–5, 297, 298, 299
  of relation 3, 5
  of relevance 4
  speaker's 4, 325
  of truth 74
Mdec 121, 122, 124
memories 142
mental models 17, 59
metaphor 10, 14–15, 16
  automaticity 73, 80–1, 87, 88
  class-inclusive 76
  comparison theory of 74
  comprehension 199–201
  comprehension, theories of 74–81
  implicit similes 78–81
  stimulus driven 75–6
  understanding via comparison 76–8
  electrophysiology 187
  familiar 200
  interference effect 76
  interpretation generation 81–3
  nominal 77
  predicative 77
  scrambled 75–6
  unfamiliar 200
metonymies 279, 280
micro-analysis 113
might 303, 306
M-implicature 258, 259, 260
minimal meaning 264
Minimalists 11
modal auxiliary 209
modifier 83, 84
modular view 174
Modus Ponens 97, 234–7, 240–6, 248–9
Modus Tollens 97, 103–4, 234–7, 242–3, 250
monotonicity 120–1, 122, 135
  downward 125–6
Moses illusion 279
MP3 technology 202, 204
M-principle 4

Subject Index 345

N400 component 190–3, 195–6, 198–201, 203, 317
negation 117–19, 124–6
Negative Polarity Item 287
Neo-Gricean 6, 7, 8, 18, 269, 288, 302
neuropsychological measures 316–18
neutral context 218, 219, 222, 223, 224
NLQs 121, 124, 125
Nonevident felicity condition 52–3, 54
non-linguistic cues 202–4
nonmodular systems 174
nonmonotonic effects 103–5
non-words 326, 329, 330
noun
 features 84, 85, 86
 head 81, 82, 83, 84, 85
 modifier 81, 82
 -noun 81, 82, 83, 87
 see also adjective-noun; noun phrase
noun phrase 36, 39, 90, 279
 attributive use 55, 56, 57, 58, 59
 elementary 37
 episodic 37
 expanded 37–8
 holder 38
 instalment 37
 invited 38
 referential use 55, 56, 57, 58, 59
 self-repaired 37
 trial 38
null set inclusion judgements 122–4

Obligation felicity condition 51, 52, 53, 54
observation *see* field method
obstacle hypothesis 60–1, 62
one-step process 26
'or' 301, 302, 303
'or': competence in children and adults 283–99
 children's understanding 292–4
 children's pragmatic knowledge 294–6
 felicity judgement task 297–9
 problem 284–5
 scalar implicature computation: semantics and pragmatics interaction 285–92

P300 191, 194
P600 194, 203
participatory act 34, 36, 41, 43, 45
particularized conversational implicatures 263
past tense 219, 220, 221, 222, 223, 224
phrase features 83–4, 85, 86
pioneering approaches 14–15
placing-for 42, 43
pointing 44–5
poising 43
polarity 15, 190
possession utterances 61
post-Gricean approaches 302–4
post-task sentence verification methodologies 250
practical reasoning 99
pragmatic
 condition 313
 delay hypothesis 297
 instructions 314
 interpretation 316
 responses 314–15, 317, 318
 predictions 175–6, 208, 270–2
 predictive-assertion statements 212, 214, 215
premises, credibility of 100–3
preparatory condition 208, 210–11, 213–20, 222–5
present tense 209
presumption of optimal relevance 6
presupposition 27n
priming methodology 250
priming tasks 69
principle 5, 6, 8
probabilistic judgement studies 106–9
probes 177, 179, 182
process 90
process models of comprehension 72–4
processing effort 6, 165
Processing Limitation hypothesis 298–9
promises 16, 17, 68, 69
 making and understanding 51–4
 see also speech acts in children: promises
promise-to-act statements 211–12, 213, 214, 215
pronouns 42, 117
proper name 58

property attribution   81, 82, 83
propositional content   208, 283
propositions   95–6, 97, 230
  affirmative   124
  reasoning and judgement   103
psycholinguistic experiments and
    linguistic pragmatics   50–70
  definite descriptions   54–9
  indirect speech acts   59–62
  literal meaning and what speakers
    say   62–8
  promises, making and understanding
    of   51–4
psychological accounts   247–9
psychology of reasoning   12–13

Q-heuristic   303, 304
Q-implicature   258, 331
Q-principle   4, 261, 262, 265, 268, 325
quantifiers   15, 116–35
  communication   130–4
  deduction studies   96–7
  denial: judgements of tags   124–8
  existential   306–7, 308–9, 311–16
  expectations   128–30
  focus and negation   117–19
  inferences and logical terms   306
  negative   119, 122, 134, 290
  positive   119, 122
  reasoning: 'even if'   229, 230–1, 232,
    238, 244, 245, 246
  reasoning and judgement   97
  truth of inferences   119–24
  universal   290, 306
question   94–5

R-implicature   258
reaction-time data   317
reading-time experiments   62, 68, 69
realistic content   233
reasoning   12–13, 15
  'even if'   16, 17, 228–52
    experiment 1   233–7
    experiment 2   237–43
      exam scenario   239
      train scenario   239
    linguistic accounts   243–7
    psychological accounts   247–9
    formal   99
    and judgement   94–113
      class inclusion and
        categorization   109–12
      deduction studies   96–105
      hypothesis testing   105–6
      probabilistic judgement
        studies   106–9
      syllogistic   135
    tasks, relational   146–56
    see also conditional reasoning
reassurance   53
recipient corollary   4, 325
reference   14, 25–46
  as cooperative process   26–33
  as interactive process   33–41
  with language and gestures   41–5
  primary aspect   55–6
  secondary aspect   56
referent, topical   182
referring, ahistorical   27
reflexive belief   30
relational
  linking   81, 82, 83
  problems   147–8, 149
  reasoning   16
  reasoning tasks   146–56
relatively unfamiliar   233
relevance   12, 141–69
  automaticity   85, 86
  maxim of   4
  optimal   6, 145, 146
  principle   74
  see also Cognitive Principle of
    Relevance; Communicative
    Principle of Relevance;
    implicature, relevance and
    default pragmatic inference;
    Relevance Theory
Relevance Theory   5–8, 12,
    16–19, 141–6, 168
  automaticity   73, 78, 90
  cognitive principle   143–4
  communicative principle   144
  conversational implicatures   323
  electrophysiology   201
  implicature, relevance and
    inference   261, 267, 269
  inferences and logical terms   302–4,
    307, 311, 313, 315–17, 319
  optimal relevance   144–5
  psycholinguistic experiments   68–9

relevance of an input to an individual 142–3
relevance-guided comprehension procedure 145–6
relevance-guided comprehension heuristic 6–7
representation 90
 of the task 95, 109
representativeness theory 108

said/implied different condition 66
said/implied identical condition 65, 66
salience against current common ground 31
saying 64
scalar implicature 17–19, 180, 258, 259, 268, 269, 278
 automaticity 87–8
 competence in comprehension of 'or' 283, 286, 293, 294, 296, 297, 298–9
 computation: semantics and pragmatics interaction 285–92
 inferences and logical terms 302
 reasoning and judgement 97
scalar inferences 303, 304, 307, 308–9, 311, 313–14, 318–19
scalars 270–2
scalp distribution 190
secure referent hypothesis 56
Selection Task 15, 106
 see also Wason Selection Task
Semantic Core Model 18, 283–4, 289, 296, 299
semantics 83, 110, 246, 283, 317
 automaticity 72, 73, 75, 76, 81, 82, 87, 90, 91
 conversational implicatures 322, 323–4, 325, 326
 electrophysiology 195, 199, 203
 implicature, relevance and inference 261, 262, 263, 267–8, 270, 279, 280
 quantifiers 121, 122–3, 124, 125, 127, 130, 134, 135
 scalar implicature computation 285–92
sentence meaning 2, 73
shape information 89
Short condition 315–16

signal 41
 composite 42
similes 76, 77, 78–81
simple future tense 209–10, 218–24
sincerity condition 208, 210, 212, 215, 224, 225
situation models 59
slow cortical potentials 194–5
some 301, 302, 303
sometimes 302
speaker's maxims 4, 325
speaker's meaning 2, 3, 4, 73
Speech Act Theory 17
speech acts in children: promises 207–25
 experiment 1 210–15
  method 210–13
  results 213–15
 experiment 2 215–24
  method 217–21
  results 221–3
speech acts theory 51
speech production task 165–9
square of opposition 96
State-of-World obstacles 61
statements of fact 53
statements of future acts 53
subalterns 96
subcontraries 96
supermaxim 3
surprise 231
syllogistic reasoning 135
syntactic positive shift see P600
syntax 270, 322
 automaticity 72, 73, 75, 76, 81, 82, 87, 90, 91

tacit contract 28
tacit knowledge 52
tags 124–8
task complexity 307
tense see future; immediate future; past; present; simple future; verb tense
theories 7–12
think 302
threats 52
time prepositions 209, 216
time, telling of 16
token 56–7, 58

348 Subject Index

truth conditions 55, 73, 88, 283, 285, 316
truth of inferences 119–24
  downward-entailment judgements 120–2
  null set inclusion judgements 122–4
Truth Value Judgement Task 291, 294
truth-table tasks 99
2-4-6 problem 15

underinformative items 312–13, 314, 315, 316, 317, 318
Underspecification Model 18, 261–9, 272, 274–80
understanding 16

unexpectedness 231, 238
United States 72
universal sentence 97

valid arguments 100–5
verb 83
  infinitive 209, 211
  tense 209, 216, 219
visual information 203–4

walk 31
Wason Selection Task 15, 16, 146, 156–64, 251
'what is implicated' 63
'what is said' 63
word frequency 269